T0325172

Intermetallic-Based Alloys—Science, Technology and Applications

MATERIALS RESEARCH SOCIETY
SYMPOSIUM PROCEEDINGS VOLUME 1516

Intermetallic-Based Alloys—Science, Technology and Applications

Symposium held November 25–30, 2012, Boston, Massachusetts, U.S.A.

EDITORS

Ian Baker

Dartmouth College
Hanover, New Hampshire, U.S.A.

Martin Heilmaier

*Karlsruhe Institute of
Technology (KIT)*
Karlsruhe, Germany

Sharvan Kumar

Brown University
Providence, Rhode Island, U.S.A.

Kyosuke Yoshimi

Tohoku University
Sendai, Miyagi, Japan

Materials Research Society
Warrendale, Pennsylvania

CAMBRIDGE
UNIVERSITY PRESS

CAMBRIDGE
UNIVERSITY PRESS

Shaftesbury Road, Cambridge CB2 8EA, United Kingdom

One Liberty Plaza, 20th Floor, New York, NY 10006, USA

477 Williamstown Road, Port Melbourne, VIC 3207, Australia

314–321, 3rd Floor, Plot 3, Splendor Forum, Jasola District Centre, New Delhi – 110025, India

103 Penang Road, #05–06/07, Visioncrest Commercial, Singapore 238467

Cambridge University Press is part of Cambridge University Press & Assessment, a department of the University of Cambridge.

We share the University's mission to contribute to society through the pursuit of education, learning and research at the highest international levels of excellence.

www.cambridge.org
Information on this title: www.cambridge.org/9781605114934

Materials Research Society
506 Keystone Drive, Warrendale, PA 15086
http://www.mrs.org

First published 2013

CODEN: MRSPDH

A catalogue record for this publication is available from the British Library

ISBN   978-1-605-11493-4   Hardback

CONTENTS

TITANIUM ALUMINIDES I

*Invited Paper

TITANIUM ALUMINIDES II

*Invited Paper

POSTER SESSION

SHAPE MEMORY ALLOYS

$L1_2$, B2 AND DO_3 COMPOUNDS

B2, $L1_0$ AND LAVES PHASE

SILICIDES AND LPSO PHASES

*Invited Paper

PREFACE

Symposium JJ on "Intermetallics-Based Alloys: Science, Technology and Applications," held November 26th through November 28th at the 2012 MRS Fall Meeting in Boston, Massachusetts, was the fifteenth in a series of symposia on intermetallic phases held every two years at the MRS Fall Meetings. It encompassed six oral sessions and one poster session, attracted about 150 participants from all over the world, and made it one of the best attended regular meetings on intermetallic phases.

This symposium focused on new understanding and developments within the realm of microstructure, processing, and properties of intermetallic compounds and multiphase alloys where intermetallic compounds are the major constituents. Topics which were dealt with included fundamentals of phase stability and their effects on microstructural design and microstructural degradation in extreme environments/service, physical and mechanical response to various loading conditions and the role of defects in influencing them, developments in innovative processing from the solid, liquid and vapor states, and technological considerations for successful commercial applications. Intermetallic phases of interest included aluminides, silicides, Laves phases and Heusler phases, and various other geometrically- and topologically-close-packed compounds. An entire oral session was devoted to shape memory alloys based on intermetallic phases. From an applications perspective, insights were given into structural and functional applications for high temperature use in the aerospace and automotive industries, the fossil fuel and nuclear industries, for ferromagnetic applications, catalysis, and thermoelectric power.

This volume presents forty-six papers which together give a representative overview of the current state of research on intermetallic phases, providing an insight into the current state of development of individual intermetallic alloy systems and identifying areas for future research.

Ian Baker
Martin Heilmaier
Sharvan Kumar
Kyosuke Yoshimi

March 2013

ACKNOWLEDGMENTS

It is a pleasure to acknowledge financial support for this symposium from the following organizations:

Brown University, Providence, RI, USA
Deutsche Forschungsgemeinschaft (DFG), Bonn, Germany
General Electric Global Research Center, Niskayuna, NY, USA
Tohoku University, Sendai, Japan

The organizers of this symposium would like to thank all of the participants, authors, reviewers and the MRS staff for their efforts in preparing these proceedings.

MATERIALS RESEARCH SOCIETY SYMPOSIUM PROCEEDINGS

MATERIALS RESEARCH SOCIETY SYMPOSIUM PROCEEDINGS

MATERIALS RESEARCH SOCIETY SYMPOSIUM PROCEEDINGS

Volume 1534E — Low-Dimensional Semiconductor Structures, 2012, T. Torchyn, Y. Vorobie, Z. Horvath, ISBN 978-1-60511-511-5

Prior Materials Research Society Symposium Proceedings available by contacting Materials Research Society

Titanium Aluminides I

Mater. Res. Soc. Symp. Proc. Vol. 1516 © 2012 Materials Research Society
DOI: 10.1557/opl.2012.1654

Advanced β-Solidifying Titanium Aluminides – Development Status and Perspectives

Helmut Clemens[1], Martin Schloffer[1], Emanuel Schwaighofer[1], Robert Werner[1], Andrea Gaitzenauer[1], Boryana Rashkova[1], Thomas Schmoelzer[1], Reinhard Pippan[2], Svea Mayer[1]
[1]Department of Physical Metallurgy and Materials Testing, Montanuniversität Leoben, A-8700, Leoben, Austria
[2]Erich Schmid Institute of Materials Science, Austrian Academy of Sciences, A-8700, Leoben, Austria

ABSTRACT

After almost three decades of intensive fundamental research and development activities intermetallic titanium aluminides based on the γ-TiAl phase have found applications in automotive and aircraft engine industries. The advantages of this class of innovative high-temperature materials are their low density as well as their good strength and creep properties up to 750°C. A drawback, however, is their limited ductility at room temperature, which is reflected by a low plastic strain at fracture. This behavior can be attributed to a limited dislocation movement along with microstructural inhomogeneity. Advanced TiAl alloys, such as β-solidifying TNM™ alloys, are complex multi-phase materials which can be processed by ingot or powder metallurgy as well as precision casting methods. Each production process leads to specific microstructures which can be altered and optimized by thermo-mechanical processing and/or subsequent heat-treatments. The background of these heat-treatments is at least twofold, i.e. concurrent increase of ductility at room temperature and creep strength at elevated temperature. In order to achieve this goal the knowledge of the occurring solidification processes and phase transformation sequences is essential. Therefore, thermodynamic calculations were conducted to predict phase fraction diagrams of engineering TiAl alloys. After experimental verification, these phase diagrams provided the base for the development of heat treatments to adjust balanced mechanical properties. To determine the influence of deformation and kinetic aspects, sophisticated ex- and in-situ methods have been employed to investigate the evolution of the microstructure during thermo-mechanical processing and subsequent multi-step heat-treatments. For example, in-situ high-energy X-ray diffraction was conducted to study dynamic recovery and recrystallization processes during hot-deformation tests. Summarizing all results a consistent picture regarding microstructure formation and its impact on mechanical properties in TNM alloys can be given.

INTRODUCTION

Innovative structural materials have to withstand the highly demanding conditions in the next generation of aircraft and automotive engines, which are targeted to exhibit higher efficiency leading to reduced fuel consumption as well as significantly decreased CO_2 and NO_x emissions. Intermetallic γ-TiAl based alloys possess numerous attractive properties which meet these demands and they also represent a good example of how fundamental and applied research along with industrial development can lead to a new class of advanced engineering materials [1,2,3,4,5,6,7,8]. Although the transfer of new materials from R&D to engineering application

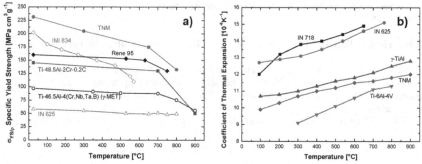

Figure 1. Variation of a) specific yield strength and b) coefficient of thermal expansion with temperature of selected structural materials in comparison with γ-TiAl based alloys [6,7,8].

is always difficult to achieve, the interest in TiAl was accelerated by the trend-setting step of General Electric, who equipped the last stage(s) of the low-pressure turbine of the so-called GEnX jet engine with cast TiAl blades [9]. Additionally, the automotive industry is eager to establish TiAl as innovative structural materials in their drive system, e.g. for the use as turbocharger turbine wheels, to meet future environmental requirements demanded by legislative authorities.

This review deals mainly with results achieved by the authors on β-solidifying TNM™ alloys in the framework of European as well as international collaborations with universities, research facilities and industries. These alloys are characterized by a high content of β-stabilizing alloying elements, such as Nb and Mo, and are based on the nominal composition of Ti-43.5Al-4Nb-1Mo-0.1B (in at.%) [10]. Since Nb and Mo represent the decisive alloying elements, this γ-TiAl based alloy family has been named "TNM alloys" in order to distinguish them from the well-known high-strength "TNB alloys", which rely on a high Nb concentration [7,11]. A more comprehensive account of the authors' work, which also contains references to other world-wide activities, can be found in a current review paper [8]. The advantage of γ-TiAl based alloys is mainly seen in low density (3.9 - 4.2 g/cm³, depending on composition and constitution), high specific yield strength, high specific stiffness, good oxidation resistance, resistance against "titanium fire", and good creep properties up to high temperatures. The variation of the specific yield strength of various γ-TiAl based alloys in comparison with Ni-base and Ti-base alloys is shown in Figure 1a. As an example of thermo-physical properties the coefficient of thermal expansion of the TNM alloy, also in comparison with conventional materials, is depicted in Figure 1b. At room temperature strength levels > 1000 MPa can be achieved in TNM alloys by appropriate thermo-mechanical processing and subsequent heat treatments. It is important to note that also high temperature properties, such as creep resistance, can be considerably improved, e.g. by implementation of precipitation hardening. Advanced TiAl alloys are mostly complex multi-phase alloys which can be processed by ingot or powder metallurgy as well as precision casting methods. Each process leads to specific microstructures which can be altered and optimized by thermo-mechanical processing and/or subsequent heat treatments. In order to further increase the economic feasibility of wrought processing for manufacturing of components, alloys are needed which can be hot-forged under isothermal conditions, but also near-conventionally, which means that a conventional forging equipment with minor and inexpensive modifications can be used [12]. Generally, the background of conducting heat treatments is manifold. Frequently the motivation is to acquire chemical and microstructural homogeneity as well as to increase the ductility at room temperature and creep strength at elevated temperature. In order to achieve this goal, the knowledge of the occurring solidification processes along with the following phase transformation sequences are essential. Therefore, thermodynamic calculations are helpful to predict the phase diagram of engineering TiAl alloys. After verification with experimental methods, such as short and long-term heat treatments combined with quantitative metallography, differential scanning calorimetry (DSC), conventional X-ray diffraction (XRD) and in-situ high-energy X-ray diffraction (HEXRD), these phase diagrams provide the basis for the development of effective heat treatments. To account for the influence of deformation and kinetic aspects sophisticated ex- and in-situ methods have been employed to investigate the evolution of both microstructure and texture during thermo-mechanical processing and subsequent multi-step heat treatments [13]. For example, HEXRD was conducted to study dynamic recovery and recrystallization processes in TNM alloys during hot-compression testing. However, while XRD diffraction gives high intensities for the main reflections of the occurring phases, ordering phenomena are difficult to detect. Due to the peculiar scattering behavior of TiAl alloys, neutron diffraction gives high intensities for the superstructure peaks of ordered phases, while the intensities of the main peaks are very small. Consequently, in-situ neutron diffraction experiments were employed to determine the ordering temperature of the α_2- and β_o-phase in the TiAl-Nb-Mo system [13]. Summarizing all results a consistent picture regarding microstructure formation and its impact on mechanical properties in advanced β-solidifying TiAl alloys can be given.

DESIGN CONCEPT OF TNM ALLOYS

In order to increase the economic feasibility of wrought processing for TiAl components alloys are needed which can be processed near conventionally. Here, the major challenge is forging of small parts, e.g. turbine blade pre-forms, or rolling of thin sheets which, in comparison to large ingots, possess only a small amount of stored heat energy. In addition, a fine-grained and isotropic casting microstructure is favorable to both ingot breakdown and secondary forming operations. The alloys should be designed to allow robust industrial heat treatments, i.e. the material must tolerate a specified, and technologically realistic, variation of the constituting elements, without pronounced changes of phase transition temperatures and phase volume fractions as well as related variations in mechanical properties. The desired microstructures should be achieved by simple heat treatments which can be conducted at an industrial scale and which are almost independent from the geometrical dimension of the final part. Furthermore, the microstructure must show a resistance against microstructural changes during long-term exposure to service temperatures in the range of 600 to 750°C. In addition, the machinability of the material by a wide variety of techniques must be guaranteed. Finally, the alloy should show balanced mechanical properties and acceptable oxidation resistance.

From these requirements the following demands on the alloy are deduced: (a) after casting and solidification the alloy should possess a refined equiaxed microstructure showing no significant casting texture; (b) the composition of the alloy must be well defined to ensure a solidification path according to $L \rightarrow L + \beta \rightarrow \beta \rightarrow ...$, instead of a peritectic solidification, i.e.,

L → L + β → α → ..., which is prone to segregation [7]; (c) during ingot breakdown, which might be obsolete in case of a homogeneous fine-grained and isotropic casting microstructure, as well as secondary hot-forming operations, a significant volume fraction of disordered bcc β-phase should be present providing a sufficient number of independent slip systems. Thus, it may improve the deformability at elevated temperature, where, for example, processes such as extrusion, forging and rolling are carried out. A number of authors, see [8] and references therein, have demonstrated that by stabilizing the β-phase through alloying with Nb, Ta, Mo, V or other elements, a significant improvement in hot-workability can be achieved. Takeyama and co-workers, who were the first pointing out the use of disordered β-phase for the development of wrought TiAl alloys, demonstrated that ingot breakdown of a β-phase containing TiAl is possible on a conventional forge with the associated high deformation rates, e.g. see [14]. At service temperature, however, the volume fraction of the β_o-phase, which then shows an ordered B2 structure, should be low in order not to deteriorate creep properties [3,15]; (d) the alloys should provide a well-balanced phase ratio of γ-TiAl, α_2-Ti$_3$Al and β_o-TiAl which can be converted in designed microstructures by means of simple and reproducible heat treatments. In order to select an alloy which fulfils the demands as defined in a) to d) a computer-aided method was applied for the design of an engineering γ-TiAl based alloy which finally - after further optimization - became the so-called TNM alloy. To this end, thermodynamic calculations based on the CALPHAD method were conducted for the prediction of the constituent phases and the related phase transition temperatures [10]. The software package ThermoCalc® was applied using a commercial TiAl database developed by Saunders [16]. Whereas the database is an excellent prediction tool for low alloyed TiAl alloys, e.g. Ti-48Al-2Cr-2Nb, it was found to poorly describe the transition temperatures and phase proportions in high Nb bearing γ-TiAl based alloys as reported in ref. [17]. Therefore, it should be pointed out that the calculations used for alloy design were conducted to study alloying trends rather than to obtain absolute values regarding phase fractions and transition temperatures. Figure 2 shows the calculated phase fractions as a function of temperature for three different β-stabilized γ-TiAl based alloys. The three alloys have a constant Al concentration (43 at.%), but different contents of β-stabilizing alloying elements (Nb, Mo). The concentration of B is kept constant with 0.1 at.%. Before Figure 2 will be discussed in detail, the role and the main effects of the selected alloying elements are addressed. Nb stabilizes the bcc β-phase and thus the β-phase field region is shifted to higher Al contents. Additionally, the eutectoid temperature (T_{eu}), where the order/disorder reaction $\alpha \leftrightarrow \alpha_2$ occurs, is increased significantly. Nb also lowers the stacking fault energy and thus increases the ductility of the alloys at room temperature by increasing the propensity for mechanical twinning [7]. Moreover, Nb retards diffusion processes and modifies the structure of the oxidation layer [6,18,19]. Like Nb, Mo is also a β-stabilizing element in the Ti-Al system, which raises the activation energy of diffusion in both γ-TiAl and α_2-Ti$_3$Al, but exhibits a much higher partition coefficient $k_{\beta\alpha}$ than Nb [20]. It must be taken into account that phases, when stabilized by such slow-diffusing elements, are expected to exhibit sluggish dissolution behavior. The B content of 0.1 at.% was selected to ensure a grain refining effect during solidification [21,22]. Boron, which tends to form very stable borides, is also beneficial in case of heat treatments conducted at high temperatures. Here, the borides retard grain coarsening by pinning the grain boundaries. In addition, the borides favor the formation of the lamellar

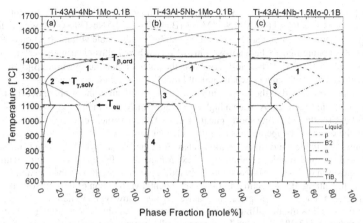

Figure 2. Calculated phase fractions as a function of temperature for three different β-stabilized Ti-43Al-Nb-Mo-0.1B alloys including 450 mass-ppm O [10]. For the calculation the database of Saunders [16] was used (see text).

microstructure (α→ α + γ) over the massive transformation (α → γ_M) by heterogeneous nucleation of γ-TiAl lamellae [21].

Figures 2a-c demonstrate that all chosen alloys solidify entirely via the β-phase. Figure 2 predicts for TNM alloys an ordering reaction β → β_o(B2) at about 1410°C to 1420°C (T_β,ord), depending on the actual alloy composition, which contradicts the excellent hot-working behavior observed at significantly lower temperatures [8,12]. In order to determine the ordering temperature of the β-phase (T_β,ord), in-situ neutron diffraction measurements were conducted in the temperature range from 900°C to 1450°C. It was found that in a Ti-43Al-4Nb-1Mo-0.1B alloy ordering of the β-phase occurs at around 1210°C. Furthermore, T_eu, which is connected with the α → α_2 ordering reaction as well as the dissolution of the γ-phase at the so-called γ-solvus temperature (T_γ,solv) were obtained by neutron diffraction [23]. From comparison of the phase fractions of the three different compositions shown in Figure 2 and additional experimental investigations it was found that Mo is a 3.9 times stronger β-stabilizer than Nb, which agrees with the results of other studies [24,25,26,27]. For all three alloys the thermodynamic calculation shows a minimum of the mole fraction of the β-phase at around 1275°C. Above this temperature the fraction of β-phase strongly increases with temperature (see "1" in Figure 2). However, at the temperature where the minimum of β-phase appears, Ti-43Al-4Nb-1Mo-0.1B shows the smallest fraction of β-phase and only this alloy variant exhibits a pronounced "C-shaped" course of the β-phase (see "2" in Figure 2a). By increasing the amount of Nb or Mo by 1 at.% or 0.5 at.%, respectively, the mole fraction of the β-phase in its minimum at 1275°C is considerably increased (see "3" in Figures 2b,c). Below the temperature of its minimum the mole fraction of the β-phase slightly increases (Figure 2a) or shows an approximately constant value (Figures 2b,c). It should be emphasized that the magnitude of the volume fraction of the β-phase at the minimum is of particular interest as far as heat treatments subsequent to the forging process are concerned. Below the eutectoid temperature (T_eu ~ 1115°C) the ordered β_o-phase fraction decreases with decreasing temperature and seems to vanish at about 600°C in case of alloys Ti-43Al-4Nb-1Mo-0.1B and Ti-43Al-5Nb-1Mo-0.1B, see "4" in Figures 2a,b. It should be noted that all three alloys exhibit an adjustable volume fraction of β-phase in the temperature range where hot-working of γ-TiAl based alloys is usually performed. However, only alloy Ti-43Al-4Nb-1Mo-0.1B shows the potential that the appearing β-phase can almost be removed by a subsequent heat treatment [10,12]. To this end, the annealing temperature must correspond to the temperature where the minimum of the β-phase occurs, i.e. to temperature "2" in Figure 2a. At such a high temperature the dissolution kinetics of the β-phase are rather fast, thus allowing short annealing times [12]. In principle, a removal of the β-phase below the eutectoid temperature is also possible, but would require long, and thus costly, annealing times due to the sluggish dissolution behavior of the Mo and Nb containing β-phase. From additional thermodynamic calculations and experimental verification a TNM alloy with nominal composition of Ti-43.5Al-4Nb-1Mo-0.1B shows the best match with the demands defined in the beginning of this section. For the experimental verification of the course of the phase fractions as a function of temperature, heat treatments following quantitative metallography as well as in-situ diffraction measurements were conducted employing synchrotron radiation [13,28]. Whereas $T_{\gamma,solv}$ agrees well with experimental data derived from quantitative analysis of heat-treated specimens and in-situ HEXRD diffraction experiments, T_{eu} is significantly underestimated by the thermodynamic calculation. The course of the experimental phase fractions around $T_{\gamma,solv}$ is in agreement with the calculated data, but differs increasingly with increasing distance to the γ-solvus temperature. For more detailed information concerning the determination of phase fraction diagrams and the establishment of phase diagrams the reader is referred to [8,29]. It should be pointed out that annealing experiments to determine phase volume fractions are rather time consuming and large numbers of samples are required, whereas in-situ HEXRD investigations can be conducted and evaluated considerably faster [13,28].

PROCESSING

The industrial scale processing routes established for wrought γ-TiAl based alloys during the last two decades are summarized in Figure 3a, where the focus was put on ingot metallurgy (IM) and forging [6,7,8]. TiAl alloy ingots can be produced either by vacuum arc re-melting (VAR) [30] or plasma arc melting (PAM) [31]. Recently, GfE - Metalle und Materialien GmbH, Germany, has introduced a production technique for TiAl alloys, which combines VAR skull melting with centrifugal casting in permanent moulds [32], as schematically shown in Figure 3b. The advantage of this process is the possibility of producing adjusted feed stock materials for both hot-working and investment casting directly from the ingot with a remarkably improved materials yield. The small ingots exhibit a satisfying microstructural and chemical homogeneity as well as impurity concentrations (O, N) well below 700 wt-ppm. For example, the as-cast microstructure of a TNM ingot as produced by VAR skull melting and subsequent centrifugal casting is shown in Figure 4a. In conclusion, a significant progress in γ-TiAl ingot processing on industrial scale has been achieved during the last decade. However, a particular challenge for the future is to develop cost savings strategies which will certainly include the implementation of recycling concepts for TiAl alloys.

8

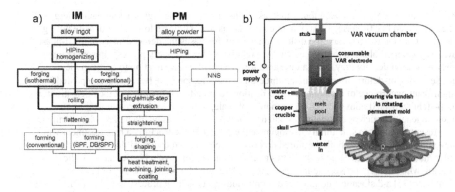

Figure 3. a) Manufacturing and processing routes established for wrought γ-TiAl based alloys on an industrial scale. The bold lines highlight the routes developed for TNM alloys so far. Corresponding microstructures are shown in Figure 4. IM: ingot metallurgy; PM: powder metallurgy; HIP: hot-isostatic pressing; DB: diffusion bonding; SPF: superplastic forming; NNS: near-net shape; b) Schematic illustration of the VAR skull melting process employed by GfE Metalle und Materialien GmbH (Germany) for the production of advanced TiAl alloys [32]. Subsequently, the melt is processed by means of centrifugal casting in permanent moulds.

In general, hot-working of TiAl alloys is conducted exclusively above their brittle-to-ductile transition temperature (BDTT > 700°C) and can be divided into primary and secondary hot-working steps. The aim of primary hot-working of cast ingots is to convert the coarse-grained microstructure into a fine-grained and uniform one suitable for subsequent wrought processing or heat treatments. This is accomplished by employing hot-working parameters at which dynamic recrystallization is prevalent, and macroscopic as well as microscopic damage can be neglected, i.e. at temperatures between the eutectoid and the γ-solvus temperature and applying relatively low deformation rates. During the last decade several wrought γ-TiAl alloys with complex alloy compositions have been developed [1,2,3,4,5,7]. These alloys exhibit excellent mechanical properties, but show narrow processing windows. Therefore, they can only be forged under isothermal conditions. However, isothermal forging of γ-TiAl based alloys must be performed at high temperatures, requiring special dies and environmental conditions which increase manufacturing costs. Consequently, TNM alloys have been developed which are equally suited for hot-die forging under near conventional conditions [10,12] and isothermal forging [33]. In addition, the homogeneous and fine-grained microstructure obtained after solidification offers the opportunity to omit the cost-intensive ingot breakdown. The processing routes which were established for TNM alloys on an industrial scale, with the exception of sheet rolling which was conducted on a laboratory roller mill, are depicted in Figure 3 (bold lines). In order to fundamentally study the hot-working behavior of TNM alloys in real time, and, thus, to optimize industrial hot-forming processes, in-situ HEXRD experiments were conducted during hot-compression tests in the (α + β) phase field [13]. A novel technique was employed to evaluate the diffraction data with respect to the microstructural processes occurring during deformation [34]. Additional electron back-scatter diffraction (EBSD) investigations have been carried out which allowed to distinguish between the mechanisms of recovery and recrystallization as reported in [13,35]. It has been found that, as a consequence of the high stacking fault energy of the disordered bcc β-phase, a fast dynamic recovery is maintaining a low defect density during deformation. On the contrary, the dislocation activity increases the defect density in the α-phase which is counterbalanced to some extent by a relatively slow recovery process. In some regions the dislocation density may cross a threshold for dynamic recrystallization to occur in the α-phase [35]. Hot-deformation within the three-phase field region ($\alpha + \beta + \gamma$) is accompanied by significant occurrence of dynamic recrystallization, leading to a refinement of the microstructure. In β-stabilized TiAl alloys forged on an industrial scale, however, microstructural inhomogeneities such as elongated β_o- and α_2-grains as well as residual lamellar colonies might be present. In order to structurally homogenize the forged microstructure, a heat treatment was developed which utilizes both the stored deformation energy and chemical driving force present after the forging process. Figure 4b shows a representative image of a heat-treated hot-die forging which clearly proves the increase in microstructural homogeneity. According to Schloffer [36] a dislocation-dominated discontinuous precipitation process is accountable for the observed microstructural improvement. Such a microstructure as shown in Figure 4b offers a good starting point for a further heat treatment which then determines the mechanical properties of the fabricated component.

MICROSTRUCTURE AND MECHANICAL PROPERTIES

Figures 4c-e show three different microstructures which were adjusted in a TNM alloy by means of a two-step heat treatment as described in [12,37]. The starting material was hot-die forged and subsequently subjected to a homogenization heat treatment (Figure 4b). For establishing a so-called nearly lamellar γ (NL+γ) microstructure (Figure 4c) the first heat treatment step is conducted in the ($\alpha + \beta + \gamma$)-phase field region, followed by air cooling. After this step the microstructure consists of a small volume fraction of globular β_o- and γ-grains as well as supersaturated α_2-grains with a grain size well below 100 μm. The volume fractions of the globular β_o- and γ-grains within the final microstructure are indicated in Figure 4c. The second heat treatment step, which in fact is an ageing treatment within the ($\alpha_2 + \beta_o + \gamma$)-phase field region, has a strong effect on the mechanical properties because ultra-fine lamellar γ/α_2-colonies are formed by decomposition of the supersaturated α_2-grains. In-situ TEM and HEXRD experiments were carried out to study nucleation and precipitation of very fine γ-lamellae from supersaturated α_2-grains during the second heat treatment step [38]. The mean interface spacing, including α_2/γ- and γ/γ-interfaces, is in the range of 15 nm to 30 nm and depends primarily on the temperature at which the second heat treatment is performed. For adjusting the nearly lamellar β variants (NL+β1 and NL+β2, see Figures 4d,e) the temperature of the first heat treatment step was set within the ($\alpha + \beta$)-phase field region, again followed by air cooling. At room temperature the microstructure consists of supersaturated α_2- and globular β_o-grains (for volume fractions see inset in Figures 4d,e), but no γ grains are present. For the NL+β1 microstructure the temperature of the second heat treatment step was identical to that selected for NL+γ. In case of NL+β2 (Figure 4e), however, the ageing treatment was conducted at a considerably higher temperature in order to induce the occurrence of a cellular reaction at the boundaries of the lamellar γ/α_2-colonies as evidenced by the TEM image shown in Figure 4f. The cellular reaction transforms the very fine γ and α_2-laths to a zone which consists of irregularly shaped γ, α_2 and β_o (+ ω)-grains. Detailed information on the three different microstructures is given in [37].

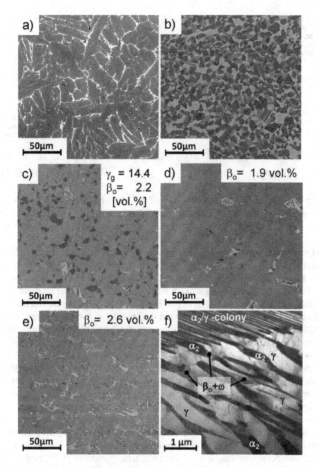

Figure 4. Microstructures of a TNM alloy with a nominal composition of Ti-43.5Al-4Nb-1Mo-0.1B. a) as-cast microstructure; b) microstructure after hot-die forging and subsequent homogenization heat treatment; c) nearly lamellar γ (NL+γ) microstructure (vol.% of globular β_o- and γ-grains are indicated); d) nearly lamellar β1 (NL+β1) microstructure; e) NL+β2 microstructure. a) - e) light-optical microscope images. f) Bright-field TEM image: the upper parts shows a lamellar γ/α2-colony which was partly transformed in cellular structure [39].

The tensile properties of TNM material with NL+γ microstructure have thoroughly been described in ref. [12]. The strength properties are dominated by the spacing between the α_2/γ as well as γ/γ-interfaces within the lamellar colonies. The observed behavior of the yield strength can be explained by a modified Hall-Petch relationship where the dominant structural parameter

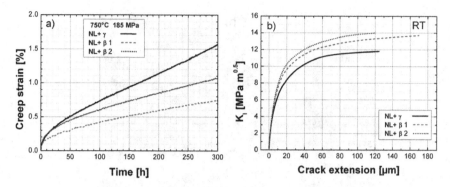

Figure 5. a) Creep behavior at 750°C and 185 MPa of the TNM microstructures shown in Figures 4c-e; b) Fracture resistance curves based on K_I for the same microstructures at room temperature [37].

is the mean interface spacing. The stress required to propagate dislocations through the interfaces is inversely proportional to the square root of the mean interface spacing [7,40]. Here, the α_2/γ and γ/γ-interfaces act as glide obstacles and lead to dislocation pile-ups as observed by TEM investigations [7,41,42]. The plastic elongation at fracture of TNM alloys at ambient temperatures, which is in the range of 1-2%, depends critically on the volume fraction of the globular γ- and β_o-grains which are arranged around the lamellar α_2/γ-colonies, because β_o-phase tends to embrittle the material, whereas γ-phase increases the ductility. Additionally, the average lamellar spacing and thus the volume fraction of γ-phase within the colonies contributes to the overall deformation behavior.

The creep curves recorded at 750°C and 185 MPa for TNM material, exhibiting the microstructures shown in Figures 4c-e, are depicted in Figure 5a. As in case of the tensile properties the creep strength is controlled by the average lamellar spacing of the lamellar colonies and the volume fraction of the globular γ- and β_o-grains. Evidently, the presence of globular γ-grains as well as the appearance of the cellular reaction, which leads to a local refinement of the nearly lamellar microstructure (see Figure 4e), have a negative impact on creep strength. The stress exponent of a TNM alloy with a NL+γ microstructure comparable to that shown in Figure 4c was determined to be n ~ 3 for a test temperature of 750°C and a stress range of 150 to 250 MPa. The corresponding activation energy for creep is about 370 kJ/mol [37,43], which agrees well with the result derived from internal friction measurements on TNM specimens [44] and those reported for other β-containing TiAl alloys [4,45]. A comprehensive analysis of the influence of the microstructural constituents (arrangement, morphology and phase fraction) on the creep properties, however, is the topic of ongoing research and will be reported in a forthcoming paper [37]. Finally, Figure 5b shows fracture (K) resistance curves based on averaged stress intensity factor for the three microstructures shown in Figures 4c-e at room temperature. Single-edge notch tension (SENT) specimens with an initial crack length of about 6 mm were used and the tests were performed displacement controlled with a loading rate of 0.15 mm/min. The propagation of the crack during loading was measured continuously by applying the DC-potential drop technique. All three microstructures show a pronounced R-curve behavior and the individual curves do not differ much among each other. The examination of fractured specimens has indicated that the colony size, their relative orientation to the propagating crack and lamellar spacing as well as crack bridging, nano-cracking and mechanical twinning contribute to the crack growth resistance. In addition, the arrangement of the γ and β_o-phase along the colony boundaries, see Figures 4c-e, plays an important role and is responsible for the small differences between the R-curves shown in Figure 5b [37].

APPLICATIONS

The world-wide drive to reduce fuel consumption, emissions (soot, CO_2) and noise in motor vehicles is steadily increasing. For example, regulations which limit emissions from mid- to large size diesel engines will soon be enforced in Europe and nations which show a lasting commitment to pollution control. One strategy to meet this obligation is downsizing of conventional combustion engines. Additionally, efficiency and engine performance will be enhanced by increasing combustion gas temperatures up to 850°C (diesel engine) and 1050°C (gasoline engine) where gas pressures and engine revolutions per minute (rpm) are simultaneously rising. By the use of light-weight TiAl alloys for turbocharger turbine wheels the following benefits are expected to be gained. First of all, a reduction of emissions, especially soot, as a result of the quicker charging of fresh air for the combustion process. An improvement of the response behavior ("turbo-lag") increases the agility of the vehicle's responsiveness. An appreciable reduction of noise and vibration is caused by the shift of the resonance frequencies to higher levels. Furthermore, the foundations are laid for an increase in the maximum rpm of the turbine rotor, and consequently, the turbocharger and engine efficiency. An additional beneficial effect of the lower rotor mass is a lower necessary protection against busting of the turbine wheel, which leads to lower housing wall thicknesses. Moreover, bearing friction is reduced and thus the whole turbocharger system can achieve higher efficiency and longer durability. Already in 1999 Mitsubishi implemented TiAl turbocharger wheels in their Lancer 6 sports car [46]. The turbine wheels were produced by means of the Levicast process, a modification of the lost wax precision casting method, as well as by centrifugal casting. Researchers of DAIDO in cooperation with universities have developed advanced TiAl alloys which contain high additions of Nb to improve the high-temperature properties and other alloying elements to ensure good castability [47]. In Europe, research projects are aimed at the development of cost-effective production routes for turbocharger wheels and the qualification of advanced TiAl alloys. However, it must be critically stated that the use of - even advanced - TiAl alloys is currently restricted to diesel engines. In order to implement this innovative class of materials in gasoline engines further alloy development is required to improve both creep strength and oxidation resistance.

The next generation of aircraft engines is targeted to exhibit higher efficiency which leads to reduced kerosene consumption, significant lower emission of CO_2 and NO_x as well as to a noticeable reduction of engine noise. There are at least three major payoff areas for γ-TiAl based alloys in advanced aero-engines [1,2,5,48,49,50,51]: (a) γ-TiAl has a specific elastic stiffness 50% greater than structural materials commonly used in aircraft engines. The higher specific stiffness (E/ρ) also shifts acoustically excited vibrations towards higher frequencies, which is usually beneficial for structural components, e.g., compressor and turbine blades and parts within the exhaust nozzle area. (b) The good creep resistance of advanced γ-TiAl based alloys in the temperature regime of 600 to 800°C enables the substitution of heavy Ni-based alloys for certain applications. (c) The high fire resistance of γ-TiAl based alloys (nearly as resistant as Ni-based alloys) can enable the substitution of heavy and expensive fire-resistant designed Ti-based alloys in some components. Only few years ago, General Electric announced the initiation of investment cast blades of a so-called 2nd generation TiAl alloy in the low-pressure turbine (LPT) because this section of the engine offers the highest weight reduction potential. Since then certification and flight tests with TiAl equipped engines have been conducted successfully and in the meantime the regular service for carriage of freight and passenger transport has begun already [9]. Meanwhile also other major aero-engine manufacturers have announced in press releases and their homepages the future use of TiAl in their next jet engine generation. Depending on engine type 100 to 150 LPT blades per stage are required [52]. Different engine concepts as reported in [9,48,53,54], however, involve higher mechanical loading of the LPT blades, thus the use of high-strength advanced γ-TiAl based alloys is required.

SUMMARY

Intermetallic γ-TiAl based alloys are considered as important candidate material for advanced applications in aerospace, automotive and related industries. World-wide research and development on γ-TiAl alloys have led to a better understanding of the fundamental influence of alloy composition and microstructure on processing behavior and mechanical properties. During the last years industry has started to implement this new class of light-weight high-temperature materials. In particular, all major aircraft and automotive engine manufacturers are advancing the qualification and introduction of γ-TiAl components. Engineering γ-TiAl based alloys, such as TNM alloys, can be processed using advanced metallurgical methods - a factor, which is decisive for these specific materials to be economically competitive with other state-of-the-art materials. In current γ-TiAl based alloys a good balance of properties like room temperature ductility, fracture toughness, high-temperature strength, creep, and oxidation resistance can be achieved. In order to promote further the use of these innovative intermetallic materials more applications in jet and automotive engines as well as other areas must be identified.

ACKNOWLEDGMENTS

A part of the work presented in this review paper was conducted within the framework of the following projects: FFG project 830381 "fAusT" Österreichisches Luftfahrtprogramm TAKE OFF, Austria; FFG project 832040 "energy-drive", Research Studios Austria, Austria; BMBF project O3X3530A. Research activities performed at DESY have received funding from the European Community's Seventh Framework Programme (FP7/2007-2013) under grant agreement n°226716. Finally, we thank our project partners for many years of fruitful cooperation.

REFERENCES

1.   *Structural Intermetallics 2001*, edited by K. J. Hemker, D. M. Dimiduk, H. Clemens, R.

Darolia, H. Iui, J. M. Larson, V. K. Sikka, M. Thomas, and J. D. Whittenberger, (TMS, Warrendale, PA, 2001).

2. *Gamma Titanium Aluminides 2003*, edited by Y-W. Kim, H. Clemens, and A. H. Rosenberger, (TMS, Warrendale, PA, USA, 2003).

3. *Advanced Intermetallics-Based Alloys*, edited by J. Wiezorek, C.-L. Fu, M. Takeyama, D. Morris, and H. Clemens, (Mater. Res. Soc. Symp. Proc. **980**, Pittsburgh, PA, 2007).

4. *Structural Aluminides for Elevated Temperature Applications*, edited by Y-W. Kim, D. Morris, R. Yang and C. Leyens, (TMS, Warrendale, PA, 2008).

5. *Intermetallic-Based Alloys for Structural and Functional Applications*, edited by M. Palm, B. P. Bewlay, K. S. Kumar, and K. Yoshimi, (Mater. Res. Soc. Symp. Proc. **1295**, Pittsburgh, PA, 2011).

6. *Titanium and Titanium Alloys*, edited by C. Leyens and M. Peters, (WILEY- VCH, Weinheim, Germany, 2003).

7. F. Appel, J. D. H. Paul and M. Oehring, *Gamma Titanium Aluminide Alloys - Science and Technology*, (WILEY- VCH, Weinheim, Germany, 2011).

8. H. Clemens and S. Mayer, *Adv. Eng. Mater.*, DOI 10.1002/adem.201200231.

9. M. Weimer, B. Bewlay and T. Schubert, paper presented at the "4th International Workshop on Titanium Aluminides", Nuremberg, Germany (September 14-16, 2011).

10. H. Clemens, W. Wallgram, S. Kremmer, V. Güther, A. Otto, and A. Bartels, *Adv. Eng. Mater.* **10**, 707 (2008).

11. F. Appel, M. Oehring, R. Wagner, *Intermetallics* **8**, 1283 (2000).

12. W. Wallgram, T. Schmoelzer, L. Cha, G. Das, V. Güther, and H. Clemens, *Int. J. Mat. Res.* **100**, 1021 (2009).

13. T. Schmoelzer, K.-D. Liss, P. Staron, S. Mayer, and H. Clemens, *Adv. Eng. Mater.* **13**, 685 (2011).

14. T. Tetsui, K. Shindo, S. Kobayashi, M. Takeyama, *Scripta Materialia* **47**, 399 (2002).

15. F. Appel, J. D. H. Paul, M. Oehring, U. Fröbl, and U. Lorenz, *Metall. Mater. Trans. A* **34**, 2149 (2003).

16. N. Saunders, in *Gamma Titanium Aluminides 1999*, edited by Y-W. Kim, D. M. Dimiduk, M. H. Loretto, (TMS, Warrendale, PA, 1999) pp. 183-188.

17. H. F. Chladil, H. Clemens, G. A. Zickler, M. Takeyama, E. Kozeschnik, A. Bartels, R. Gerling, S. Kremmer, L. Yeoh, and K.-D. Liss, *Int. J. Mat. Res.* **98**, 1131 (2007).

18. C. Herzig, T. Przeorski, M. Friesel, F. Hirsker, and S. Divinski, *Intermetallics* **9**, 461 (2001).

19. Y. Mishin, Chr. Herzig, *Acta Materialia* **48**, 589 (2000).

20. R. Kainuma, Y. Fujita, H. Mitsui, I. Ohnuma, and K. Ishida, *Intermetallics* **8**, 855 (2000).

21. Z. Zhang, K. J. Leonard, D. M. Dimiduk, and V. K. Vasudevan, *Structural Intermetallics 2001*, edited by K. H. Hemker et al., (TMS, Warrendale, PA, 2001), p. 515.

22. U. Hecht, V. Witusiewicz, A. Drevermann, and J. Zollinger, *Intermetallics* **16**, 969 (2008).

23. I. J. Watson, K.-D.Liss, H. Clemens, W. Wallgram, T. Schmoelzer, T. C. Hansen, and M. Reid, *Adv. Eng. Mater.* **11**, 932 (2009).

24. M. Takeyama, S. Kobayashi, *Intermetallics* **13**, 989 (2005).

25. F. S. Sun, C. X. Cao, S. E. Kim, Y. T. Lee, M. and G. Yan, *Met. Mat. Trans. A* **32**, 1573 (2001).

26.   D. R. Johnson, H. Inui, S. Muto, Y. Omiya, and T. Yamanaka, *Acta Mat.* **54**, 1077 (2006).
27.   R. M Imayev, V. M. Imayev, M. Oehring, and F. Appel, *Intermetallics* **15**, 451 (2007).
28.   T. Schmoelzer, K.-D. Liss, G. A. Zickler, I. J. Watson, L. M. Droessler, W. Wallgram, T. Buslaps, A. Studer, and H. Clemens, *Intermetallics* **18**, 1544 (2010).
29.   R. Werner, M. Schloffer, E. Schwaighofer, H. Clemens, and S. Mayer, these proceedings.
30.   V. Güther, A. Otto, J. Klose, C. Rothe, H. Clemens, W. Kachler, S. Winter, and S. Kremmer, *Structural Aluminides for Elevated Temperature Applications*, edited by Y-W. Kim, D. Morris, R. Yang, and C. Leyens, (TMS, Warrendale, PA, 2008) pp. 249-256.
31.   J. R. Wood, in [2], pp. 227-232.
32.   M. Achtermann, V. Güther, J. Klose, and H.-P. Nicolei, paper presented at the "4th International Workshop on Titanium Aluminides", Nuremberg, Germany (September 14-16, 2011).
33.   N. Rizzi, presentation at the Symposium "Structural Aluminides for Elevated Temperature Applications", TMS 2008 Annual Meeting, New Orleans, LA, USA (March 9-13, 2008).
34.   K. Liss, T. Schmoelzer, K. Yan, M. Reid, M. Peel, R. Dippenaar, and H. Clemens, *J. Appl. Phys.* **106**, 113526 (2009).
35.   T. Schmoelzer, K.-D. Liss, C. Kirchlechner, S. Mayer, A. Stark, M. Peel, H. Clemens, *Intermetallics*, Submitted for publication (2012).
36.   M. Schloffer, Diploma thesis, Montanuniversität Leoben, Austria (2010).
37.   M. Schloffer, T. Leitner, H. Clemens, S. Mayer, and R. Pippan, *Intermetallics*, in preparation.
38.   L. Cha, T. Schmoelzer, Z. Zhang, S. Mayer, H. Clemens, P. Staron, and G. Dehm, *Adv. Eng. Mater.* **14**, 299 (2012).
39.   L. Cha, H. Clemens and G. Dehm, *Int. J. Mat. Res.* **102**, 703 (2011).
40.   G. Cao, L. Fu, J. Lin, Y. Zhang, and C. Chen, *Intermetallics* **8**, 647 (2000).
41.   F. Appel and R. Wagner, *Mat. Sci. Eng.* **R22**, 187 (1998).
42.   A. Chatterjee, H. Clemens, H. Mecking, G. Dehm, and E. Arzt, *Z. Metallkd.* **92**, 1001 (2001).
43.   A. Gaitzenauer, M. Müller, H. Clemens, R. Hempel, P. Voigt, and S. Mayer, *Berg- und Hüttenmännische Monatshefte (BHM)*, submitted for publication (2012).
44.   P. Simas, T. Schmoelzer, M.L. No, H. Clemens, and J. S. Juan, in [5], pp. 139-144.
45.   J. G. Wang and T. G. Nieh, *Intermetallics* **8**, 737 (2000).
46.   T. Tetsui, *Adv. Eng. Mater.* **3**, 307 (2001).
47.   T. Noda, *Intermetallics* **6**, 709 (1998).
48.   W. Smarsly, H. Baur, G. Glitz, H. Clemens, T. Khan, and M. Thomas, in [1], pp. 25-34.
49.   E. A. Loria, *Intermetallics* **8**, 1339 (2000).
50.   X. Wu, *Intermetallics* **14**, 1114 (2006).
51.   A. Lasalmonie, *Intermetallics* **14**, 1123 (2006).
52.   J. Aguliar, O. Kättlitz, T. Stoyanov, paper presented at the 4th European Conference on Materials and Structures in Aerospace, Hamburg, Germany, (February 7-8, 2012).
53.   R. Martens, *VDI Nachrichten*, June 17, 2011.
54.   G. Das, W. Smarsly, F. Heutling, C. Kunze, D. Helm, paper presented at the "4th International Workshop on Titanium Aluminides", Nuremberg, Germany (September 14-16, 2011).

Mater. Res. Soc. Symp. Proc. Vol. 1516 © 2012 Materials Research Society
DOI: 10.1557/opl.2012.1664

First Investigations on a TNM TiAl Alloy Processed by Spark Plasma Sintering

Thomas Voisin[1], Jean-Philippe Monchoux[1], Helmut Clemens[2] and Alain Couret[1]

[1]CEMES/CNRS, 29 Rue J. Marvig, BP 94347, 31055 Toulouse Cedex 4, France

[2]Department of Physical Metallurgy and Materials Testing, Montanuniversitaet Leoben, A-8700, Leoben, Austria.

ABSTRACT

The processing of a TNM powder (Ti-43.9Al-4Nb-0.95Mo-0.1B, in at.%) by Spark Plasma Sintering (SPS) was investigated for the first time. SPS experiments were performed at varying temperatures. The microstructures of the products were analyzed and interpreted in reference to the available phase diagram. Results of tensile tests conducted at room temperature are also presented.

INTRODUCTION

Among TiAl alloys, TNM alloys, which are containing niobium and molybdenum as alloying elements, attract a great deal of interest because of an excellent hot-workability and balanced mechanical properties [1,2]. The high workability is attributed to the presence of β-phase which provides independent slip systems at elevated temperatures. This is of primary interest for forging, especially when forging of small parts is concerned. However, this β-phase is suspected to be detrimental to the creep resistance at service temperature and to the ductility at room temperature.

The phase diagram and the solidification sequence of TNM alloys have been largely investigated for a wide range of chemical compositions [3,4]. Here, attention will be focused on alloys with the chemical composition of the powder used in the present work: Ti-43.9Al-4Nb-0.95Mo-0.1B (Figure 1). Consistently with the phase diagram, this alloy solidifies via the β phase following the sequence [4]:

$$L \rightarrow L + \beta \rightarrow \beta \rightarrow \beta + \alpha \rightarrow \alpha \rightarrow \alpha + \gamma \rightarrow \alpha + \gamma + \beta \rightarrow \alpha + \alpha_2 + \gamma + \beta_0 \rightarrow \alpha_2 + \gamma + \beta_0$$

The existence of the single α-phase field region was experimentally confirmed in a previous study by heat-treating a sample at 1265 °C for 1h [5]. The addition of Nb and Mo leads to the formation of a four-phase field region corresponding to the ordering of the α phase into α_2 phase [3].

During the last few years, several TiAl alloys were successfully sintered by SPS [6,7]. SPS is a powder metallurgy technique for which the densification is due to the simultaneous application of a pulsed direct current and an uniaxial pressure. Because it involves strong high temperature deformation as in the forging process, SPS is assumed to be adapted for processing TNM alloys. This paper presents a first investigation concerning the consolidation of a TNM powder by SPS. In particular, it will be studied whether the β solidification route allows to adjust refined microstructures by means of sintering.

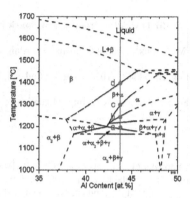

Figure 1. Phase diagram of the Ti-(35-50)Al-4Nb-1Mo alloy [3]. Points a to d situate the dwell temperatures of the SPS densification performed in the present work.

EXPERIMENTAL DETAILS

SPS experiments were conducted on a Ti-43.9Al-4Nb-0.95Mo-0.1B powder produced by argon gas atomization using the Electrode Induction Melting Gas Atomization (EIGA) technique [8]. The diameter of the powder particles is smaller than 180μm. A previous study conducted on identical powder has evidenced a fine dendritic solidification structure and a non-equilibrium phase distribution of 73 vol.% of α/α_2- phase and 27 vol.% of β/β_0-phase [5]. Using this powder, samples were sintered at varying temperatures, from 1200°C to 1400°C. Details about SPS as well as the standard cycle which was used in the present work can be found elsewhere [7]. Cylindrical samples with diameters of 8mm and 36 mm (from now on termed Ø8 and Ø36 samples) were sintered using a constant pressure of 100 MPa and an holding time of 2 min at dwell temperature. All temperatures given in the present paper are those of the samples, which are calculated by adding 25°C for Ø8 samples and 60°C for Ø36 samples to the measured external temperature as reported in Ref [7]. The microstructures were studied in a scanning electron microscope (SEM) using back-scattered electron imaging (BSE). The mechanical properties were measured by means of tensile tests at room temperature. The yield stress is measured at 0.2% of plastic strain.

EXPERIMENTAL RESULTS

Microstructures

Ø8 billets were sintered at varying dwell temperatures to study the evolution of the microstructure with temperature and Figure 2 presents SEM micrographs for four of them. BSE mode was used for phase identification by utilizing the chemical contrast. At 1200°C (Fig. 2a) the microstructure consists of γ grains (black contrast), α_2 grains (grey) and small β_0 grains (bright). In this work, the hexagonal and cubic centred structures are assumed to be disordered at high temperature (α and β) and ordered at ambient temperature (α_2 - DO19 and β_0 - B2). Information on the ordering behaviour of TNM alloys is given in [9].The size of α_2 and γ grains is of the order of a few micrometers. The largest β_0 grains are often situated at boundaries between γ grains, whereas fine β_0 zones are situated at the periphery of α_2 grains. Figure 3 shows the same microstructure at a lower magnification. Obviously, the α_2 grains are not homogeneously distributed. Firstly, they are concentrated at the former particles surface probably because of a higher oxygen amount [10]. Secondly, they are aggregated in small groups, mostly situated at the former β dendrites of the powder which might be attributed to the lower Al content and a segregation of Nb and Mo.

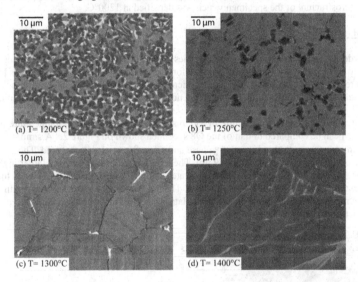

Figure 2. SEM BSE images of the four microstructures formed in Ø8 specimens during SPS.

Figure 2b shows the microstructure after sintering at 1250 °C. It is formed of lamellar grains (α_2 +γ) with an average grain size of 15 µm and small γ grains. A few thin plates of β_0 phase can be detected at the periphery of the γ grains. At 1300°C (Fig. 2c), these γ grains were dissolved, which is in accordance with the phase diagram. However, a very thin film of γ phase (in dark) can be seen along the boundaries of the lamellar colonies. β_0-phase in plate-like shape (in bright) can also be found at the colony boundaries. The average size of the lamellar colonies is 30 µm. It is probable that the small γ grains and the thin films of γ phase evidenced at 1250°C and 1300°C were formed during the cooling of the sample.

At 1400 °C the microstructure is the typical result of a SPS treatment in the vicinity of the $\beta_{transus}$ temperature, followed by the transformation of the high-temperature β-phase into the hexagonal α-phase upon cooling which eventually undergoes a lamellar transformation. At room temperature the microstructure consists of lamellar colonies and fine layers of β-phase at their boundaries.

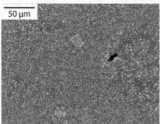

Figure 3. Microstructure of the specimen which was densified at 1200°C.

Mechanical properties

In order to measure the mechanical properties of SPS material, it is necessary to densify Ø36 billets to be able to machine samples suitable for tensile testing and creep experiments [11]. Even if it is relatively easy to adjust comparable microstructures in Ø8 and Ø36 specimens, however, slight modifications mainly due to differences in the cooling rate must be taken into account.

Figure 4a shows the microstructure of a Ø36 sample sintered at a dwell temperature of 1304 °C which can be compared to that of the Ø8 specimen shown on Figure 2c. A similar lamellar microstructure with thin layers of γ-phase and a few β-particles is observed. The average size of lamellar colonies is about 19 μm. The lamellae width is apparently larger in this sample, probably due to a lower effective cooling rate. Figure 4b shows two tensile curves for specimens extracted of this 1304 °C- Ø36 sample. The tensile properties are summarized in Table 1. The material exhibits promising strength data (both yield and ultimate stress) and a moderate ductility.

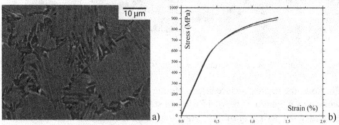

Figure 4. SEM BSE image (a) and tensile stress-strain curves (b) of a Ø36 sample sintered at 1304 °C

Table I. Tensile properties of two samples extracted from the Ø36 billet sintered at 1304 °C.

	YS (MPa)	UTS (MPa)	Plastic fracture strain (%)
Sample 1	763	911	0,75
Sample 2	752	897	0,73

DISCUSSION

When processing a material via spark plasma sintering, the situation is different from melt solidification because the sample is heated from room temperature to the dwell temperature and cooled down. Consequently, the SPS microstructures are probably more similar to those obtained by a thermal treatment. Generally, the initial gas atomized TiAl powder is in a strong non-equilibrium condition. In case of G4-TiAl powder containing tungsten and rhenium [12] and of a high niobium containing powder (TNB) [13], the return to phase equilibrium is achieved at 975°C, whereas the densification occurs between 900°C and 1100°C.

At 1200°C (Fig. 2a and Fig. 3), the observed microstructure results from a SPS treatment in the $\alpha + \gamma + \beta$ phase field region, which, according to the phase diagram (Fig. 1), extends from 1190 °C to 1220°C. The same three phases were observed after a heat treatments of 1h at 1240 °C [1] and at 1230°C [5] but with lower proportion of β_0 and γ-phases. The retention of the non-equilibrium β_0-phase probably explains the large amount of this phase observed in the present study after SPS at 1200°C. However, the volume fraction of the γ-phase is very close to that observed in the alloy with the same chemical composition at 1230°C [5].

At 1250°C, a duplex microstructure, formed by lamellar colonies and globular γ-grains, is evidenced. At 1300°C, a lamellar + β_0 microstructure similar to that observed at 1290°C [5] is observed. The decrease of the volume fraction of the γ-phase is also consistent with this experimental study and with the phase diagram, which indicates that the α-transus temperature is situated between 1250°C and 1295°C. Interestingly, in these two last microstructures some fine plate-like β-phase is observed at the boundaries of the former α grains which possess a fine grain size. This confirms that residual β-phase, lying at grain boundaries, impedes α grain growth [5]. For the case of SPS alloys, the β-phase proportion is higher at 1300°C than at 1250°C, indicating a rapid β precipitation when the temperature reaches the $\alpha \rightarrow \beta + \alpha$ transition temperature. At 1250 °C, it is unclear whether the few β plates result from the starting of this β precipitation or if they result from the retention of the initial non-equilibrium phase. Note that Imayev et al. [2] have shown that at 1100°C the β transformation is activated due to the high contents of Mo and Nb which act as β-stabilizing elements. Anyway, the present study demonstrates that the presence of residual β phase allows limiting the size of the lamellar colonies in alloys processed by SPS. A beneficial role of the 0.2 at. % of B could be also possible, because a recent study has demonstrated a refining effect in a GE alloy sintered by SPS [15].

At 1400 °C the observed microstructure results from a SPS treatment close to the $\beta_{transus}$ temperature. More generally, all the observed microstructures are consistent with the diagram shown in Figure 1, indicating that during SPS a few minutes are enough to activate transformation kinetics.

The first tensile tests confirm a good reproducibility of the mechanical properties of TiAl alloys processed by spark plasma sintering [6]. This is due to the powder metallurgy route, which provides predominantly homogeneous microstructures. The shape of the stress-strain curves, with a continuous and pronounced decrease of the work hardening, is correlated to lamellar or duplex microstructures. In comparison, near γ or $\gamma + \alpha_2$ double-phase microstructures exhibit very low work hardening (see for instance [6,11,13]). The present TNM-SPS alloy exhibits a higher strength and a lower ductility than the G4 alloy [11] and quite similar properties than the best SPS-TNB alloys [13], which possess a double-phase microstructure. Creep tests are in progress to check if the fine lamellar microstructure of the SPS-TNM alloy leads to better high

21

temperature mechanical properties. Even if it is limited, the ductility is nevertheless interesting for a lamellar microstructure, probably in reason of the control of the grain size by retained β_o-phase. The ductility might be improved by increasing the volume fraction of globular γ-phase situated at the colony boundaries.

CONCLUSIONS

The first experiments on a TNM powder demonstrates that this alloy can be easily processed by Spark Plasma Sintering. The microstructures are approximately consistent with the phase diagram. It is shown that the retention and/or the precipitation of β-phase allow refining the lamellar microstructure. Promising mechanical properties have been measured at room temperature. For a better understanding of the microstructure and the related mechanical properties investigations by transmission electron microscopy will be performed in a forthcoming study.

REFERENCES

1. H. Clemens, W. Wallgram, S. Kremmer, V. Güther, A. Otto, and A. Bartels, *Advanced Engineering Materials* **10**, 707 (2008).
2. V.M. Imaev, R.M. Imaev, T.I. Oleneva, and T. G. Khismatullin, *The Physics of Metals and Metallography* **106**, 641 (2008).
3. H. Clemens, B. Boeck, W. Wallgram, T. Schmoelzer, L. M. Droessler, G. A. Zicker, H. Leitner, and A. Otto, *Mater. Res. Soc. Symp. Proc.* **1128**, U03 (2009).
4. T. Schmoelzer, K. D. Liss, G. A. Zicker, I. J. Watson, L. M. Droessler, W. Wallgram, T. Buslap, A. Studer, and H. Clemens, *Intermetallics*, **18**, 1544 (2010).
5. M. Schloffer, F. Iqbal, H. Gabrisch, A. Schwaighofer, F. P. Schimansky, S. Mayer, A. Stark, T. Lippmann, M. Göken, F. Pyczak, and H. Clemens, *Intermetallics*, **22**, 231 (2012).
6. A. Couret, G. Molénat, J. Galy, and M. Thomas, *Intermetallics* **16**, 1134 (2008).
7. T. Voisin, L. Durand, N. Karnatak, S. Le Gallet, M. Thomas, Y. Le Berre, J.F. Castagné, and A. Couret, *Journal of Materials Processing Technology*, **213**,269 (2012).
8. R. Gerling, H. Clemens H, and F.P. Schimansky. *Advanced Engineering Materials*, **6**, 23 (2004).
9. I. J. Watson, K.-D.Liss, H. Clemens, W. Wallgram, T. Schmoelzer, T. C. Hansen, and M. Reid, *Adv. Eng. Mater.* **11**, 932 (2009).
10. W. Lefebvre, A. Loiseau, M. Thomas, and A. Menand, *Phil. Mag. A.*, **11**, 2341(2002).
11. H. Jabbar, J.P. Monchoux, M. Thomas, and A. Couret, *Acta Materialia*, **59**, 7574 (2011).
12. H. Jabbar, A. Couret, L. Durand, and J.P. Monchoux, *Journal of Alloys and Compounds*, **509**: 9826(2011).
13. H. Jabbar, J.P. Monchoux, F. Houdellier, M. Dolle, F.P. Schimansky, F. Pyczak, M. Thomas, and A. Couret, *Intermetallics*, **18**, 2312 (2010).
14 R.M. Imayev, V.M. Imayev, M. Oehring, and F. Appel, *Intermetallics*, **15**, 451 (2007).
15. J. S. Luo, T. Voisin, J.P. Monchoux, and A. Couret, submitted to Intermetallics (2012).

Mater. Res. Soc. Symp. Proc. Vol. 1516 © 2012 Materials Research Society
DOI: 10.1557/opl.2012.1665

Near Conventional Forging of an Advanced TiAl Alloy

Daniel Huber[1], Helmut Clemens[2] and Martin Stockinger[1]
[1] Business Development, Research & Innovation, Böhler Schmiedetechnik GmbH & Co KG,
Mariazellerstrasse 25, A-8605 Kapfenberg, Austria
[2] Department Physical Metallurgy and Materials Testing, Montanuniversität Leoben, A-8700
Leoben, Austria

ABSTRACT

Balanced mechanical properties are needed for TiAl low pressure turbine blades envisaged for use in new generation aircraft engines. However, thermomechanical processing of γ-TiAl based alloys is a challenging task due to a small "processing window". Isothermal forging, as state of the art process for this class of material, results in high productions costs and lower productivity. Due to these facts Bohler Schmiedetechnik GmbH & Co KG has developed a higher efficient "near conventional" thermomechanical processing technology. Lower die temperature and processing at standard atmosphere as well as the use of standard hydraulic presses with higher ram speed result in a highly economical process. Subsequent heat treatment strategies can be used to tailor microstructure and, therefore, mechanical properties according to customer needs. The paper summarizes our effort to establish a near conventional forging route for the fabrication of TiAl components for aerospace industry.

INTRODUCTION

The strong demand for higher efficiency, reduction of fuel consumption, CO_2 and NO_x emissions as well as weight reduction in aircraft engines lead to a substitution of presently used materials by novel light-weight, high-temperature materials like γ-TiAl based alloys. Turbine blades are engine parts that are subjected to high mechanical and thermal loading. Thus alloys are required which provide high strength and good fatigue properties at elevated temperatures as well as a sufficient ductility at room temperature. Presently, Ni-base alloys are state of the art. World-wide fundamental research conducted over the last two decades has clearly shown that balanced material properties can be obtained by hot-working and subsequent heat treatment of TiAl alloys. Due to a small "deformation window" hot-working of TiAl alloys is a complex and challenging task and, therefore, isothermal forming processes are favored so far [1]. In order to expand the process window, a novel Nb and Mo containing γ-TiAl based alloy (TNM™ alloy) was developed [2]. As a result of a high volume fraction of disordered bcc β-phase at elevated temperatures this alloy can be hot-die forged under near conventional conditions, which entails that conventional forging equipment with minor and inexpensive modifications can be used. By means of subsequent heat treatment a significant reduction of β-phase can be achieved and mechanical properties can be tailored to customer requirements. The paper summarizes our effort to establish a near conventional forging route for the fabrication of TiAl components for aerospace industry. The path from lab scale compression tests for material data generation via finite element modeling to industrial scale forging trials and mechanical properties evaluation is shown.

EXPERIMENT

The investigated γ-TiAl alloy had a nominal composition of Ti-43.5Al-4Nb-1Mo-0.1B (in atomic percent). This composition is commonly known as TNM^{TM} alloy. Primary ingots were produced via a double vacuum arc remelting (VAR) process to achieve good chemical homogeneity. These primary ingots were then remelted via VAR skull melting and casted into small billets via spin casting in permanent molds [3]. In order to close residual porosity the billets were hot-isostatically pressed (HIP). Casting diameter varied between 40mm (blade forgings) and 60mm (trial forgings). After HIP the billets were machined to get a smooth surface. The final billet diameters were 35mm and 55mm. To investigate the influence of varying chemistry within specification range two alloys with border chemistry in terms of β-stabilizing elements (Nb, Mo) were produced. These alloys, hereinafter referred to as high-β and low-β alloy, had a chemical composition of Ti-43.5Al-4.3Nb-1.2Mo-0.1B and Ti-44.4Al-3.8Nb-0.9Mo-0.1B, respectively. For phase fraction analysis samples were heat treated in a Carbolite furnace Type RHF 16/15 for two hours followed by oil quenching. Volume phase fractions were analyzed by means of quantitative metallography using optical microscopy. Compression tests for flow stress data generation were performed on a Gleeble® 3800 at different temperatures, varying between 1100 °C and 1350 °C, and strain rates from 0.05s^{-1} to 0.5s^{-1}. Cylindrical samples with a diameter of 10mm and a height of 12mm were taken from cast + HIP billet material. Industrial hot-die forging trials were conducted on a 1000 metric tons hydraulic press with special die heating system and adapted control system to guarantee a constant deformation speed. For these trials billets with 55mm in diameter (pancake forging perpendicular to casting direction) and billets with 35mm in diameter (turbine blade forgings), with special preforming along the casting direction and final forging perpendicular to casting direction, were used. The billets were covered with a special coating to prevent heat loss during forging process. Prior to forging the billets were heated in an electric furnace and held for a defined period of time. The dies were pre-heated to a temperature between 400 °C and 800 °C below billet temperature. Subsequent heat treatments were conducted in an industrial furnace. Mechanical testing was performed at a certified test lab. For creep tests samples were taken from forged and heat treated pancakes. The creep tests were conducted according to ASTM E139.

RESULTS AND DISCUSSION

Lab scale trials

As microstructure and mechanical properties are highly linked, it is important to understand the evolution of microstructure from cast + HIP condition to forged and forged + heat treated condition. To study the influence of thermomechanical treatment on microstructure and mechanical properties several lab scale compression tests as well as industrial trial forgings were conducted. Starting with cast + HIP microstructure (Figure 1a) samples were forged at different ram speeds to a global strain of about 0.8. The influence of ram speed on microstructure evolution is depicted in Figures 1b,c, where Figure 1c shows a higher ram speed. Both microstructures are taken from a position with about 0.8 local strain. The higher ram speed was more than twice as high as the lower ram speed. Detailed process parameters, however, cannot be given due to confidentially reasons. A subsequent homogenization annealing treatment with a special time – temperature profile [4] leads to recrystallization and therefore a fine globular homogeneous microstructure, which is shown in Figure 1d. After homogenization annealing a two-step heat treatment was applied to adjust the final microstructure. The second annealing step is used to define phase fractions of globular β and γ. This high temperature annealing step can be conducted in (α+β+γ) or (α+β) phase field region, depending on the desired microstructure. The amount of phases and their arrangement in the microstructure defined by this heat treatment step highly influences the mechanical properties of the final component. The microstructure depicted in Figure 1e was developed by annealing below the solution temperature of the γ-phase, $T_{\gamma, solv}$, in the (α+β+γ)-phase field region for 1h and subsequent air cooling. The third annealing step is the so-called stabilization treatment which is conducted below the eutectoid temperature and is used to relief residual stresses and to adjust lamellar spacing of the α_2/γ-colonies. Details on the two-step heat treatment and the evolution of microstructure is reported in [2].

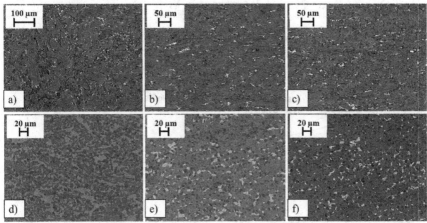

Figure 1. Microstructure after different processing steps. (a) Cast + HIP; (b) As forged with low ram speed; (c) As forged with high ram speed; (d) After forging + homogenization annealing; (e) After forging + homogenization annealing + high temperature annealing; (f) After forging + homogenization annealing + high temperature annealing + stabilization annealing. Note: β-phase shows a white contrast; globular γ is dark and α_2/γ-colonies are grey.

To study the influence of varying chemistry within the specification range, two alloys (high-β and low-β) were investigated. At first, volume phase fraction diagrams were established by quenching experiments and subsequent metallographic examination. To this end, samples were heated to specified temperatures and held for two hours. Subsequent oil quenching was used to conserve microstructure present at annealing temperature [5]. The phase fraction diagrams for high-β and low-β alloy are shown in Figures 2a,b. Additionally, calculated phase fraction diagrams are depicted. These diagrams were calculated using the software package ©MatCalc, as described in [6, 7]. More information on the volume phase fraction diagrams of TNM alloys is given in [2, 5-7]. For finite element simulation of forging processes flow stress curves for both high-β and low-β alloy were measured. Figure 2c shows, as an example, a comparison between high-β and low-β flow stress curves at a strain rate of 0.05s^{-1} and two different temperatures

(1180 °C and 1260 °C). From Figures 2a,b the volume phase fractions for high-β and low-β alloy at 1180 °C and 1260 °C were determined (Table I). The β-phase acts as "lubricant" for better malleability, because it provides good deformation behavior, which is attributed to a sufficient number of slip systems [7, 8]. It is obvious that flow stress values are higher in case of the low-β alloy.

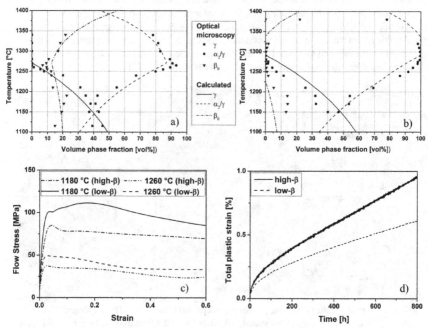

Figure 2. (a) Volume phase fraction diagram of the high-β TNM alloy; (b) Volume phase fraction diagram of the low-β TNM alloy; (c) Flow stress curves of high-β and low-β alloys at two different temperatures and a strain rate of $0.05s^{-1}$; (d) Creep curve at 750 °C and 150 MPa obtained for high-β and low-β TNM alloys.

Table I. Phase fractions for both high-β and low-β TNM alloy at 1180 °C and 1260 °C (data taken from Figures 2a,b).

	Temperature [°C]	γ [vol%]	$α_2/γ$ [vol%]	$β_0$ [vol%]
high-β	1180	34.9	42.5	22.6
low-β		32.0	54.0	14.0
high-β	1260	1.2	89.4	9.4
low-β		4.4	95.0	0.6

Varying chemistry has not only an impact on flow stress and, therefore, hot working behavior, but also on heat treatment strategies. The knowledge of phase transformation temperatures for each heat lot is essential to guarantee a specified microstructure after heat treatment. To show the effect of chemistry on microstructure and mechanical properties, one identical heat treatment was performed on both high-β and low-β alloy. To this end, the forged and homogenization annealed material was subjected to a two-step heat treatment consisting of 1300 °C / 30min / spray cooling + 850 °C / 6 hours / air cooling. As shown in Figures 2a,b 1300 °C is well above $T_{\gamma, solv}$ for both alloys, i.e. the annealing took place in the (α+β) phase field region. Also important is the fact, that at 1300 °C the amount of β-phase in the high-β and low-β alloy was 17.0 vol% and 0.4 vol%, respectively. As β-phase reduces grain coarsening of the α-grains [7], which then transform to α_2/γ colonies, grain coarsening was observed in the low-β alloy showing a mean α_2/γ colony diameter of 180µm, compared to 120µm in high-β alloy. After the final heat treatment step the volume phase fraction for β-phase of the low-β and the high-β alloy variant was about 1 vol% and 20 vol%, respectively. The influence of microstructure on the creep properties is depicted in Figure 2d. From Figure 2d it is evident that the volume fraction of β-phase has an appreciable influence on creep. During creep only a small change of the phase fraction took place. A detailed analysis will be reported in a forthcoming paper. However, our fundamental study on the property-microstructure relationship can be used to identify the optimum heat treatment parameters, which lead to a microstructure with balanced mechanical properties, i.e. high strength at elevated temperature and sufficient elongation to fracture at ambient temperature.

Industrial scale forging trials

After lab scale trials and finite element simulations industrial scale forging trials were performed. For confidentially reasons no specific details on simulation as well as industrial forging parameters can be given. The results of finite element simulation studies for preforming and final forging step are shown in Figure 3a. Cylindrical pre-material of nominal composition was first preformed and then forged to turbine blade geometry on a 1000 metric tons hydraulic press. For the final forging step a so called near conventional hot-die forging process was conducted using an ordinary hydraulic press, see "Experiment" for details.

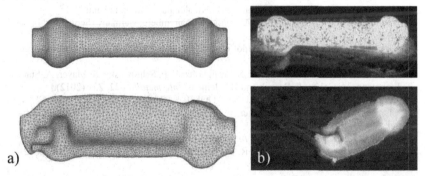

a)   b)

Figure 3. (a) Preform and final forged part as modeled with Deform™3D; (b) Preform at forging temperature and forged turbine blade with flash.

Prior to forging, the preforms were heated in an electric furnace and held for a defined period of time. The heated preform and the final forged turbine blade with flash are shown in Figure 3b. Evidently, the TNM alloy shows an excellent hot-workability. In addition, from comparison of Figure 3a with 3b it is obvious that the flow and die-filling characteristics of the alloy can be predicted.

SUMMARY

Near conventional hot-die forging of advanced γ-TiAl alloys is a complex and challenging task due to a small "processing window". The TNM alloy system is suitable for forging at high deformation rates due to the presence of disordered bcc β-phase at processing temperature. Extensive lab scale trial forgings and material characterization was performed by Bohler Schmiedetechnik GmbH & Co KG and its partners in various research projects. The outcome was the development of a robust hot-die forging process with short process times and high yield rates. Microstructure can be tailored by subsequent heat treatments leading to balanced mechanical properties.

ACKNOWLEDGMENTS

Part of the work presented in this paper was conducted within the framework of the FFG project 826989 "ProStTiAl" Basisprogramm, Austria. We would like to thank our project partners for fruitful cooperation.

REFERENCES

1. S. Kremmer, H. F. Chladil, H. Clemens, A. Otto and V. Güther, in *Ti-2007 Science and Technology*, edited by M. Niinomi, S. Akiyama, M. Hagiwari, M. Ikeda and K. Maruyama (The Japan Institute of Metals, Sendai, 2007), pp. 989–992.
2. W. Wallgram, T. Schmölzer, L. Cha, G. Das, V. Güther, H. Clemens, *Int. J. Mat. Res.* **100**, 1021 (2009).
3. M. Achtermann, V. Güther, J. Klose and H.-P. Nicolei, paper presented at the „4th International Workshop on Titanium Aluminides", Nuremberg, Germany (September 14-16, 2011).
4. H. Clemens, W. Wallgram and M. Schloffer, U.S. Patent No. US 2011/0277891 A1 (17 November 2011).
5. M. Schloffer, F. Iqbal, H. Gabrisch, E. Schwaighofer, F. P. Schimansky, S. Mayer, A. Stark, T. Lippmann, M. Göken, F. Pyczak and H. Clemens, *Intermetallics* **22**, 231 (2012).
6. H. Clemens and S. Mayer, *Adv. Eng. Mater.*, DOI 10.1002/adem.201200231.
7. H. Clemens, W. Wallgram, S. Kremmer, V. Güther, A. Otto and A. Bartels, *Adv. Eng. Mater.* **10**, 707, (2008).
8. F. Appel, H. Kestler and H. Clemens, *Intermetallic compounds – principles and practice, Vol. 3*, edited by J. H. Westbrook and R. L. Fleischer (WILEY-VCH, Chicester, UK, 2002).

Mater. Res. Soc. Symp. Proc. Vol. 1516 © 2012 Materials Research Society
DOI: 10.1557/opl.2012.1561

Optimized Hot-forming of an Intermetallic Multi-phase γ-TiAl Based Alloy

Andrea Gaitzenauer[1], Martin Müller[1], Helmut Clemens[1], Patrick Voigt[2], Robert Hempel[2], Svea Mayer[1]
[1]Department of Physical Metallurgy and Materials Testing, Montanuniversität Leoben, A-8700 Leoben, Austria
[2]Titanium Solutions GmbH, D-28195 Bremen, Germany

ABSTRACT

A robust processing route at low cost is an essential requirement for high-temperature materials used in automotive engines. Because of their excellent high-temperature properties, their low density, high elastic modulus as well as high specific strength, intermetallic γ-TiAl based alloys are potential candidates for application in advanced automotive turbochargers. So-called 3rd generation alloys, such as TNM[TM] alloys with a nominal composition of Ti-43.5Al-4Nb-1Mo-0.1B (in at%), are multi-phase alloys consisting of γ-TiAl, $α_2$-Ti_3Al and a low volume fraction of $β_o$-TiAl phase. In this paper a novel hot-processing route, which is a combination of a one-shot hot-forging step and a controlled cooling treatment, leads to mechanical properties required for turbocharger turbine wheels. The observed strength can be attributed to the small lamellar spacing within the $α_2$/γ colonies of the nearly lamellar microstructure. In order to analyze the microstructure and the prevailing phase fractions microscopic examinations and X-ray diffraction measurements were conducted. The mechanical properties were determined by hardness measurements as well as tensile and creep tests. The evolution of the microstructure during the hot-forming process is described and its relation to the obtained mechanical properties.

INTRODUCTION

Because of the continuously increasing automotive traffic as well as decreasing oil resources the demand for reduced fuel consumption, emissions (soot, CO_2) and noise in motor vehicles has become a world-wide issue. For example, in Europe a strict regulation, such as Euro 6, will be implemented in 2014 aimed for a significant reduction in both fuel consumption and harmful emissions [1]. The automotive industry is meeting this challenge by further downsizing their conventional combustion engines. One strategy is to equip the engines with efficient turbochargers. A turbocharger is a device in which the energy (heat) of the exhaust gas of the engine propels a turbine along with a compressor on the same shaft, such that the inflow air is pressurized by the compressor and supplied to the cylinder of the engine, so that the combustion efficiency of the engine is improved [2]. Additionally, efficiency and engine performance will be enhanced by increasing the combustion gas temperatures up to 850 °C (Diesel engine) and 1050 °C (Otto engine) respectively, while gas pressures and engine revolutions per minute are simultaneously rising. Consequently, the requirements for rotating parts operating at high temperatures are steadily increasing. Thus, new light-weight, high-temperature materials and cost-effective production techniques must be developed and applied. In this context γ-TiAl based alloys exhibit a promising combination of low density (half of Ni base alloys), high stiffness and high strength at temperatures up to 800 °C [2-5]. While the turbocharger turbine inlet temperature is lower for Diesel engines than for Otto engines, the requirements are still severe for TiAl alloys. However, only outer areas of the blades reach the indicated temperature (which are subjected to relatively low rotational stress levels), while other regions of the turbine wheel remain at temperatures that are lowered by more than 200 °C, but are subjected to high rotational stresses (such as the back side of the wheel) [3,6]. That means that at least for Diesel engine application creep problems are not a decisive issue, while oxidation of the blades would be a serious type of damage that might occur in TiAl turbine wheels used in turbochargers [6]. Already in 1999 the first commercial application of γ-TiAl based alloys was announced. Mitsubishi has utilized TiAl turbocharger wheels in their Lancer Evolution VI sports car [2]. The turbine wheels were produced using the so-called LEVICAST process, a modification of the lost wax precision casting method, as well as centrifugal casting. DAIDO in cooperation with universities have developed TiAl alloys which contain a high content of Nb to improve the high-temperature properties and other alloying elements to ensure a good castability [7]. Additionally, the entire joining technology of the TiAl turbine wheel to the steel shaft has been developed [2,8].

Another possibility to produce TiAl turbine wheels is to machine them from the bulk material. In contrast to casting, thinner blade geometries can be achieved. However, for cost-effective processing a pre-shaped starting material is required. To this end, a novel hot-processing route, comprising a combination of a one-shot hot-forging step and a controlled cooling treatment, was developed [9]. The alloy used in this study was a so-called TNM^{TM} alloy which contains a balanced concentration of the β-stabilizers Nb and Mo as well as a small content of B [4,5]. At elevated temperatures this alloy system possesses a large amount of disordered β-phase with bcc lattice which improves hot-workability. In previous studies it was demonstrated that TNM^{TM} alloys can be forged using a conventional forging process [10,11]. In order to analyze the microstructure after hot-forging and creep testing scanning and transmission electron microscopy (SEM, TEM) were conducted. To evaluate the mechanical properties tensile and creep tests were carried out.

EXPERIMENTAL DETAILS

The investigated TNM^{TM} alloy has a nominal composition of Ti-43.5Al-4Nb-1Mo-0.1B (in at.%). More information regarding this β-solidifying TiAl alloy, including phase diagram, hot-forming characteristics, properties, and applications, is given in [4,5,12]. First, small ingots ("slugs") were produced by a production technique which combines vacuum arc skull melting and centrifugal casting in permanent moulds [4]. The advantage of this process is the possibility to produce adjusted feed stock materials for hot-working as well as investment casting directly from a large ingot with a remarkably improved materials yield. The slugs exhibit a satisfying microstructural homogeneity as well as a homogeneous alloying element distribution. In order to close residual casting porosity the slugs were subsequent hot-isostatically pressed (HIPed) at 1200 °C for 4 hours at a pressure of 200 MPa. After HIPing the slugs were cut in dimensions suitable for the following hot-forging process. Forging was conducted in the upper temperature range of the (α + β) phase field region, where the volume fraction of the β-phase is considerably high. The one-shot forging process was conducted on an industrial forging facility at a deformation rate which is significantly higher than normally used for hot-die forging [10] and isothermal forging [13] of TiAl alloys. Due to confidentiality reasons, however, no specific value regarding the deformation rate can be given. Subsequently, the forging was controlled cooled to room temperature. A representative TNM™ forging is shown in Figure 1a. Since the dimension of this part is too small for providing material for the preparation of tensile and creep test specimens, additional forging trials were conducted. Here, starting material was forged to pancakes with a sufficient dimension to take out specimens for mechanical testing. In order to achieve a microstructure comparable to that in the forging shown in Figure 1a the forging ratio and the cooling procedure were adapted correspondingly. For SEM investigations the specimens were metallographically prepared according to [14], whereas the procedure for TEM specimens is reported in [15]. For SEM and TEM examinations an EVO 50 from Zeiss and a Philips CM 12 (120 kV) were used, respectively. In order to determine the prevailing volume fractions of the constituting phases X-ray diffraction (XRD) measurements were conducted on a Bruker AXS D8 Advance Diffractometer in Bragg-Brentano arrangement using a Cu-$K_{\alpha 1}$ radiation. For Rietveld analysis the commercial software package DIFFRAC^{plus} TOPAS was used. Vickers (HV 10) macro-hardness measurements were performed on an Emcotest M4C 025 G3M. Uniaxial tensile and creep tests were conducted according to ASTM-E8 and ASTM-E139.

RESULTS AND DISCUSSION

The microstructure after the one-shot hot-forging process and subsequent controlled cooling to room temperature is shown in Figure 1b. The obtained hardness is indicated in the image. The microstructure consists mainly of lamellar α_2/γ-colonies. The ordered β_o-phase is located along the boundaries of the colonies. In the following this type of microstructure is termed "nearly lamellar β" (NL + β). In some β_o-grains the presence of small lens-shaped γ-particles can be seen. These particles have their origin from a $\beta/\beta_o \rightarrow \gamma_{Pr}$ precipitation reaction during the cooling sequence. From Rietveld analysis of the corresponding XRD pattern the following volume phase fractions were determined: γ 57.9 %, α_2 37.3 % and β_o 4.8 %. The evolution of the microstructure and the transformation of the involved phases during cooling from forging temperature can be summarized as $\alpha + \beta \rightarrow (\alpha + \gamma)_{colonies} + \beta + \gamma_{Pr} \rightarrow (\alpha + \gamma)_{colonies} + \beta_o + \gamma_{Pr} \rightarrow (\alpha_2 + \gamma)_{colonies} + \beta_o + \gamma_{Pr}$. The ordering temperature of the β_o- and α_2-phase as well as the formation temperature of the γ-phase in the TNM™ alloy was investigated by means of in-situ neutron diffraction experiments. The results are reported in [16]. The lamellar spacing within the colonies is controlled by the cooling rate. Figure 1c shows a TEM image of a lamellar α_2/γ-colony. The mean interface spacing, including α_2/γ- and γ/γ-interfaces, is about 40 nm.

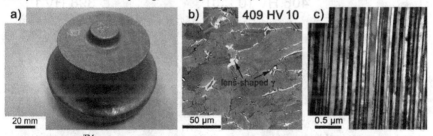

Figure 1. a) TNM™ part which was made in a single-shot hot-forging process; b) Microstructure of the forged material. The β_o-phase is situated along the lamellar α_2/γ-colonies. Because of the high local concentration of Nb and Mo in the β_o-phase it shows the brightest contrast. SEM image taken in back-scattered electron mode; c) TEM image of a lamellar α_2/γ colony taken in [2-1-10]α_2 direction.

The tensile strength is controlled by the mean interface spacing within the lamellar colonies as well as the volume fraction of β_o-phase and, if present, γ-phase [11,17]. The observed behavior of the yield strength can be explained by a modified Hall-Petch relationship where the dominant structural parameter is the mean interface spacing. The stress required to propagate dislocations through the interfaces is inversely proportional to the square root of the mean interface spacing [13,18]. Here, the α_2/γ and γ/γ-interfaces act as glide obstacles and lead to dislocation pile-ups as observed by TEM investigations [13,19,20]. Figures 2a,b show the dependence of the specific elastic modulus and the specific tensile properties as a function of test temperature. For comparison the data of a Ni base alloy (IN 713C), conventionally used for turbine wheels, are indicated [6].

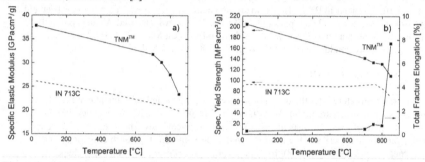

Figure 2. a) Specific elastic modulus, b) Specific yield strength and total fracture elongation as function of temperature of a TNM^{TM} alloy with NL + β microstructure (see Figure 3a). For comparison the corresponding values for IN 713C are shown [21].

The microstructure of the "pancake-material" which were used for tensile and creep tests is shown in Figure 3a. From comparison with Figure 1b it is evident that the material used for the determination of the mechanical properties is almost identical. The tensile properties of the TNM^{TM} alloy depicted in Figure 2 are typical for nearly lamellar TiAl alloys with a high content of β-stabilizing alloying elements (see [5] and references therein). Due to the small average

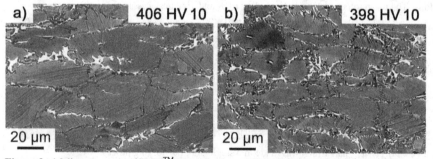

Figure 3. a) Microstructure of TNM^{TM} "pancake material" used for mechanical testing. Note that the microstructure is comparable to that of the forged part shown in Figure 1b; b) Microstructure after a 300 hours creep test at 750 °C and 200 MPa.

lamellar spacing the specific yield strength at room temperature is about 205 MPa and decreases to about 110 MPa at 850 °C. Below the brittle-to-ductile transition temperature the total fracture elongation shows a small value, but increases for test temperatures higher than 800 °C as observed also for other TiAl alloys [4,5,13,19]. First results of creep tests on TNM^{TM} material with a NL + β microstructure are summarized in Table I. From Table I the stress exponent and the apparent activation energy for creep can roughly be estimated, leading to a stress exponent of about 3 and an activation energy of 370 kJ/mol. The obtained stress exponent indicates that the dominating creep mechanism is a diffusion-assisted climb process of dislocations, which is commonly found in γ-TiAl based alloys with lamellar microstructures [13]. The determined activation energy agrees well with the results derived from internal friction measurements on TNM^{TM} specimens with a nearly lamellar microstructure [22] and those reported for other β-containing TiAl alloys [23]. Finally, Figure 3b shows the microstructure of a creep specimen which was tested at 750 °C and 200 MPa. After 300 hours and a creep strain of 3.7 % the test was terminated. From comparison of Figure 3b with 3a it is obvious that no significant change of the microstructure took place during the creep test. Also phase fraction analysis by means of XRD and Rietveld evaluation indicated no significant difference between starting material and crept samples.

Table I. Results of creep tests conducted on TNM^{TM} material with NL + β microstructure (see Figure 3a).

Temperature [°C]	Load [MPa]	Time [h]	1 % Creep strain [h]	Creep strain after 300 h [%]	Creep rate x 10^{-8} [s^{-1}]
700	200	300	---	0.8	0.28
725	200	300	196.6	1.3	0.82
750	200	300	48.8	3.7	2.6
750	140	300	152.8	1.5	0.96
750	235	300	25.0	8.6	4.59
750	250	300	18.9	18.0	6.24

SUMMARY

For the intermetallic TNM^{TM} alloy with a nominal composition of Ti-43.5Al-4Nb-1Mo-0.1B a simple and cost-effective hot-working route was developed for the manufacture of forgings which can further processed to turbocharger turbine wheels. The hot-working process is a combination of a one-shot hot-forging step and a controlled cooling treatment, leading to a nearly lamellar microstructure exhibiting a small lamellar spacing. The obtained mechanical properties should fulfill the requirements demanded by the manufacturers of turbocharger devices used in Diesel engines.

ACKNOWLEDGMENTS

A part of this research was conducted in the framework of the FFG project 832040 "energy-drive", Research Studios Austria. The authors thank Boryana Rashkova for TEM investigations.

REFERENCES

1. Reduction of pollutant emissions from light vehicles, http://europa.eu/legislation_summaries/environment/air_pollution/l28186_en.htm.
2. T. Tetsui, *Advanced Engineering Materials* **3**, 307 (2001).
3. H. Baur, D.B. Wortberg and H. Clemens, in *Gamma Titanium Aluminides 2003*, edited by Y.-W. Kim, D. M. Dimiduk and M. H. Loretto, (TMS, Warrendale, PA, 2003), pp. 23–31.
4. H. Clemens, M. Schloffer, E. Schwaighofer, R. Werner, A. Gaitzenauer, B. Rashkova, T. Schmoelzer, R. Pippan, and S. Mayer, these proceedings.
5. H. Clemens and S. Mayer, *Advanced Engineering Materials*, DOI: 10.10002/adem.201200231.
6. T. Tetsui, in *Gamma Titanium Aluminides 1999*, edited by Y.-W. Kim, D. M. Dimiduk, M. H. Loretto, (TMS, Warrendale, PA, 1999), pp. 15–23.
7. T. Noda, Intermetallics **6**, 709 (1998).
8. T. Tetsui, Intermetallics **9**, 253 (2001).
9. A. Gaitzenauer, M. Müller, H. Clemens, P. Voigt, R. Hempel, and S. Mayer, *Berg- und Hüttenmännische Monatshefte* **8-9**, 319 (2012).
10. S. Kremmer, H. Chladil, H. Clemens, A. Otto, and V. Güther, in *Ti-2007 Science and Technology*, (JIM, Sendai, Japan, 2008) pp. 989–992.
11. W. Wallgram, T. Schmoelzer, L. Cha, G. Das, V. Güther, and H. Clemens, *International Journal of Materials Research* **100**, 1021 (2009).
12. H. Clemens, W. Wallgram, S. Kremmer, V. Güther, A. Otto, and A. Bartels, *Advanced Engineering Materials* **10**, 707 (2008).
13. F. Appel, J.D.H. Paul and M. Oehring, *"Gamma Titanium Aluminides Alloys- Science and Technology"*, (WILEY-VCH Verlag, Weinheim, 2011).
14. M. Schloffer, T. Schmoelzer, S. Mayer, E. Schwaighofer, G. Hawranek, F.-P. Schimansky, F. Pyczak, and H. Clemens, *Practical Metallography* **48**, 594 (2011).
15. L. Cha, H. Clemens and G. Dehm, *International Journal of Materials Research* **102**, 703 (2011).
16. I.J. Watson, K.-D. Liss, H. Clemens, W. Wallgram, T. Schmoelzer, T. C. Hansen, and M. Reid, *Advanced Engineering Materials* **11**, 932 (2009).
17. M. Schloffer, T. Leitner, H. Clemens, S. Mayer, and R. Pippan, *Intermetallics*, in preparation.
18. G. Cao, L. Fu, J. Lin, Y. Zhang, and C. Chen, *Intermetallics* **8**, 647 (2000).
19. F. Appel and R. Wagner, *Materials Science and Engineering: R: Reports* **22**, 187 (1998).
20. A. Chatterjee, H. Clemens, H. Mecking, G. Dehm, and E. Arzt, *International Journal of Materials Research (formerly Zeitschrift für Metallkunde)* **92**, 1000 (2001).
21. Engineering Properties of Alloy 713C http://www.nickelinstitute.org/~/media/Files/TechnicalLiterature/Alloy713C_337_.ashx
22. P. Simas, T. Schmoelzer, M.L. Nó, H. Clemens, and J. San Juan, in *Materials Research Society Symposium Proceedings 2011*, edited by M. Palm, B. P. Bewlay, K. S. Kumar, and K. Yoshimi (MRS, Warrendale, PA, 2011), pp. 139–144.
23. J.G. Wang and T.G. Nieh, *Intermetallics* **8**, 737 (2000).

Mater. Res. Soc. Symp. Proc. Vol. 1516 © 2012 Materials Research Society
DOI: 10.1557/opl.2012.1575

TiAlNb-alloy with a modulated B19 containing constituent produced by powder metallurgy

Heike Gabrisch, Uwe Lorenz, Michael Oehring, Jonathan Paul, Florian Pyczak, Marcus Rackel, Frank-Peter Schimansky, Andreas Stark

Helmholtz-Zentrum Geesthacht, Max-Planck Str.1, 21502 Geesthacht, Germany

ABSTRACT

Intermetallic TiAl alloys are of interest to the aero engine industry because of their light weight, corrosion resistance and excellent high temperature strength. This justifies the continued effort to improve properties and processing of these alloys.

A critical parameter that limits the practical implementation of Ti aluminides is their low ductility at room temperature. Recently, a new class of TiAl alloys based on a modulated lath structure has been introduced that exhibit an excellent combination of ductility and strength. A key component in this alloy is the orthorhombic phase B19 that is attributed to alloying with high amounts of niobium. The driving forces and mechanisms that lead to the observed modulated structures involving the B19 phase are not fully understood yet. As a first step to a better understanding we present a study of the thermal stability range of the phases involved.

INTRODUCTION

Structural applications in aero-engines require good high-temperature strength in combination with low density. These demands are met by intermetallic TiAl alloys that are considered as replacement for Ti alloys and Ni base superalloys in aero-engines and in the automotive industry [1-3].

The microstructure of TiAl alloys varies with chemical composition and processing parameters [4]. While the microstructure can be adjusted for optimized strength, a general drawback of all TiAl alloys is their low ductility at room temperature [2]. The best room temperature ductility is provided by near γ or duplex microstructures, whereas lamellar $\alpha_2 + \gamma$ microstructures have better high-temperature strength [5]. For practical applications alloys are needed that possess sufficient room temperature ductility as well as excellent strength at elevated temperatures.

Such a promising combination of mechanical strength and ductility has been reported for a recently developed TiAl alloy with high amounts of Nb addition [6]. Its microstructure consists of lamellar $(\alpha_2 + \gamma)$ colonies and pearlite-like regions of $B19/\beta_0$ and γ laths. The typical feature of this alloy is the closely spaced co-existence of β_0 and B19 phases within $B19/\beta_0$ laths. In the TEM this structure is recognized from contrast modulations within the laths that are caused by the elastic strain between the co-existing β_0 and B19 phases. While these nano-laminates are thought to improve the mechanical properties of the alloy, their origin is not known yet. First principles calculations indicate that the B19 phase may form through decomposition of β/β_0 phase [7]. On the other hand, according to experimental findings, B19 is reported to form in α_2 phase after fast cooling from the α regime [8].

The present study is motivated by the expected improvement of the mechanical properties resulting from incorporation of the orthorhombic B19 phase into a lamellar γ-TiAl alloy. Below we explore the feasibility of a powder metallurgical processing route for this type of nano-laminated microstructure and its evolution during different heat treatments.

EXPERIMENT

Ti-42Al-8.5Nb was produced in house following a powder metallurgical (PM) route starting from pure elements that are molten into buttons by plasma arc melting [9]. The buttons were re-molten four times to ensure chemical homogeneity and subsequently cast into a rod. Powder was produced from rods by the EIGA (Electrode Induction Melting Gas Atomization) technique and powder particles with diameters 45- 180 μm were compacted by HIPing at 1250 °C under 200 MPa in Ar for 2 h. From the HIPed compacts specimens were prepared for heat treatments.

Table 1: Heat treatments performed after HIPing

specimen	temperature	annealing time	cooling
A1	1250°C	2h	air cool
A2	A1 + 1030°C	2h	furnace cool
B, C, D	600°C	30min, 1h, 2 h	oil quench
E (in-situ)	20-1000°C	heating rate 5K/min	air cool

Heat treatments were conducted between 600 and 1250°C as listed in table 1. The heat treatment A2 corresponds to the two step heat treatment of the reference alloy [6]. Specimens for observations by TEM and SEM were prepared from the annealed samples. In-situ heating high-energy x-ray diffraction (HEXRD) experiments were performed in the HZG synchrotron beam line HEMS at the Deutsches Elektronen-Synchrotron (DESY).

The annealed specimens were investigated by HEXRD to monitor the occurrence of the B19 phase. Samples were measured in transmission mode using a beam cross section of 0.2 mm x 0.2 mm (rotation angle of ± 30°). High-energy x-rays having a photon energy of 87.1 keV (λ = 0.14235 Å) were used in the investigation. The resulting Debye-Scherrer diffraction rings were recorded on a Mar345 image plate detector with an exposure time of 10-40 seconds. Conventional diffraction patterns were generated by an azimuthal integration of the Debye-Scherrer rings using the fit2D software package [10]. Lattice parameters and phase fractions were determined by Rietveld analysis using PowderCell version 2.3 [11].

RESULTS AND DISCUSSION

High temperature anneal
Fig. 1 illustrates the development of the alloy's microstructure from the HIPed state over the intermediate state after anneal at 1250°C to the final state after heat treatment A2. After HIPing a near lamellar microstructure is observed with small amounts of β_o, α_2 and γ phase at the grain boundaries (Fig. 1a). After annealing at 1250°C followed by air cool the microstructure consists of α_2 grains with a minor phase at the grain boundaries (Fig. 1b). The crystal structure of

Fig. 1: Evolution of the alloy's microstructure during high temperature anneal. **a.)** HIPed state, **b.)** after heat treatment A1 (1250°C,2h, air cool), **c.)** after two step heat treatment A2 (A1+1030°C,2,furnace cool).

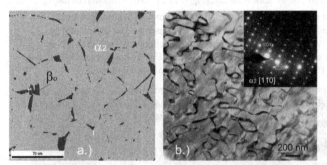

Fig. 2 : Microstructure after heat treatment A1. **a.)** EBSD image illustrating the distribution of α_2 (grey) β_0 phases (black). **b.)** TEM image showing thermal APBs within one α_2 grain.

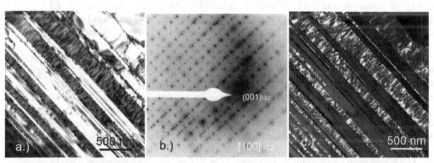

Fig. 3: a.) TEM image after heat treatment A2 showing contrast modulations in a lamellar colony. **b.)** Diffraction pattern of a., c. where reflection used for imaging in c. is marked. **c.)** Dark field image illustrating that contrast modulations are observed in α_2 phase.

37

the minor constituent was determined by EBSD as β_o phase as is illustrated in Fig. 2a. TEM imaging of this alloy in Fig. 2b shows that the α_2 grains are pervaded by networks of anti phase boundaries. These are formed during fast cooling from the $(\alpha + \gamma)$ phase field when the ordering of Ti and Al atoms takes place during the $\alpha \rightarrow \alpha_2$ transition (thermal APBs) [12]. The fast cooling to RT also suppresses the formation of γ phase and leads to a super-saturation of α_2 phase with Al. During the subsequent heat treatment at 1030°C fine γ lamellae form within the supersaturated α_2 grains as can be seen in Fig. 1c. Imaging of this microstructure in the TEM in Fig. 3a shows that a modulated contrast is observed within the lamellar colonies, similar to that described in [6] and unlike the appearance of lamellar colonies in other TiAlNb alloys, e.g. Ti-45Al-7.5Nb [13]. This contrast modulation is considered the indicator for the presence of multiple phases. The corresponding diffraction pattern in Fig. 3b is the superposition of contributions from α_2 in [100] orientation and two variants of γ in [110] orientation (Blackburn orientation relationship [14]). In Fig. 3c dark field imaging with a ($\bar{1}$10) reflection of the α_2 phase shows that the contrast modulation is restricted to the α_2 lamellae. It should be noted that the diffraction patterns of the B19 phase and α_2 are identical in this zone axis orientation (α_2 (100)// B19 (010); (α_2 [001])// B19 [100], [6]). Therefore a distinction between the two phases by electron diffraction is not straightforward. The bright phase in Fig. 3c may therefore be either α_2 or B19 or a coexistence of both (which would agree with the contrast modulation).

Fig. 4 : High resolution image taken from a modulated lamella after heat treatment A2. **a.)** Regular patterns of bright atom columns at the left and right side of the image. **b.)** Image simulation comparing the appearance of corresponding orientations of the B19 and α_2 phase.

In Fig. 4a the high resolution image of a modulated lamella shows that some atom columns have a distinct bright contrast and form a regular pattern. A magnification of this pattern is shown in the inset at the bottom. Such an arrangement is not usually observed in high resolution images of α_2 phase in [100] orientation, see for example [15]. In Fig. 4b a simulated high resolution image of the B19 phase in [010] orientation is shown that reproduces this regular pattern (a magnification of this pattern is shown in the small inset at the bottom). For comparison a simulated image of α_2 in [100] orientation is shown in the large inset where this pattern is also present, but very faint. These observations suggest that the areas where the regular pattern is formed by bright atom columns might be identified as B19 phase. The simulations were however performed with unit cells containing Ti and Al atoms only. The outcome is different when Nb atoms are considered that preferentially replace Ti atoms [16, 17]. A simulation of α_2 phase with Nb atoms on Ti sites reproduces the strong contrast seen in the image simulation of the B19 phase. This indicates that the presence of Nb obscures the distinction between B19 and α_2, at least in some zone axis orientations. In the present case investigation by HEXRD could not confirm presence of B19 phase in this heat treated state. Regarding uncertainties in the detection limit of the HEXRD experiments and in the interpretation of the high resolution images (Fig. 4) we cannot conclude on the presence or absence of the B19 phase after heat treatment A2. The strain that leads to contrast modulations in Fig. 3 may result from chemical inhomogeneities or lattice mismatch of co-existing phases.

HEXRD studies

The high brilliance of the incident beam in HEXRD experiments enables the distinction between the closely related hexagonal α_2 and the orthorhombic B19 phases in the synchrotron. Representative results of the HEXRD experiments are shown in Fig. 5 where reference peak positions of the phases are listed below the diffractograms. The close relationship between the α_2 and the B19 phase results in largely superimposing peak positions. A distinct difference is the occurrence of two diffraction peaks in B19 at the position of the single (200) and (201) peaks of the α_2 phase. This peak splitting originates from the orthorhombic lattice of B19 where atom positions in the close packed B19 (100) planes are shifted with respect to those in to the corresponding α_2 (001) planes [18].

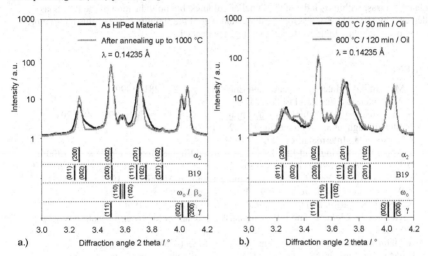

Fig. 5 : Results of HEXRD experiments. **a.)** Comparing the microstructure before and after heating the HIPed material to 1000°C. **b.)** Formation of B19 phase during annealing at 600°C.

In Fig. 5a the results of the in-situ heating experiment E of the HIPed alloy are shown. The darker curve of the HIPed specimen exhibits small shoulders near the α_2 (200) and α_2 (201) peaks that stem from a small amount of B19 phase. After heating to 1000°C these shoulders have disappeared, indicating that the B19 phase is no longer present (or is present in amounts below the detection limit). The doublet peak at 2 theta = 3.58° corresponds to ω_0 phase that convert to

β_0 phase during annealing [19]. Analysis of the in-situ heating experiment showed that the B19 phase is stable up to approximately 695°C and has a peak volume fraction around 600- 650°C.

Based on these results low temperature heat treatments at 600°C were performed as listed in table 1. Investigation of the alloys after these heat treatments showed that the highest amount of B19 phase (in the range of 30 vol.-%) is present after annealing at 600°C for 1 and 2 hours. The HEXRD spectra in Fig. 5b give a comparison of the microstructure after annealing at 600°C for 30 min and 2 hours. It can be seen that the small shoulders present after 30 min (darker curve) grow during the two hours anneal at 600°C corresponding to an increased amount of B19 phase.

CONCLUSIONS

TiAl alloys with a modulated microstructure have been produced by powder metallurgical methods and subsequent heat treatments. The optimum temperature for the formation of the B19 phase was found to be 600°C. At temperatures above 700°C this phase starts to dissolve which limits the beneficial effect of the B19 phase for practical applications to the temperature regime below 700°C. Alloys that were annealed at 1030°C exhibited a modulated contrast within α_2 laths of the lamellar colonies, but no indication of the B19 phase was detected by HEXRD. For a complete understanding of the stability range of this phase more investigations, including high resolution imaging, are needed.

REFERENCES

1. D.M. Dimiduk, Materials Science and Engineering A 263 281 (1999).
2. X. Wu, Intermetallics 14 1114 (2006).
3. G. Electric, http://www.flightglobal.com/news/articles/power-house-207148/ (2012).
4. C. McCullough, et al., Acta metall. 37 1321 (1989).
5. Z.C. Liu, et al., Intermetallics 10 653 (2002).
6. F. Appel, M. Oehring, and J.D.H. Paul, Advanced Engineering Materials 8 371 (2006).
7. D. Nguyen-Manh and D.G. Pettifor, eds. *Origin of Phase and Pseudo-Twinning in Ti-Al-Nb Alloys: a First-Principles Study.* Gamma Titanium Aluminides 1999, ed. D.M.D. Y.W. Kim, and M.H. Loretto. 1999, The Minerals, Metals & Materials Society.
8. E. Abe, T. Kumagai, and M. Nakamura, Intermetallics 4 327 (1996).
9. R. Gerling, H. Clemens, and F.P. Schimansky, Advanced Engineering Materials 6 23 (2004).
10. A. Hammersley, http://www.esrf.eu/computing/scientific/FIT2D/index.html
11. B.f.M.u. -prüfung, http://www.bam.de/de/service/publikationen/powder_cell.htm
12. S.A. Jones, et al., Scripta Metallurgica 22 1235 (1988).
13. L. Cha, et al., Intermetallics 16 868 (2008).
14. M.J. Blackburn, Science, Technology, and Application of Titanium 633 (1970).
15. S.R. Singh and J.M. Howe, Philosophical Magazine A 66 739 (1992).
16. Y.L. Hao, et al., Acta Materialia 47 1129 (1999).
17. D.G. Konitzer, I.P. Jones, and F. H.L, Scripta Metallurgical 20 265 (1986).
18. T. Schmoelzer, et al., Advanced Engineering Materials 14 445 (2012).
19. A. Stark, et al., Advanced Engineering Materials 13 700 (2011).

Mater. Res. Soc. Symp. Proc. Vol. 1516 © 2012 Materials Research Society
DOI: 10.1557/opl.2012.1576

Relaxation Processes at High Temperature in TiAl-Nb-Mo Intermetallics

Pablo Simas[1], Thomas Schmoelzer[2], Svea Mayer[2], Maria L. Nó[3], Helmut Clemens[2] and Jose San Juan[1]

[1]Física Materia Condensada, Facultad de Ciencia y Tecnología, Universidad del País Vasco, Apdo. 644, 48080 Bilbao, Spain.
[2]Department of Physical Metallurgy and Materials Testing, Montanuniversitaet Leoben, Franz-Josef-Str. 18, A-8700 Leoben, Austria.
[3]Física Aplicada II, Facultad de Ciencia y Tecnología, Universidad del País Vasco, Apdo. 644, 48080 Bilbao, Spain.

ABSTRACT

In the last decades there was a growing interest in developing new light-weight intermetallic alloys, which are able to substitute the heavy superalloys at a certain temperature range. At present a new Ti-Al-Nb-Mo family, called TNM™ alloys, is being optimized to fulfill the challenging requirements.

The aim of the present work was to study the microscopic mechanisms of defect mobility at high temperature in TNM alloys in order to contribute to the understanding of their influence on the mechanical properties and hence to promote the further optimization of these alloys.

Mechanical spectroscopy has been used to study the internal friction and the dynamic modulus up to 1460 K of a TNM alloy under different thermal treatments. These measurements allow to follow the microstructural evolution during in-situ thermal treatments. A relaxation process has been observed at about 1050 K and was characterized as a function of temperature and frequency in order to obtain the activation parameters of the responsible mechanism. In particular, the activation enthalpy has been determined to be H= 3 eV. The results are discussed and an atomic mechanism is proposed to explain the observed relaxation process.

INTRODUCTION

Among the different families of intermetallics, γ-TiAl based alloys were soon envisioned as good candidates for advanced automotive and jet engines [1-3]. In the eighties of the last century intensive work on the fundamental aspects of the first generation of binary TiAl alloys was performed, followed by the development of a second generation of more complex ternary alloys Ti-(45-48)Al-(1-3)Cr-(2-5)Nb,Ta,Mo (at%). During the last 15 years a third generation with a high niobium content Ti-45Al-(5-10)Nb was developed to improve both room temperature ductility and high temperature creep resistance [4,5]. Parallel, basic thermal treatment parameters were defined to adjust the microstructure from fully lamellar to duplex and near-gamma [5]. In spite of all this effort, they were not able to fulfill an important requirement for an industrial scale development, namely hot-working at reasonable costs. The challenge was to design an alloy fulfilling that particular requirement, leading to the development of a new type of TiAl alloy which contains both Nb and Mo in well balanced quantities (a so-called TNM alloy). The basic strategy was to develop a β-stabilized γ-TiAl based alloy, which allows a near conventional processing and the adjustment of a creep-resistant microstructure suitable for long-term service up to 750° C [6]. To achieve this goal a combination of Nb, which e.g. slows down diffusion processes [7,8], with a stronger β-stabilizer such as Mo [9,10] and a small amount of boron as grain refiner [11,12] was precisely adjusted. The phase and phase fraction diagrams [6,13,14], the α-α$_2$ and β-β$_0$ ordering reactions [6,15] for such compositions as well as the thermal treatments and decomposition kinetics were studied [16, 17]. However, optimizing this new designed TNM alloy requires a precise control of the microstructure and a deep comprehension of the microstructural phenomena contributing not only to the precipitation processes taking place during processing and thermal treatments, but also an understanding of the atomic mechanisms susceptible of being activated during further high-temperature service conditions.

So, the aim of the present work was to study the processes involving the mobility of atomic defects (point defects, dislocations etc.) taking place during the thermal treatments carried out to adjust the microstructure in a very large temperature range up to 1460 K. To study the mobility of the defects up to this high temperature we have applied an innovative approach, like the measurement of the internal friction and dynamic modulus variation by using mechanical spectroscopy due to the high sensitivity of this technique to atomic relaxation processes [18-20].

RESULTS AND DISCUSSION

The investigated TNM alloy with a nominal composition of Ti-43Al-4Nb-1Mo-0.1B (at. %) was produced by vacuum arc remelting (VAR) in order to obtain a good chemical homogeneity and less than 1000 ppm of interstitial impurities (O, C, N, H) [21]. The ingot of 200 mm diameter was extruded in a steel can down to 55 mm diameter, being further thermally treated at 980°C for 4 hours in Ar atmosphere for stress relaxations [22]. Then testing samples were annealed for 1 hour in the (α+β+γ) phase field region at 1230°C, with subsequent air-cooling (sample named TNM1230). However, due to the relatively fast air-cooling the resulting microstructure consists of supersaturated α$_2$ grains as well as β$_0$ and γ phase and is far from thermodynamic equilibrium. The second heat-treatment, which in fact is an ageing treatment to precipitate fine γ-laths from the supersaturated α$_2$ phase, was conducted at two different temperatures below the eutectoid temperature, like 950°C (sample named TNM950) and 850°C (sample named TNM850), for 6 hours followed by furnace cooling. In the present work we studied these three specimens by means of mechanical spectroscopy. In case of the TNM1230 sample the maximum temperature was 1460 K, whereas the two aged sample variants were measured up to the temperature corresponding to their second heat-treatment in order to avoid any microstructural changes during testing and thus to obtain reproducible results. Internal friction and dynamical modulus variation measurements have been carried out in a high-temperature sub-resonant torsion pendulum [23, 24] between 600 K and 1460 K at a heating rate of 1.5 K/min, under high vacuum (5×10^{-6} mbar), with a maximum oscillating stress amplitude of $\varepsilon = 10^{-5}$, and at different frequencies from 0.003 Hz to 3 Hz.

In order to study the initial state just after high temperature annealing and fast cooling, the internal friction (IF) and dynamic modulus variation (DMV) have been measured in sample TNM1230 up to 1460 K and the results are shown in Figure 1a. A strong IF peak is observed at about 1050 K and a fast growing high temperature background (HTB) extends up to 1460 K. At the position of the IF peak the dynamic modulus undergoes an initial decrease with subsequent hardening before going down in the temperature range of the HTB. As it was expected, this microstructure would not be stable, because of the presence of supersaturated α_2 grains in the starting microstructure due to the fast cooling. Therefore a second heating run was performed just after the previous one showed in Figure 1a. Indeed, the instability of the microstructure is confirmed by the comparison of both measurements as shown in Figure 1b. Obviously, at 1050 K the IF peak decreases and the hardening stage in the DMV curve disappears. The hardening stage of the DMV curve is associated to the precipitation of the γ-laths inside the supersaturated α_2 grains as it has been confirmed by High Energy X-Ray Diffraction [25,26], while the initial decrease of the modulus as well as the low temperature side of the IF peak seems to be related to a relaxation process [26, 27].

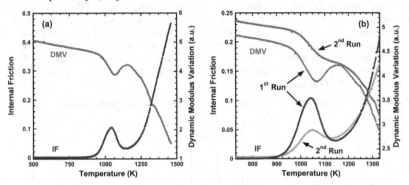

Figure 1. (a) Internal friction spectrum and dynamic modulus variation curve (at 1 Hz) during the first heating run of sample TNM1230 and **(b)** comparison of the previous result with the one measured during the subsequently heating run, evidencing the evolution of the microstructure.

After the second heating run up to 1460 K the IF peak in Fig. 1b is strongly annealed leaving only a remnant peak that is also present in both microstructural conditions after aging at 950°C and 850°C, as can be seen in Figure 2, where the IF spectra and the DMV curves measured at 0.1 Hz are shown for samples TNM950 and TNM 850.

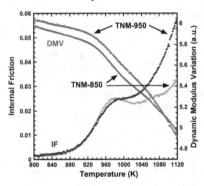

Figure 2. Internal friction spectra and dynamic modulus variation measured at 0.1 Hz, in samples TNM950 and TNM850.

The relaxation peak, at about 980 K because of the lower frequency, seems to be equivalent in both samples. That means that the microstructure becomes stable after aging, and the main difference of the behavior exhibited by these two samples is basically due to the HTB, as was already discussed previously [27]. However, there arises the question whether it is the same peak than the one measured before ageing (see Figure 1) or is it associated to another different mechanism appearing after the ageing treatment, or the peak already exists during the first run because of the much higher strength of the transitory peak. This is an interesting point because a relaxation peak at around 1050 K for 1 Hz has been previously reported in literature for several γ-TiAl intermetallics [28-32] and was attributed to different mechanisms such as reorientation of structural point defects or movement of dislocations. At present the interpretation of this relaxation is still open and object of controversy. Therefore the interpretation of this peak was a matter of interest. In order to elucidate the nature of this peak its activation parameters have been measured. In a previous work [27] a first estimation was done by measuring the IF spectrum as a function of temperature at different frequencies, but it could be argued that the method is prone to some error, because of the strong influence of the HTB. Consequently, we have measured the spectra in frequency in isothermal conditions at different temperatures, as shown in Figure 3a and 3b for samples TNM950 and TNM850, respectively.

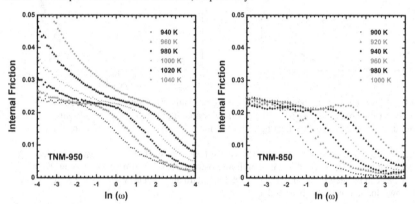

Figure 3. Samples TNM950 and TNM850: Internal friction spectra measured in isothermal conditions as a function of frequency for different temperatures.

As reported in [27] we notice a shift of the HTB to lower frequencies. The more pronounced shift to lower frequencies (longer times) in case of sample TNM850 is in agreement with the higher creep resistance observed for this sample [33].

From the curves shown in Figure 3, we can also determine the activation enthalpy H_{act} of the microscopic mechanism responsible of this relaxation by using [20]:

$$\ln (\omega_2/\omega_1) = (H_{act}/k_B)(T_1^{-1} - T_2^{-1}) \tag{1}$$

Equation (1) can be applied to the high frequency lateral side, provided that the peaks have the same strength, which is practically the case. From the mean shift of the frequency curves we have obtained the activation enthalpies of H_{act}(TNM950) = 2.9±0.1 eV and H_{act}(TNM850) = 3.1±0.1 eV, respectively. The obtained activation enthalpy H_{peak} = 2.9 - 3.1 eV agrees well with previous results from temperature spectra and correlates with the activation enthalpy for Ti self-diffusion (H_{Ti}=2.99 eV) in α_2-Ti$_3$Al [34, 35]. Therefore, it is tempting to suggest that the 1050 K peak is attributed to a Zener peak involving Ti atom jumps in the α_2-Ti$_3$Al phase from supersaturated α_2 grains or from the lamellar structure ($\alpha_2 + \gamma$). This interpretation agrees with the decreasing strength of the peak during the precipitation of the γ-lamellae in the α_2 grains. A deeper analysis and an atomic model for this mechanism, exceeding the scope of the present work, will be presented elsewhere [26].

SUMMARY

An advanced γ-TiAl-Nb-Mo (TNM) alloy has been studied by mechanical spectroscopy in order to obtain knowledge on the microscopic mechanism of the mobility of defects at high temperature. A relaxation peak that becomes stable after ageing has been observed, with a similar behavior and activation enthalpy of H_{act}=3.0±0.1 eV in both microstructures prevailing after aging of supersaturated α_2 grains at 950°C and 850°C. This relaxation is attributed to a Zener-like mechanism of Ti atoms in the lattice of the α_2 phase.

ACKNOWLEDGMENTS

The authors thank the financial support as Consolidated Research Group IT-10-310 from the Education Department and by the project ETORTEK ACTIMAT-10 from the Industry Department of the Basque Government. A part of this work was also conducted within the framework of the FFG project 832040 "energy drive", Research Studios Austria, Austria.

REFERENCES

1. *Gamma Titanium Alumindes 1999*, edited by Y.W. Kim, D.M. Dimiduk, and M.H. Loretto (TMS, Warrendale, PA, 1999).
2. *Titanium and Titanium alloys*, edited by C. Leyens, M. Peters (Wiley-VCH, Weinheim, Germany, 2003).
3. *Gamma Titanium Alumindes 2003*, edited by Y.W. Kim, H. Clemens and A.H. Rosemberg (TMS, Warrendale, PA, 2003).
4. F. Appel and R. Wagner, *Mater. Sci. Eng. R* **22**, 187 (1998).
5. H. Kestler and H. Clemens, in Ref. (2), (2003), p 351-392.
6. H. Clemens, W. Wallgram, S. Kremmer, V. Güther, A. Otto and A. Bartels, *Adv. Eng. Mater.* **10**, 707 (2008).
6. F. Appel and M. Oehring, in Ref. (2), (2003) p 89-152.
8. C. Herzig, T. Przeorski, M. Friesel, F. Hisker and S. Divinski, *Intermetallics* **9**, 461 (2001).
9. M. Takeyama and S. Kobayashi, *Intermetallics* **13**, 989 (2005).
10. R.M. Imayev, V.M. Imayev, M. Oehring and F. Appel, *Intermetallics* **15**, 451 (2007).
11. Z. Zhang, K.J. Leonard, D.M. Dimiduk and V.K. Vasudevan, *Structural Intermetallics 2001*. (TMS, Warrendale, PA, 2001) p. 515.

12. Y.W. Kim and D.M. Dimiduk, *Structural Intermetallics 2001*. (TMS, Warrendale, PA, 2001) p. 625.
13. H. Clemens, B. Boeck, W. Wallgram, T. Schmoelzer, L.M. Droessler, G.A. Zickler, H. Leitner and A. Otto, (*Mater. Res. Soc. Symp. Proc.* **Volume 1128**, Warrendale, PA, 2009) p.115.
14. I.J. Watson, K.D. Liss, H. Clemens, W. Wallgram, T. Schmoelzer, T.C. Hansen and M. Reid, *Adv. Eng. Mater.* **11**, 932 (2009).
15. H. Clemens, H.F. Chladil, W. Wallgram, G.A. Zickler, R. Gerlig, K.D. Liss, S. Kemmer, V. Güther and W. Smarsly, *Intermetallics* **16**, 827 (2008).
16. L.M. Droessler, T. Schmoelzer, W. Wallgram, L. Cha, G. Das and H. Clemens, (*Mater. Res. Soc. Symp. Proc.* **Volume 1128**, Warrendale, PA, 2009) p. U03-08.
17. T. Schmoelzer, K.D. Liss, G.A. Zickler, I.J. Watson, L.M. Droessler, W. Wallgram, T. Buslaps, A. Studer and H. Clemens, *Intermetallics* **18**, 1544 (2010).
18. A.S. Nowick and B.S. Berry, *Anelastic Relaxation in Crystalline Solids*. (Academic Press, New York, 1972).
19. *Mechanical Spectroscopy Q^{-1} 2001*, edited by R. Schaller, G. Fantozzi and G. Gremaud G. (Trans Tech Publications, Uetikon-Zuerich (SW), 2001).
20. J. San Juan, *Mater. Sci. Forum* **366-368**, 32 (2001).
21. V. Güther, J. Otto, J. Klose, C. Rothe, H. Clemens, W. Kachler, S. Winter and S. Kremmer, *Structural Intermetallics for Elevated Temperature Applications*, edited by Y.W. Kim, D. Morris, R. Yang and C. Leyens. (TMS, Warrendale, PA, 2008), p. 249.
22. V. Güther, C. Rothe, S. Vinter and H. Clemens, *BHM* **155**, 325 (2010).
23. P. Simas, J. San Juan, R. Schaller and M. L. Nó, *Key. Eng. Mat.* **423**, 89 (2009).
24. P. Simas, PhD Thesis, University of the Basque Country, (2012).
25. L. Cha, T. Schmoelzer, Z. Zhang, S. Mayer, H. Clemens, P. Staron and G. Dehm, *Adv. Eng. Mater.* **14**, 299 (2012).
26. J. San Juan, P. Simas, T. Schmoelzer, S. Mayer, H. Clemens and M.L. Nó, to be published.
27. P. Simas, T. Schmoelzer, M.L. Nó, H. Clemens and J. San Juan, (*Mater. Res. Soc. Symp. Proc.* **Volume 1295**, Warrendale, PA, 2011), p. 139.
28. M. Weller, G. Haneczok, H. Kestler and H. Clemens , *Mater. Sci. Eng.* A **370**, 234 (2004).
29. M. Perez-Bravo, M.L. Nó, I. Madariaga, K. Ostolaza and J. San Juan, in *Gamma Titanium Aluminides 2003*, edited by Y.W. Kim, H. Clemens and A.H. Rosemberg. (TMS, Warrendale, PA, USA, 2003) p 451.
30. M. Perez-Bravo, M.L. Nó, I. Madariaga, K. Ostolaza and J. San Juan, *Mater. Sci. Eng.* A **370**, 240 (2004).
31. M. Weller, H. Clemens, G. Haneczok, G. Dehm, A. Bartels, S. Bystrzanowski, R. Gerling and E. Arzt, *Phil. Mag. Letters* **84**, 383 (2004).
32. M. Weller, H. Clemens and G. Haneczok, *Mater. Sci. Eng.* A **442**, 138 (2006).
33. W. Wallgram, T. Schmoelzer, L. Cha, G. Das, V. Güther, and H. Clemens, *Int. J. Mat. Res.* **100**, 1021 (2009).
34. J. Rusing and C. Herzig, *Intermetallics* **4**, 647 (1996).
35. Y. Mishin and C. Herzig, *Acta Mater.* **48**, 589 (2000).

Titanium Aluminides II

Mater. Res. Soc. Symp. Proc. Vol. 1516 © 2013 Materials Research Society
DOI: 10.1557/opl.2013.44

The Science, Technology, and Implementation of TiAl Alloys in Commercial Aircraft Engines

B. P. Bewlay[1], M. Weimer[2], T. Kelly[2], A. Suzuki[1], and P.R. Subramanian[1]

[1] GE Global Research, Niskayuna, NY, United States.

[2] GE Aviation, Cincinnati, OH, United States.

ABSTRACT

The present article will describe the science and technology of titanium aluminide (TiAl) alloys and the engineering development of TiAl for commercial aircraft engine applications. The GEnx[TM] engine is the first commercial aircraft engine that is flying titanium aluminide (alloy 4822) blades and it represents a major advance in propulsion efficiency, realizing a 20% reduction in fuel consumption, a 50% reduction in noise, and an 80% reduction in NOx emissions compared with prior engines in its class. The GEnx[TM] uses the latest materials and design processes to reduce weight, improve performance, and reduce maintenance costs.

GE's TiAl low-pressure turbine blade production status will be discussed along with the history of implementation. In 2006, GE began to explore near net shape casting as an alternative to the initial overstock conventional gravity casting plus machining approach. To date, more than 40,000 TiAl low-pressure turbine blades have been manufactured for the GEnx[TM] 1B (Boeing 787) and the GEnx[TM] 2B (Boeing 747-8) applications. The implementation of TiAl in other GE and non-GE engines will also be discussed.

INTRODUCTION

Titanium aluminide (TiAl)-based alloys have received considerable attention for high-performance applications in the aerospace and automotive industries. There have been recent articles by Clemens [1], Wu [2], and Lasalmonie [3] that describe the state of the art of TiAl alloys and the challenges to implementation. In the last two years, TiAl has been introduced into service by GE in commercial aircraft engines as a new lightweight low-pressure turbine (LPT) blade material. Figure 1(a) shows a photograph of a TiAl LPT blade as used in the GEnx[TM] engine, and Figure 1(b) shows a photograph of the assembly of TiAl LPT blades as used in an early GE CF6 test engine; the data provided by this CF6 test established the foundations for the use of TiAl in the GEnx[TM] engine.

From the early TiAl research [1-5] it became evident that the principal advantages of TiAl-based alloys are low density (3.9–4.2 g/cm³, depending on composition), high specific strength, high specific stiffness and improved creep, oxidation resistance, and burn resistance (in comparison with conventional titanium alloys) properties up to 800°C. At temperatures between 600 and 800°C, TiAl alloys have higher specific strength than conventional titanium alloys.

Figure 2 shows the specific tensile strength of second- and third-generation TiAl alloys from room temperature to 800°C [1, 4]. Figure 2 shows that TiAl has a higher specific strength than alloy steels or Ni-based super-alloys. While the specific strength of TiAl is higher than that of competing materials, the room-temperature ductility is poor, typically ~1%. The low ductility has been considered to be a major barrier to the application of TiAl-based alloys as structural components [1, 2].

(a)   (b)

Figure 1. (a) a photograph TiAl LPT blade as used in the GEnx engine, and
(b) a photograph of the assembly of TiAl LPT blades as used in the GE CF6 test engine.

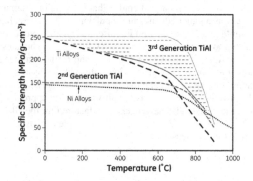

Figure 2. Graph showing the specific tensile strength of second and third generation TiAl alloys as a function of temperature [1, 4]. The graph shows the specific strength benefits of TiAl alloys in comparison with conventional titanium alloys and nickel-based alloys.

In the past 20 years there have been several large programs in Europe, the United States, and Japan to develop TiAl-based alloys for commercial applications. These programs were stimulated by almost 20 prior years of research on alloy development and processing of a wide range of TiAl-based alloys. One of the largest TiAl programs in the last two decades was the IMPRESS Project [6], which was coordinated by the European Space Agency (ESA). The objective of the project was to understand the strategic link between the material processing, the structure, and the final properties of new intermetallic alloys, such as TiAl, that has many different applications ranging from aerospace components to power generation systems. The most recent technical activities [1-3] have also revealed the tremendous importance of processing technology in the manufacture of the defect-sensitive TiAl-based alloys.

The first commercial application of TiAl (alloy 4822) low-pressure turbine (LPT) blades was announced by GE in 2006 for the GEnx™ engine. The GEnx™ implementation decision for TiAl was made after many years of diligent material and component design, extensive and disciplined development, and rigorous testing. Figure 3(a) shows the GEnx™ -1B aircraft engine as used on the Boeing 787, and Figure 3(b) shows a cut-away of the complete turbine section. The turbine consists of two high-pressure stages and seven low-pressure stages, including two stages of TiAl low-pressure turbine blades. More than 600,000 pounds of TiAl (alloy 4822) have been produced, and more than 40,000 blades have been produced. At present, there are approximately 20,000 TiAl blades flying daily throughout the World on nineteen Boeing 787s and thirty one Boeing 747-8s. It has required the skills and dedication of many people in the design, engineering, and supply chain departments to drive the implementation of TiAl 4822.

The objective of present article is to review the commercial status of TiAl alloys, and to describe the key technical activities relevant to the implementation of TiAl alloys in aircraft engines. The present article will also describe some of the engineering and manufacturing considerations regarding the scale-up of TiAl 4822 to the manufacture of 30-cm long LPT blades at substantial production volumes (more than 30,000 blades per year). This article will not present any information on component processing, material property curves, and design practices.

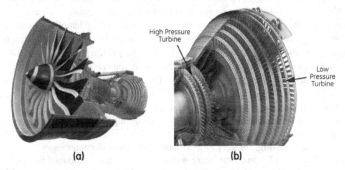

Figure 3. Renditions of the GEnx-1B aircraft engine as used on the Boeing 787. (a) shows the complete engine, and (b) shows the turbine section. The turbine consists of two high-pressure stages and seven low-pressure stages, including two stages of low-pressure TiAl turbine blades.

PRINCIPLES OF TiAl-BASED ALLOYS

The early TiAl engineering alloys have been classified as *first-generation* and *second-generation* TiAl alloys [1,4,5], and the early focus was on improvements in room-temperature ductility, oxidation performance, and creep. The second-generation alloys have received the most serious consideration for engineering applications. Most second-generation alloys were developed in the 1990s, including the GE alloy Ti–48Al–2Cr–2Nb, the ABB alloy Ti–47Al–

2W–0.5Si, and the Howmet alloys [7], which are similar to the GE composition but with Mn replacing Cr. The XD alloys which are reinforced by boride particles [1, 7] can also be considered in the second-generation category. The composition of second-generation TiAl alloys have been described by Clemens and Smarsly [1] using the following expression in atom per cent: Al : 45-48: Cr, Mn, V : 1-3: Nb, Ta, W, Mo :2-5 : Si, B, C : <1.

A range of alloys with higher temperature capability have been developed containing higher levels of Nb and Mo; these alloys have become known as third-generation TiAl alloys [1, 8, 9]. The TNB and TNM alloys fall into this category [1]. The third generation TiAl alloys can extend the useful temperature range of TiAl alloys up to ~850 °C, as shown in Figure 2. The composition of third generation TiAl alloys have also been described by Clemens and Smarsly [1] using the following expression: Al : 42-46: Cr, Mn, Nb : 0-10: W, Mo, Hf, Zr :0-3 : B, C : 0-1 : Rare earth elements : 0–0.5. The low-temperature strength of third-generation TiAl alloys is obtained through a combination of improved chemistry and high-temperature extrusion; typically the product form is rod/billet. Blades can be produced through isothermal forging and machining, which introduces additional cost. There remain some oxidation behavior questions regarding the use of these third-generation alloys, at higher temperature.

For TiAl alloys, the composition range that has been investigated previously involves Al contents between 44 and 48 at.%, although for wrought applications Al contents as low as 42 at.% have been investigated [1, 2, 4] ; at lower Al contents the high-temperature beta phase field is extended towards lower temperatures, which makes hot working easier and, if high creep temperature properties are not required, this can be a useful approach for lower cost processing.

For most applications Al concentrations closer to 46 at.% are required for high-temperature strength. If alloys of this composition range are cooled reasonably quickly, such as at rates similar to those of thin-walled castings, from the α phase-field, a fully lamellar structure is formed as γ is precipitated on the (0001) planes of the α phase. The face centered tetragonal γ phase precipitates on the (0001) planes with the {111} of the γ phase parallel to the (0001) plane in the hexagonal α phase and the $\langle 110 \rangle$ in γ phase parallel to the $\langle 11{-}20 \rangle$ in the α phase, according to the Blackburn orientation relationship. The α grain size that forms on cooling after casting is very important, because there is only one family of (0001) planes in the α phase within the γ/α lamellae formed during cooling, and the lamellae can extend across the width of the α grains. This is important because the α grain size formed during solidification is inherited in cast products. Depending on the specific casting process, the grain size can be very small (<100 μm) in thin wall investment castings, or very large (>5 mm) in a large diameter ingot.

Typically, a fully lamellar microstructure is observed in castings produced from hypoeutectic alloys, and the heat treatment of the as-cast microstructure to generate a full balance of properties for the application requires very careful consideration. The castings are typically hot isostatically pressed, because in the as-cast condition the material typically has very low ductility. If the cooling is slightly slower than is found in thin-walled castings, a near fully lamellar structure can be formed, where some γ lamellae coarsen.

If the alloy is hot worked in the two-phase region, the cast structure can be eliminated and a duplex structure consisting of γ phase grains and lamellar grains is formed on subsequent cooling. After hot working and heat treatment, the grain size of wrought alloys with a duplex microstructure can be typically 20 μm, which is considerably smaller than the typical grain or colony size in as-cast lamellar structures where grains of several hundred micrometers are common unless grain refiners are added [7].

If the alloy is not hot worked, heat treatment practices can be employed to refine the grain size [2]. The general practice is to hot isostatically press (HIP) TiAl castings to remove residual micro and macro porosity. The TiAl castings are also heat treated to convert the fully lamellar structure to a near fully lamellar or duplex structure, depending on the balance of properties that is required. The design of the suitable heat treatment window that can provide acceptable microstructure throughout the full section of the blade is extremely important. The selection of suitable heating rates, hold temperatures and times, and cooling rate are all parameters to which TiAl alloys can have strong sensitivity [10].

PROCESSING OF TiAl-BASED ALLOYS

The major challenges for the implementation of TiAl are low ductility and processing. In addition, the supply chain infrastructure is not well established. A range of processing routes has been explored including, casting [2], extrusion, and forging [1, 11]. Powder metallurgy and additive manufacturing routes have been explored but they are limited by the availability of acceptable quality powder at a competitive cost [2, 3]. Powder processing is potentially important, especially for larger products where segregation can limit the homogeneity of products and the uniformity of properties, but this process route has not been explored extensively [2, 3]. Casting offers an attractive cost-effective route for the production of complex TiAl shapes because the infrastructure for investment casting of conventional titanium alloys is well established [12]; these manufacturing methods have typically been employed for structural castings.

While the manufacture of engineering components from TiAl-based alloys is limited by their processing difficulties, significant progress has been made in the last ten years. Extrusion and forging have been used to produce compressor blades for engine testing. Small compressor blades have been produced using this process route by Thyssen, GfE Gesellschaft für Elektrometallurgie mbH, Leistritz and GKSS for Rolls-Royce [2] using an alloy with 45Al–8Nb and small amounts of carbon and boron to increase the high-temperature strength. Even though these blades are only a few centimeters long, this is a significant engineering and supply chain accomplishment.

In principle, casting offers the most cost-effective route for the production of complex shapes, since investment casting is a well-established process. The casting difficulties arise mainly from the high reactivity of molten TiAl-based alloys with ceramic crucibles and ceramic molds, and casting defects, such as shrinkage and porosity. TiAl alloys can present some difficult shrinkage challenges, but TiAl 4822 is an alloy that generally possesses high fluidity and good castability. The high reactivity of molten TiAl-based alloys with ceramic crucibles [13] has resulted in the use of cold wall crucibles, such as in induction skull melting and vacuum arc melting operations, but these are thermally inefficient, and induction skull melting can only offer a maximum melt superheat capability of ~60°C. The lower superheat leads to the requirement of pre-heated molds in order to improve filling for airfoil geometries, and this leads to slow cooling and to the production of components with large α grain sizes and thus long lamellae. The solidification and porosity related defects are harder to resolve, and they can be manifested as surface dimples after HIPping investment cast turbine blades [2]. Despite these challenges, engineering solutions have been developed, and turbine blades are now produced in both overstock and net-shape casting forms. The control of alloy chemistry and component microstructure is important to ensure consistent performance.

Figure 4(a) shows a photograph of an oversize gravity cast GEnx™ LPT blade with the gating removed, and Figure 4(b) shows a photograph of a centrifugally cast GEnx™ net-shape LPT blade with the gating still attached. Substantial efficiency improvements can be achieved through the use of net-shaped casting approaches, such as centrifugal casting, but with TiAl alloys net-shape casting is very complicated, and requires sophisticated techniques. Gravity casting of TiAl benefits from the existing supply chain infrastructure for conventional titanium alloy structural castings, but it is limited in its ability to produce thin-wall airfoils with long aspect ratios and wide chords.

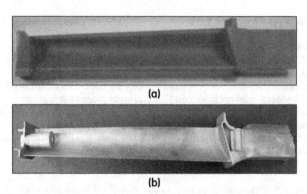

(a)

(b)

Figure 4. (a) Photograph of an oversize gravity-cast GEnx™ LPT blade with the gating removed, and (b) Photograph of a GE spin cast GEnx™ net-shape LPT blade with gating.

There has been significant interest in centrifugal casting of TiAl, such as the activity at ACCESS [14, 15]. Both horizontal and vertical forms of centrifugal casting have been explored. Simulation of horizontal centrifugal casting has been performed in order to consider fluid flow, mold fill, heat transfer, solidification, and solidification shrinkage. Simulations of mold filling, heat transfer, and solidification have been performed and validated at subscale (one-third-size) and full scale. Figure 5 shows an example of subscale simulation of mold filling. The subscale geometries were designed to simulate the most difficult elements of the full size LPT blade geometry, such as the shroud, seal teeth, platform, under platform pocket, and dovetail at their actual sizes; only the airfoil section of the blade was simulated on a reduced length basis but with the same chord, and chord profile. The geometry of the wax models used for the sub-scale casting trials is shown in Figure 5. These tools can be used to reduce the process development time, and to improve the properties of the casting. Centrifugal casting of TiAl requires the most advanced casting ceramics. For melting crucibles, yttria has benefits, in principle, in terms of reduced interstitial contamination; there has been significant work in several laboratories in Europe and Japan in the last five years to evaluate ceramic melting crucible approaches for TiAl [13]. For centrifugal casting the mold must be strong and it must not react with the molten alloy; typical mold systems for nickel alloys do not meet the requirements for centrifugal casting of TiAl alloys.

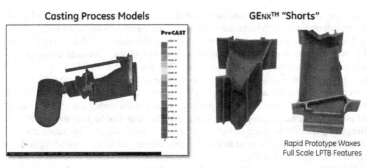

Figure 5. Simulations of a centrifugally cast TiAl blade. Simulations of mold filling, heat transfer, and solidification have been performed and validated on a subscale (one-third-size) and full scale.

TiAl IMPLEMENTATION IN GE ENGINES

GE announced in 2006 that TiAl would be used in LPT blades for the GEnx™ engine [16]. The LPT section of the GEnx™ engine employs approximately 200 TiAl blades per engine. The GEnx™ aircraft engine is the premium product in its class, offering substantial improvements in emissions, noise, reliability, and cost of ownership. The FAA certification for the GEnx™ engine was issued in April 2011. Throughout the certification testing the TiAl blades performed according to the design expectations.

At the time of writing the present article, more than 600,000 pounds of TiAl alloy 4822 have been produced, more than eighteen separate aircraft engine part configurations have been cast, and more than 40,000 blades have been produced. While these numbers are impressive, the volume of TiAl alloy production is still low in comparison with, for example, commercial titanium alloy grades. There are approximately 20,000 TiAl blades flying daily on nineteen Boeing 787s and thirty-one Boeing 747-8s. A total of 162 engines with approximately 20,000 TiAl blades have accumulated more than 270,000 hours and 50,000 cycles of operation. The decision to use TiAl in the GEnx™ LPT was made after many years of diligent, disciplined, and extensive development. More than 1300 GEnx™ engines that have been ordered, so the following years will see continuous large-volume production of TiAl LPT blades. More recently, CFM, a 50/50 joint company between GE-Snecma has announced that TiAl will be used in the LEAP-X engine family [17].

BROADER IMPLEMENTATION OF TiAl-BASED ALLOYS

TiAl-based products are currently used in automotive and aircraft engine applications. The most successful of the automotive applications has been the development in Japan of TiAl turbochargers that were fitted in 1998 to top-of-the-range Lancer cars; the initial success led to more than 20,000 cars being equipped with second-generation turbochargers in 2003. Several other companies are now manufacturing TiAl-based turbochargers. Thermo-mechanically processed valves have been used for many years in Formula One cars but Formula One regulations no longer allow their use. Attempts to develop cast TiAl-based valves have been successful in terms of performance but are still far too expensive for passenger vehicles.

In the aircraft engine industry, there has been substantial interest in the application of TiAl-based alloys in aircraft engines, with technical activities being reported by most engine manufacturers, but to date the only successful commercial application of TiAl has been by GE in 2006 for the GEnx[TM] engine. At present there are more than 160 GEnx[TM] engines flying in commercial service throughout the world on the Boeing 787 and the Boeing 747-8.

There have also been demonstration programs in the United States, such as the CAESAR program [3] for compressor blades and shrouds, blade retainers (Pratt and Whitney), and turbine dampers (Volvo). Snecma and Turbomeca also performed several feasibility studies, including the use of TiAl powder to make a high-pressure compressor casting. For large components, powder metallurgy is unavoidable to get uniform and reliable properties, and this raises the issues of high cost and defect management in terms of processing and component lifing. Most of these programs were successful technically.

Rolls-Royce has announced on its website that: "TiAl manufacturing routes for both forging and centrifugal casting have been established. A current temperature capability of approximately 750°C, which in future development may be extended to 850°C. The first application of forged materials will be in static and rotating compressor airfoils while the biggest goal is centrifugally cast γ titanium aluminide for low-pressure turbine blades of the next-generation Trent XWB" [18]. The Rolls-Royce development reinforces the commitment of the aircraft engine industry to TiAl technology for high by-pass ratio turbofan engines, such as the GEnx[TM] and the Trent XWB engines.

In 2011 a partnership between Access, an institute at the RWTH Aachen, and TITAL GmbH, the specialist investment casting company based in Bestwig, was announced for the production of TiAl blades [19]. The commissioning of a new melting and casting plant for production of TiAl blades in Aachen was announced. Series production of these turbine parts (overall size, or edge length, ranging up to 300 mm) is planned to start in 2013. The purpose of the TITAL and Access partnership is to develop an efficient process for the casting aircraft engine components in TiAl alloys. A new concept for a casting plant that can provide cost-effective production has been developed in a further partnership with ALD Vacuum Technologies. Further details have been provided by Aguilar et al. [14, 15].

In a series of announcements [17, 20], Pratt and Whitney and MTU have indicated that they will use a TiAl alloy as a low-pressure turbine blade material for the last stage of the new geared turbofan (GTF) engine for the Airbus 320neo. The selection of TiAl turbine blades for a lower thrust class of engines than the GEnx[TM] engine is significant. No further details have been provided to date.

CONCLUSIONS

TiAl alloys have become an important class of new materials for high-performance applications in the aerospace and automotive industries. TiAl alloys offer substantial advantages compared to conventional technologies, but the cost to introduce them into production is very high. Although TiAl alloy components can be manufactured with conventional processing equipment, such as that which exists in casting, forging, and heat treatment operations, specific processes conditions and controls have to be employed for TiAl alloys. The present state of the art for TiAl technology has been derived through several decades of substantial research and development in both academia and industry throughout the World.

GE currently leads the aircraft engine industry with TiAl LPT blade implementation on the GEnx™ engine, and the other aircraft engine companies have all announced their intentions for future engines. As TiAl alloy science and technology develops, and the cost of TiAl components becomes more competitive with conventional components, the full potential of TiAl alloys will start to be realized. Alloy design and component processing to achieve the balance of application properties required for TiAl components will continue to be a very important challenge.

The future opportunities for TiAl technology lie in the areas of manufacturing efficiency and alloys with higher temperature capability. Three generations of TiAl alloys have been developed. The present temperature capability of TiAl alloys is limited to ~750°C for second-generation TiAl alloys. The third-generation alloys have the potential to provide temperature capability up to ~850°C; maturation of third-generation alloys will significantly increase the application space for TiAl alloys.

ACKNOWLEDGMENTS

The authors would like to acknowledge the numerous colleagues at GE Aviation and GE Global Research whose skills, dedication, and commitment have led to the implementation of TiAl in the GEnx™ engine. The authors would like to specifically express their gratitude to Dr Robert E. Schafrik for his continuous support and leadership.

REFERENCES

1. H. Clemens and W. Smarsly, *Advanced Materials Research* **278**, 551-556 (2011).
2. X. Wu, "Review of alloy and process development of TiAl alloys," *Intermetallics* **14**(10), 1114-1122 (2006).
3. A. Lasalmonie, *Intermetallics* **14**(10), 1123-1129 (2006).
4. Y. W. Kim, *JOM* **46** (7), pp. 30–36 (1994).
5. F. Appel, U. Brossmann, U. Christopf, S. Eggert, P. Janschek, U. Lorenz, J. Müllauer, M. Oehring, J. D. H. Paul, *Adv Eng Mater* **2** (11) pp. 699–720 (2000).
6. D. J. Jarvis and D. Voss, *Materials Science and Engineering: A* **413–414**, 583–591 (2005).
7. D. E. Larsen, *Mater Sci Eng* **A213**, pp. 128–133 (1996).
8. S. C. Deevi, W. J. Zhang, C. T. Liu, and B. V. Reddy, *Mat. Res. Symp. Proc.* **646**, pp. N.1.2.1-1.2.6 (2001).
9. W. J. Zhang, B. V. Reddy, and S. C. Deevi, *Scripta materialia*, **45**(6), 645-651 (2001).
10. A. Suzuki, R. Casey, B.P. Bewlay, GE Internal Report (2012).
11. T. Tetsui, K. Shindo, K. Satoshi, S. Kobayashi, and M. Takeyama, *Intermetallics* **13**, pp. 971–978 (2005).
12. M. Weimer, B.P. Bewlay and T. Schubert, paper presented at the "4th International Workshop on Titanium Aluminides," Nuremberg, Germany (September 14-16, 2011).
13. T. Tetsui, T. Kobayashi, T. Mori, T. Kishimoto, and H. Harada, *Materials transactions*, **51**(9), 1656 (2010).
14. J. Aguiliar, G. J. Schmitz, U. Hecht, A. Schievenbusch, R. Guntlin, G. Schuh, and C. Wesch, in *Concurrent Enterprising (ICE), 2011 17th International Conference on* (pp. 1-7). IEEE (2011, June).
15. J. Aguilar, A. Schievenbusch, and O. Kättlitz, *Intermetallics* **19**(6), 757-761 (2011).

16. G. Norris, in Flight Global June 13th, 2006.
17. G. Norris, in Aviation Week & Space Technology Nov 05, 2012 , p. 45.
18. "Low Density Materials," Rolls Royce plc, accessed December 19, 2012, http://www.rollsroyce.com/about/technology/material_tech/low_density_materials.jsp.
19. "High temperature-resistant turbine blades made from titanium aluminide," TITAL, accessed December 19, 2012, http://www.tital.de/it/news/attualit/high-temperature-resistant-turbine-blades-made-from-titanium-aluminide.html.
20. G. Das, W. Smarsly, F. Heutling, C. Kunze, D. Helm, paper presented at *4th International Workshop on Titanium Aluminides*, Nuremberg, Germany (September 14-16, 2011).

Mater. Res. Soc. Symp. Proc. Vol. 1516 © 2012 Materials Research Society
DOI: 10.1557/opl.2012.1680

Thermodynamic Calculations of Phase Equilibria and Phase Fractions of a β-Solidifying TiAl Alloy using the CALPHAD Approach

Robert Werner[1], Martin Schloffer[1], Emanuel Schwaighofer[1], Helmut Clemens[1], Svea Mayer[1]
[1]Department of Physical Metallurgy and Materials Testing, Montanuniversität Leoben, A-8700 Leoben, Austria

ABSTRACT

The CALPHAD (CALculation of PHAse Diagrams) method is widely recognized as a powerful tool in both scientific and industrial development of new materials and processes. For the implementation of consistent databases, where each phase is described separately, models are used which are based on physical principles and parameters assessed from experimental data. Such a database makes it possible to perform realistic calculations of thermodynamic properties of multi-component systems. However, a commercial available TiAl database can be applied for thermodynamic calculations to both conventional Ti-base alloys and complex intermetallic TiAl alloys to describe experimentally evaluated phase fractions as a function of temperature. In the present study calculations were done for a β-solidifying TiAl alloy with a nominal composition of Ti-43.5Al-4Nb-1Mo-0.1B (in at. %), termed TNM[TM] alloy. At room temperature this alloy consists of ordered γ-TiAl, $α_2$-Ti_3Al and $β_0$-TiAl phases. At a certain temperature $α_2$ and $β_0$ disorder to α and β, respectively. Using the commercial database the thermodynamic calculations reflect only qualitative trends of phase fractions as a function of temperature. For more exact quantitative calculations the commercial available thermodynamic database had to be improved for TiAl alloys with high Nb (and Mo) contents, as recently reported for Nb-rich γ-TiAl alloys. Therefore, the database was modified by experimentally evaluated phase fractions obtained from quantitative microstructure analysis of light-optical and scanning electron micrographs as well as conventional X-ray diffraction after long-term heat treatments and by means of in-situ high-energy X-ray diffraction experiments. Based on the CALPHAD-conform thermodynamic assessment, the optimized database can now be used to correctly predict the phase equilibria of this multi-component alloying system, which is of interest for applications in automotive and aircraft engine industry.

INTRODUCTION

With the use of the CALPHAD formalism it is possible to calculate phase diagrams and phase fraction diagrams. For this reason the required effort to determine equilibrium conditions in a multi-component system is reduced. Following the CALPHAD methodology it is possible to use the assessed excess Gibbs energies of the constituent subsystems for extrapolation to a higher component system. Details of modeling procedures can be found in the fundamental work of Kaufman and Hillert [1,2] and in the reviews of Ansara [3] and Saunders and Miodownik [4]. All types of models require an input of coefficients that uniquely describe the thermo-physical properties of the various phases of the multi-component system and these coefficients are held in databases. The main models used in the present work are the substitutional type model and the multiple sublattice model, details of each can be found in [4,5]. Both models are necessary to describe the total Gibbs energy of the existing phases in the TNM[TM] specification range [6] and will be explained shortly in the next chapter dealing with the theory behind CALPHAD modeling. The thermodynamic calculations were conducted using the thermodynamic software ThermoCalc. As far as the calculation of phase fractions as a function of temperature for the investigated TNM^{TM} alloy is concerned, three databases (A, B and C) were compared in this work. Database A is the commercial TiAl database from Saunders [7], database B is an extended, but unmodified database (using [8] for the reassessed Ti-Al-Nb system and [7] for the Mo contributions), and database C is a modified version of database B and has been optimized for calculations in the TNM^{TM} specification region.

THEORY

For the calculation of phase equilibria in a multi-component system, it is necessary to minimize the total Gibbs energy, G_{total} (as a function of pressure, temperature and composition) of all phases that take part in this equilibrium.

$$G_{total} = \sum_{p=1}^{N} n_p G_m^{\varphi_p} \rightarrow minimum, \tag{1}$$

where n_p is the number of moles, and $G_m^{\varphi_p}$ is the molar Gibbs energy of phase p.
A thermodynamic description of a system requires the assignment of thermodynamic functions for each phase. As it has been pointed out in the introduction the main models used in the present work are the substitutional type model and the multiple sublattice model. Both of these models can broadly be represented by the general equation of a phase φ_p [1]:

$$G_m^{\varphi_p}(T, x_k) = G_{m,ref}^{\varphi_p}(T, x_k) + G_{m,id}^{\varphi_p}(T, x_k) + G_{m,E}^{\varphi_p}(T, x_k), \quad with \quad k = \text{Ti, Al, Nb, Mo}, \tag{2}$$

where $G_{m,ref}^{\varphi_p}$ is the molar Gibbs energy of the phase in its pure form, $G_{m,id}^{\varphi_p}$ is the molar Gibbs energy for ideal mixing and $G_{m,E}^{\varphi_p}$ is the molar excess Gibbs energy of the phases.

Unary phases

The molar Gibbs energy function for the element k (k = Ti, Al, Nb, Mo) in the phase φ_p ($\varphi_p = \beta$-bcc_A2 $\leftrightarrow \beta_0$-bcc_B2, α-hcp_A3, L-liquid, α_2-Ti$_3$Al_D0$_{19}$ and γ-TiAl_L1$_0$) is described by Eq. (3). In the present work the molar Gibbs energy functions for each component k are taken from the Scientific Group Thermodata Europe (SGTE) compilation of Dinsdale [9].

$$^0G_{m,k}^{\varphi_p}(T) = a + bT + cT \ln T + \sum_{j \in \mathbb{Z}} d_j T^j ..., \tag{3}$$

where a, b, c, and d_j are adjustable coefficients and j are integers. To represent the pure elements, j-values are typically 2, 3, -1 and 7 or -9 [9].

Substitutional type model

The L-liquid phase is modeled as substitutional by applying the Muggianu extension of the Redlich-Kister formalism [10,11]. In general the Gibbs energy of mixing of a multi-component system for a substitutional phase can be represented by Eq. (4):

$$G_m^{\varphi_p}(T, x_k) = + \begin{cases} G_{m,\mathrm{ref}}^{\varphi_p}(T, x_k) = \sum_{i=1}^{K} x_i \, {}^0G_{m,i}^{\varphi_p}(T), \\ G_{m,\mathrm{id}}^{\varphi_p}(T, x_k) = RT \sum_{i=1}^{K} x_i \ln x_i, \\ G_{m,\mathrm{E}}^{\varphi_p}(T, x_k) = \sum_{\substack{i,j=1 \\ (i\neq j)}}^{K} x_i x_j \sum_{z=0}^{Z \in \mathrm{N}} {}^zL_{ij}^{\varphi_p}(x_i - x_j)^z + \sum_{\substack{i,j,l=1 \\ (i\neq j\neq l)}}^{K} x_i x_j x_l \sum_{r=i,j,l} {}^rL_{ijl}^{\varphi_p} x_r + ..., \end{cases}$$ (4)

where x_i, x_j and x_l are the mole fractions of component i, j and l. ${}^0G_{m,i}^{\varphi_p}$ defines the molar Gibbs energy of the phase in the pure component i, T is the temperature in Kelvin an R is the universal gas constant. The binary interaction parameters $L_{ij}^{\varphi_p} = \sum_{z=0}^{Z \in \mathrm{N}} {}^zL_{ij}^{\varphi_p}(x_i - x_j)^z$ were modeled using Redlich-Kister [11] polynomials and consist of binary interaction coefficients ${}^zL_{ij}^{\varphi_p}$ dependent on the value of z and temperature T. In practice the value for z does not usually rise above two. The ternary interaction parameters $L_{ijl}^{\varphi_p} = \sum_{r=i,j,l} {}^rL_{ijl}^{\varphi_p} x_r$ were modeled as described in [12] and those parameters are a function of the temperature dependent ternary interaction coefficients ${}^rL_{ijl}^{\varphi_p}$. In the present work only two terms were applied for the temperature dependence of the binary and ternary interaction coefficients (${}^zL_{ij}^{\varphi_p} = {}^za_{ij}^{\varphi_p} + {}^zb_{ij}^{\varphi_p}T$ and ${}^rL_{ijl}^{\varphi_p} = {}^ra_{ijl}^{\varphi_p} + {}^rb_{ijl}^{\varphi_p}T$), where a and b are parameters that were modified during the optimization process in this work too (database C).

Multiple sublattice model

The metal solid solution phases and the more complex intermetallic compounds are usually modeled using the compound energy formalism [13]. The metal solid solutions based on β-bcc_A2 and α-hcp_A3 phases are modeled as disordered solution phases using the two-sublattice model where the second sublattice is occupied by vacancies (Va). The remaining intermetallic compounds (α₂-Ti₃Al_D0₁₉ and γ-TiAl_L1₀) are treated as ordered phases by using the two-sublattice model proposed by Hillert and Staffanson [14]. The ordered α₂-Ti₃Al phase and the corresponding disordered α-hcp_A3 phase are described by separate models. The ordered β₀-bcc_B2 phase is modeled by a unified three-sublattice model, which includes the contribution from disordered β-bcc_A2 phase. Following the compound energy formalism the molar Gibbs energy of a phase φ_p has the general form:

$$G_m^{\varphi_p}(T, y_k) = + \begin{cases} G_{m,\mathrm{ref}}^{\varphi_p}(T, y_k) = \sum_{j\geq2} \left(\prod_{i=1}^{J} y_{k_i}^{s_i} \right) {}^0G_{m,(k_1:...:k_j)}^{\varphi_p}, \\ G_{m,\mathrm{id}}^{\varphi_p}(T, y_k) = RT \sum_{s=1}^{S} \eta^{\varphi_p,s} \sum_{i=1}^{K_s} y_i^s \ln\left(y_i^s\right), \\ G_{m,\mathrm{E}}^{\varphi_p}(T, y_k) = \sum_{\substack{i,j=1 \\ (i\neq j)}}^{K_1} \sum_{l=1}^{K_2} y_i^{s_1} y_j^{s_1} y_l^{s_2} \sum_{z=0}^{Z \in \mathrm{N}} {}^zL_{(i,j):l)}^{\varphi_p}\left(y_i^{s_1} - y_j^{s_1}\right)^z + ... \end{cases}$$ (5)

with $y_k^s = n_k^s / \sum_{i=1}^{K_s} n_i^s$ and $\sum_{i=1}^{K_s} y_i^s = 1$.

y_k^s is the site fraction of component k on sublattice s and n_k^s is the number of moles of constituent k on sublattice s. ${}^0G_{m,(k_1:...:k_j)}^{\varphi_p}$ is the molar Gibbs energy of formation of 'virtual compounds' or unaries where each sublattice is occupied by just one component. $\eta^{\varphi_p,s}$ is the stoichiometric coefficient for each sublattice of the phase φ_p and $G_{m,\mathrm{E}}^{\varphi_p}$ is the contribution of the molar excess Gibbs energy in the compound energy formalism, which is only depicted for a case of two sublattices in Eq. (5). The parameter $L_{(i,j:l)}^{\varphi_p} = \sum_{z=0}^{Z \in N} {}^z L_{(i,j:l)}^{\varphi_p} \left(y_i^{s_1} - y_j^{s_1} \right)^z$ describes the mutual interaction of the constituents i and j in the first sublattice, when the second sublattice is fully occupied by constituent l. This description can be extended in the same way to any number of sublattices. Building up the molar Gibbs energy terms with the use of the Eqs. (3)-(5) for all relevant phases of the TNM^{TM} system it is possible to calculate the total Gibbs energy by using Eq. (1). The basic mathematical method used for the calculation of phase fractions as a function of temperature is a constrained minimization of the total Gibbs energy (calculated from Eq. (1)) for a given temperature, pressure and overall composition of the multi-component TNM^{TM} system. For reasons of simplification of the model the low B content (0.1 at. % in the β-solidifying TiAl alloy) was not considered in the calculation of the total Gibbs energy.

EXPERIMENT

TNM^{TM} powder, produced by argon gas atomization using the Electrode Induction Melting Gas Atomization technique [15], with a particle size smaller than 180 μm in diameter, was filled into cylindrical titanium cans. Subsequently, the capsules were evacuated, sealed and hot-isostatically pressed (HIPed) at 1250 °C for 2 h at 200 MPa, followed by furnace cooling. Samples as reported in [16] were cut from the HIPed material for the subsequent isothermal heat-treatments (HT) in the temperature range of 850-1350 °C. All heat-treatments were performed in a Carbolite furnace RHF 1600 where the temperature was controlled with thermocouples, integrated in the furnace. Subsequent water quenching (WQ) was performed. The small samples were used to evaluate the predominant microstructural constituents after annealing between 1075 °C and 1285 °C with a holding time of 1 h. For the investigation of the microstructure and phase fractions, after annealing between 1290 °C and 1350 °C for 9 min up to 30 min, the larger samples were used. The holding time for the annealed samples at 850 °C and 900 °C was 6 h and details of the stepwise heat-treatments for the two lowest temperatures can be found in [16]. According to [16] the evolution of the microstructure after each individual heat-treatment step was examined by light-optical (LOM), scanning electron microscopy (SEM) as well as by conventional X-ray diffraction (XRD) and subsequent Rietveld analysis. Furthermore, the setups of the HZG beamline HARWI II at DESY in Hamburg, Germany, were used in order to perform in-situ high-energy X-ray diffraction (HEXRD) experiments with high-energy synchrotron radiation.

DISCUSSION

The experimentally evaluated phase fractions as a function of temperature are compared with the results of thermodynamic calculations using three different databases A, B and C (Figures 1a and 1b). For all three databases the calculated γ-solvus temperatures ($T_{\gamma,solv}$) agree well with experimental data derived from quantitative analysis of heat-treated specimens and in-situ HEXRD experiments. The eutectoid temperature (T_{eu}), which is connected to the $\alpha \rightarrow \alpha_2$ ordering reaction, is significantly underestimated by the thermodynamic calculations if the database A or B is used. Database C is a modified version of database B and, therefore, optimized for calculations in the TNM^{TM} specification range. In Table I the changed interaction coefficients are listed so that it is comprehensible which modifications were necessary to shift T_{eu} to a higher temperature of about 1160 °C and to obtain a better correlation with the experimental data in the (α+β)-two-phase region above $T_{\gamma,solv}$. The comparison of the changed parameters shows that consequently all *a*-values (corresponding to molar enthalpy terms) are modified in the ten listed parameters, whereas the *b*-values (corresponding to molar entropy terms) are not changed.

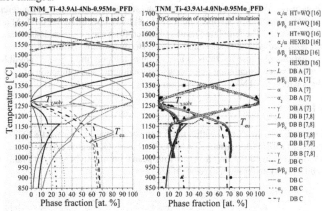

Figure 1. a) Comparison of calculated phase fractions as a function of temperature according to three different databases (A, B and C) used for the β-stabilized TiAl alloy Ti-43.9Al-4.0Nb-0.95Mo(-0.1B) without considering of boron contributions. b) Experimentally evaluated phase fractions obtained from heat treatments followed by quantitative metallography as well as in-situ HEXRD measurements are compared to the simulated results of the new established database C.

Table I. Comparison of changed interaction coefficients (databases B and C).

Phase	Changed parameter	Database B	Ref. B	Database C	Ref. C
α	$^0L^\alpha_{(Al,Mo\,:\,Va)}$	$-85570 + 25T$	[7]	$-85700 + 25T$	[this work]
	$^0L^\alpha_{(Mo,Ti\,:\,Va)}$	$+22760 - 6T$	[7]	$+18050 - 6T$	[this work]
β₀	$^0L^{\beta_0}_{(Al,Ti\,:\,Al\,:\,Va)} = {}^0L^{\beta_0}_{(Al\,:\,Al,Ti\,:\,Va)}$	$+6155$	[8]	$+900$	[this work]
	$^0L^{\beta_0}_{(Al,Ti\,:\,Ti\,:\,Va)} = {}^0L^{\beta_0}_{(Ti\,:\,Al,Ti\,:\,Va)}$	-21406	[8]	-23500	[this work]
α₂	$^0L^{\alpha_2}_{(Al,Ti\,:\,Al)}$	$-269912 + 86.338T$	[8]	$-273380 + 86.338T$	[this work]
	$^0L^{\alpha_2}_{(Ti\,:\,Al,Ti)}$	$-13704 + 5.318T$	[8]	$-44000 + 5.318T$	[this work]
γ	$^0L^{\gamma}_{(Al,Ti\,:\,Al)} = {}^0L^{\gamma}_{(Al\,:\,Al,Ti)}$	$-88993 + 41.695T$	[8]	$-90700 + 41.695T$	[this work]
	$^1L^{\gamma}_{(Al,Ti\,:\,Al)} = {}^1L^{\gamma}_{(Al\,:\,Al,Ti)}$	$+27363$	[8]	$+27950$	[this work]
	$^2L^{\gamma}_{(Al,Ti\,:\,Al)} = {}^2L^{\gamma}_{(Al\,:\,Al,Ti)}$	$+42189$	[8]	$+39000$	[this work]
	$^0L^{\gamma}_{(Al,Ti\,:\,Ti)} = {}^0L^{\gamma}_{(Ti\,:\,Al,Ti)}$	$-31963 + 6.952T$	[8]	$-33258 + 6.952T$	[this work]

CONCLUSIONS

For the evaluation of phase fractions as a function of temperature for the investigated β-solidifying TNM^{TM} alloy different heat-treated samples were examined by means of quantitative metallography and conventional XRD. The results match well with the data derived from in-situ HEXRD measurements. Additionally, the experimentally observed dependency of the phase fractions on temperature is compared with those derived from thermodynamic calculations using three different databases (A, B and C). The simulated phase fractions obtained using the databases A and B could only show an satisfying agreement with the experimental results in the vicinity of $T_{\gamma,solv}$ (Figure 1a). For a higher accuracy of quantitative calculations the commercial available thermodynamic databases from open literature [7,8] had to be improved for TiAl alloys with high Nb and Mo contents. The simulated results obtained from the new established database C are in satisfying accordance with the experimental ones. Because of the optimized thermodynamic calculations of phase fractions as a function of temperature it is now possible to obtain appropriate results when calculating quasibinary phase diagrams as a function of temperature for alloys within the TNM^{TM} specification range.

ACKNOWLEDGEMENTS

A part of this work was conducted within the framework of the FFG project 830381 "fAusT" Österreichisches Luftfahrtprogramm TAKE OFF, Austria; Research activities performed at DESY have received funding from the European Community's Seventh Framwork Programme (FP7/2007-2013) under grant agreement n°226716.

REFERENCES

1. L. Kaufman and H. Bernstein, Computer calculation of phase diagrams, Academic Press, New York (1970).
2. M. Hillert, Calcularions of phase equilibria. American Society for Metals Seminar on Phase Transformations. Metals Park, Ohio: *American Society for Metal*, 181-218 (1968).
3. I. Ansara, *Int. Met. Reviews* 22, 20-53 (1979).
4. N. Saunders and A.P. Miodownik, CALPHAD (A Comprehensive Guide), Elsevier, London (1998).
5. B. Sundman and J. Ågren, *J. Phys. Chem. Solids* 2, 227-238 (1981).
6. V. Güther, TNM^{TM} data sheet Nr. 1, GfE Metalle und Materialien GmbH, Nuremberg, Germany (2010).
7. N. Saunders, in *Gamma Titanium Aluminides*, edited by Y.-W. Kim, D. M. Dimiduk and M. H. Loretto (TMS, Warrendale, PA, 1999) p. 183-188.
8. V. T. Witusiewicz, A. A. Bondar, U. Hecht, and T. Ya. Velikanova, *Journal of Alloys and Compounds* 472, 133-161 (2009).
9. A. T. Dinsdale, *CALPHAD* 15, 317-425 (1991).
10. Y.-M. Muggianu, M. Gambino, and J.-P. Bros, *J. Chim. Phys.* 72, 83-88 (1975).
11. O. Redlich and A. Kister, *Indust. Eng. Chem.* 40, 345-348 (1948).
12. M. Hillert, *CALPHAD* 4, 1-12 (1980).
13. J.-O. Andersson, A. F. Guillermet, M. Hillert, B. Jansson, and B. Sundman, *Acta Metall.* 34, 437-445 (1986).
14. M. Hillert and L.-I. Staffansson, Acta Chem. Scand. 24, 3618-3626 (1970).
15. R. Gerling, H. Clemens, and F.-P.Schimansky, *Advanced Engineering Materials* 6, 23-38 (1996).
16. M. Schloffer, F. Iqbal, H. Gabrisch, E. Schwaighofer, F.-P. Schimansky, S. Mayer, A. Stark, T. Lippmann, M. Göken, F. Pyczak, and H. Clemens, *Intermetallics* 22, 231-240 (2012).

Mater. Res. Soc. Symp. Proc. Vol. 1516 © 2012 Materials Research Society
DOI: 10.1557/opl.2012.1583

In situ SEM Observations of the Tensile-Creep Deformation Behavior and Fracture Mechanisms of a γ-TiAl Intermetallic Alloy at Low and High Stresses.

R. Muñoz-Moreno[1,2], M. T. Pérez-Prado[1], E.M. Ruiz-Navas[2], C. J. Boehlert[1,3,4], J. Llorca[1,4]

[1] IMDEA Materials Institute, C/ Eric Kandel 2, 28906, Getafe, Madrid, Spain.
[2] Department of Materials Science and Engineering, Universidad Carlos III de Madrid, Avda. Universidad 39, 28911, Leganés, Spain.
[3] Department of Chemical Engineering and Materials Science, Michigan State University, 2527, Engineering Building, East Lansing, MI 48824, U.S.A.
[4] Department of Materials Science, Polytechnic University of Madrid, E. T. S. de Ingenieros de Caminos, 28040, Madrid, Spain.

ABSTRACT

The effect of stress on the deformation and crack nucleation and propagation mechanisms of a γ-TiAl intermetallic alloy (Ti-45Al-2Nb-2Mn (at.%) - 0.8v.%TiB$_2$) was studied by means of *in situ* tensile (constant strain rate) and tensile-creep (constant load) experiments performed at 973 K inside a scanning electron microscope (SEM). The evolution of the microstructure and the nucleation and propagation of cracks was tracked during the high temperature mechanical tests in the SEM. Colony boundary crack nucleation was found to be activated during the secondary stage in creep tests at 300 MPa and 400 MPa and during the tertiary stage of the creep tests performed at higher stresses and at constant strain rate. Interlamellar ledges were only observed during the high stress tensile-creep tests (σ>400 MPa) and during the constant strain rate test. Quantitative measurements of the nature of the crack propagation path along secondary cracks and along the primary crack were carried out. It was found that colony boundaries were preferential sites for crack propagation under all the conditions investigated. The frequency of interlamellar cracking increased with increasing stress.

INTRODUCTION

Gamma Titanium Aluminides are important intermetallics alloys targeted for high temperature aerospace applications in Low Pressure Turbines (LPT) because they can provide increased thrust-to-weight ratios and improved efficiency [1-3]. However, their high-temperature deformation and fracture behavior must be better understood in order to optimize their microstructure [4-10].

γ-TiAl alloys (γ-TiAl, tetragonal L1$_0$ structure and α$_2$-Ti$_3$Al, hexagonal DO$_{19}$ structure) exhibit various microstructures [3]. In general, fully-lamellar microstructures present enhanced creep resistance and acceptable fracture toughness, while the duplex microstructures exhibit higher elongation-to-failure [3, 11]. At temperatures above 923-973 K (T>0.4T$_m$), γ-TiAl alloys are susceptible to creep. These materials usually exhibit a limited secondary creep region, characterized by the absence of a steady-state, in which the creep rate reaches a minimum and then increases with strain leading to the tertiary stage [3, 12]. The minimum creep rate ($\dot{\epsilon}_{min}$) depends on, apart from the temperature and the stress, the alloy chemistry and microstructure [13].

Lamellar microstructures have greater fracture toughness than duplex, thanks to the crack deflection at lamellar interfaces and to the development of crack bridging ligaments and microcracking in the lamellae [14]. At high temperatures in lamellar microstructures, the fracture toughness increases [3], colony boundaries and, in particular, triple points, are favorable sites for crack nucleation [10, 15]. Colony sliding promotes the nucleation of wedge cracks which propagate along colony boundaries. Nevertheless, further work is needed to understand better the crack nucleation and propagation mechanisms.

The primary aim of this work was to investigate the deformation and failure mechanisms operative during elevated-temperature tensile-creep deformation of a Ti-45Al-2Nb-2Mn (at.%)-0.8v.% TiB$_2$ intermetallic at different stresses. A transition in the dominant crack propagation path, from interlamellar to intercolonial, was observed as the applied stress decreased [10]. Building from this knowledge, this investigation attempts to quantify the importance of intercolonial, interlamellar and translamellar cracking as a function of the stress level.

EXPERIMENTAL DETAILS

The nominal composition of the intermetallic was Ti-45Al-2Nb-2Mn (at.%)-0.8v.%TiB$_2$. The alloy was provided by GfE (Gesellschaft für Elektrometallurgie mbH) and was processed by centrifugal casting at ACCESS e. V. TechCenter (Aachen, Germany). The dimensions of the castings were 44 mm by 20 mm by 2.5 mm. The centrifugal cast specimens were hot isostatically pressed at 1458 K and 1700 bar for 4 hours in order to remove any remnant porosity in the castings. *In situ* tensile and tensile-creep experiments were conducted at 973 ± 30 K, at a constant strain rate of 10^{-3} s^{-1} and constant stresses between 300 MPa and 450 MPa, respectively. These conditions were chosen in an attempt to reproduce LPT service conditions. The *in situ* tests were performed using a screw-driven tensile stage placed inside the SEM. Temperature was controlled using a constant-voltage power supply to a 6 mm diameter tungsten-based heater located just below the gage section of the sample. Further details of this apparatus and testing technique can be found elsewhere [10, 16-19]. Backscattered electron (BSE) SEM images were taken before and after the experiments and secondary electron (SE) SEM images were taken at periodic displacements throughout tests without interrupting the experiment. The strain values were estimated from the displacement measurements taking into account the heated gage length of the samples.

RESULTS AND DISCUSSION

Understanding the high-temperature deformation and crack propagation mechanisms of γ-TiAl alloys is challenging, as the influence of microstructural parameters is still unclear and a wide variety of compositions and microstructures are possible [3]. In this study, these mechanisms are studied for a defined alloy composition with specific microstructural features. The intermetallic exhibited a nearly-lamellar polycrystalline microstructure, with a small fraction (0.18) of equiaxed α$_2$ and γ grains located at lamellar colony boundaries. The average colony size was 126 ± 52 μm, while the average α$_2$ and γ lamella thickness were, 329 and 422 nm, respectively. The B-rich phase was located both inside the colonies and at the colony boundaries and typically exhibited a lacey shape, as has been observed previously [20].

In addition, the idea to devise a general creep model for these materials is complicated since an analysis of the creep data from conventional methods, as Dorn approach, is quite controversial [3, 15]. Actually, γ-TiAl alloys only present a short secondary creep stage. However, a large range of stress exponents and activation energies are suggested from several studies based on conventional methods [21-24]. Therefore, these previous research suggest the simultaneous contribution of several deformation mechanisms, but the nature and incidence of each of them are unknown.

In order to analyze the deformation mechanisms of the intermetallic at high temperature, the evolution of the lamellar microstructure was observed during the constant strain rate test (10^{-3}s^{-1}) at different stress levels. The maximum tensile strength, σ_{max}, was 460 MPa. The micrographs show that interlamellar ledges (Fig. 1a) became more pronounced with deformation. Interlamellar ledges were first observed at a stress of 374 MPa and a strain value of $\varepsilon\sim0.8$ %. Lacey borides were observed to crack at intermediate strains close to the yield stress. Finally, ledges at colony boundaries were detected at higher strains (Fig. 1b).

(a)   (b)

Figure 1. SE SEM photomicrographs during the tensile test at: (a) $\sigma = 374$ MPa, $\varepsilon\sim0.8\%$; (b) $\sigma = 433$ MPa, $\varepsilon\sim1.4$ %. White arrows point toward interlamellar ledges, black arrows highlight reinforcement cracks and dashed arrows show the initiation of intercolony cracks; The loading axis is vertical.

During tensile-creep tests, the microstructural evolution observed at 300 MPa ($0.65\sigma_{max}$) and 400 MPa ($0.86\sigma_{max}$) was similar. Figure 2 shows representative SEM micrographs taken at different strain levels of the tensile-creep test performed at 300 MPa ($0.65\sigma_{max}$). During the secondary-creep regime, surface relief at the colony boundaries was increasingly evident, suggesting the presence of colony boundary sliding. Crack nucleation and propagation at the colony boundaries was observed during the secondary regime (Fig 2a) (at strains of above 0.4%) and continued into the tertiary creep regime (Fig 2b). The B-rich phase particles were susceptible to brittle fracture at even lower strains (Fig 2a). The primary crack path followed colony boundaries.

However, the deformation and fracture micromechanisms at 425 MPa ($0.92\sigma_{max}$) and 450 MPa ($0.98\sigma_{max}$) were different from those observed at lower stresses. The most remarkable observation is that interlamellar ledges become apparent from the beginning of the secondary stage (Fig 3a). In particular, interlamellar ledges were first detected at a strain of 0.3% and 0.1% at 425 MPa and 450 MPa, respectively. This is similar to what was observed during the tensile test. For these higher stress tensile-creep tests, colony boundary cracking was limited to the tertiary stage (Fig. 3b). B-rich phase cracking was also observed at these conditions.

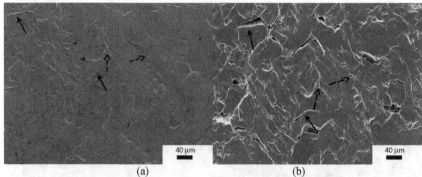

(a)　　　　　　　　　　　　　(b)

Figure 2. SE SEM photomicrographs obtained during (a) secondary stage, 50h and $\varepsilon \sim 0.4\%$, (b) tertiary stage, 76h and $\varepsilon \sim 1.5\%$. Dash arrows point colony boundary cracks and black arrow points B-rich phase cracking. The loading axis is vertical.

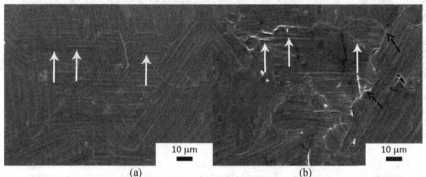

(a)　　　　　　　　　　　　　(b)

Fig 3. SE SEM photomicrographs corresponding to a test at 450 MPa illustrating interlamellar ledges: (a) secondary creep stage, 0.3 hours, $\varepsilon \sim 0.1\%$; (b) tertiary creep stage, 1 hour, $\varepsilon \sim 0.6\%$. White arrows points interlamellar ledges and dash arrows point colony boundary cracks. The loading axis is vertical.

These observations confirm the importance of colony boundary sliding (CBS) during the secondary creep stage of the Ti-45Al-2Mn-2Nb-0.8v.%TiB$_2$. Furthermore, this mechanism appears to be active at relatively high stresses, up to about 400 MPa, even when the colony size is larger than the critical size proposed previously [23]. This suggests that special care must be taken in reinforcing the CBS resistance if the creep resistance of γ-TiAl alloys under service conditions is to be improved. At stresses greater than 400 MPa, interlamellar ledges were observed during the early stages of the secondary creep regime. These were probably caused by the need to accommodate strain incompatibilities resulting from dislocation sliding between adjacent lamellae. CBS was limited to the tertiary stage. These observations are consistent with the prevalence of dislocation slip during the secondary stage at these very high creep stresses or, equivalently, during high constant strain rate tests at the quasi-static rates investigated. According to the fracture behavior, cracks formed preferentially at colony boundaries during the early stages of deformation in low-stress tensile-creep and during the later stages of deformation in high-stress tensile-creep.

The classification of the different secondary cracks (Fig. 4a), illustrated that the main damage mechanism was intercolony cracking for all the tests conditions. However, this mechanism was more prevalent at the lowest strain rates. Some interlamellar cracks were also detected but this mechanism was of less importance. Translamellar secondary cracks were rarely observed. Analysis of the primary crack path further revealed that catastrophic failure was mostly due to crack propagation along grain boundaries at low stresses (up to 400 MPa) in tensile-creep, and across the interior of grains at higher stresses in tensile-creep and upon constant strain rate loading at the quasi-static rates investigated (see Fig 4b). Interlamellar crack propagation was negligible in comparison with the other two.

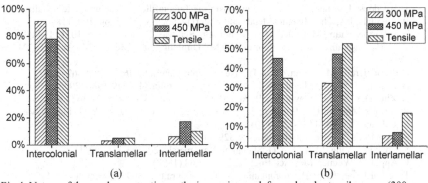

Fig 4. Nature of the crack propagation paths in specimens deformed under tensile-creep (300 MPa and 450 MPa) and tension. (a) Secondary cracks. (b) Primary crack.

CONCLUSIONS

The *in situ* SEM observations during tensile and tensile-creep test on the Ti-45Al-2Mn-2Nb-0.8v.%TiB$_2$ alloy showed that at low stresses ($\leq$ 400 MPa), colony boundary sliding was activated during the secondary creep stage. However, at higher stresses, colony boundary sliding was only observed during the tertiary stage. In addition, interlamellar ledges were only observed during the early stages of the high temperature tensile tests and the tensile-creep tests performed at the highest stresses ($\geq$425 MPa). Crack nucleation took place mostly at colony boundaries at all the testing conditions investigated. However, tensile tests and tensile-creep tests at higher stresses revealed that this phenomenon was delayed but still present. In addition, early cracking of B-rich phase reinforcements was observed. Colony boundary cracking was the main damage mechanism for all the tests conditions. Some interlamellar secondary cracks were also detected but this mechanism was of less importance. Translamellar secondary cracks were rarely observed. Translamellar crack path segments were observed only along the primary crack. It is suggested that they connected intercolonial crack segments upon catastrophic failure.

ACKNOWLEDGMENTS

Funding from the Spanish Ministry of Science and Innovation through projects (MAT2009-14547-C02-01 and MAT2009-14547-C02-02) is acknowledged. The Madrid Regional Government partially supported this project through the ESTRUMAT grant (P2009/MAT-1585). CJB acknowledges the support from the Spanish Ministry of Education for his sabbatical stay in Madrid (SAB2009-0045).

REFERENCES

1.   D.M. Dimiduk, Materials Science and Engineering A 263, 281-288 (1999).
2.   G. Lütjering, J.C. Williams in *Titanium, Enigineering Materials and Processes* edited by B. Derby (Springer-Verlag Berlin Heidelberg, Germany, 2007) pp. 337-366.
3.   F. Appel, J. D. Paul, M. Oehring in *γTitanium Aluminides*, edited by Wiley-VCH & Co (Weinheim, Germany, 2011).
4.   J. Beddoes, L. Zhao, P. Au, D. Dudzinski, J. Triantafillou, Structural Intermetallics, Warrendale, PA, The Minerals Metals and Materials Society (1997) pp. 109–118.
5.   T.A. Parthasarathy, M. Keller, M.G. Mendiratta, Scr. Metall. 38 (7), 1025–1031 (1998).
6.   T.A. Parthasarathy, M.G. Mendiratta, D.M. Dimiduk, Scr. Metall. 37 (3), 315–321 (1997).
7.   C.J. Boehlert, D.M. Dimiduk, K.J. Hemker, Scripta Materialia 46 (4), 259-267 (2002).
8.   H. Zhu, D.Y. Seo, K. Mayurama, P. Au, Scripta Materalia 54, 1979-1984 (2006).
9.   H. Zhu, D.Y. Seo, K. Mayurama, P. Au., Mat. Sc. Eng. A 483-484, 533-536 (2008).
10. R. Muñoz Moreno, C.J. Boehlert, E.M. Ruiz Navas, M.T. Pérez Prado, J. Llorca, Met. Mat. Trans. A 43 A, 1198-1208 (2012).
11. M. A. Morris, T. Lipe, Intermetallics 5, 329-337 (1997).
12. S. Liwen, L. Ying, M.A. Yue, G. Shengkai, Rare metals, 30, 323-325 (2011).
13. F. Appel, R. Wagner, Mat. Sci. Eng. R 22, 187-268 (1998).
14. T. A. Parthasarathy, P. R. Subramanian, M. G. Mendiratta and D. M. Dimiduk, Acta mater. 48, 541-551 (2000).
15. M.E. Kassner, M.T. Pérez-Prado in *Fundamentals of creep in metals and alloys* edited by Elsevier (Oxford, 2004).
16. C.J. Boehlert, S.C. Longanbach , T.R. Bieler, Phil. Mag. 88 (5), 641-664 (2008).
17. C.J. Boehlert, C.J. Cowen, S. Tamirisakandala, D.J. McEldowney, D.B. Miracle, Scr. Mater. 55, 465-468 (2006).
18. C.J. Cowen, C.J. Boehlert, Metall. Mater. Trans. A 38A, 26-34 (2007).
19. W. Chen , C. J. Boehlert, International Journal of Fatigue 32 (5), 799-807 (2010).
20. D.E. Larsen, S. Kampe, L. Christodoulou, Mater. Res. Soc. Symp. Proc. 194, 285 (1990).
21. W.J. Zhang, S.C. Deevi, Intermetallics 10, 603-611 (2002).
22. W.J. Zhang, S.C. Deevi, Intermetallics 11, 177-185 (2003).
23. J. Beddoes, W. Wallace, L. Zhao, Int. Mater. Reviews 40 (5), 197-217 (1995).
24. L. M. Hsiung, T.G. Nieh, Intermetallics 7, 821-827 (1999).

Mater. Res. Soc. Symp. Proc. Vol. 1516 © 2012 Materials Research Society
DOI: 10.1557/opl.2012.1577

In Situ High-Energy XRD Study of the Hot-Deformation Behavior
of a Novel γ-TiAl Alloy

Andreas Stark[1], Emanuel Schwaighofer[2], Svea Mayer[2], Helmut Clemens[2], Thomas Lippmann[1], Lars Lottermoser[1], Andreas Schreyer[1] and Florian Pyczak[1]

[1]Helmholtz-Zentrum Geesthacht, Institute of Materials Research, Max-Planck-Straße 1, D-21502 Geesthacht, Germany
[2]Montanuniversität Leoben, Department of Physical Metallurgy and Materials Testing, Franz-Josef-Straße 18, A-8700 Leoben, Austria

ABSTRACT

The development of suitable hot-forming processes, e.g. forging, is an important step towards the serial production of TiAl parts. Several microstructure parameters change during hot-forming. However, the underlying mechanisms can normally only be inferred from *post process* metallographic studies.

We used a deformation dilatometer modified for working in the HZG synchrotron beam-lines at DESY for hot-deformation experiments. This setup enables the *in situ* monitoring of the interaction and evolution of microstructure parameters during processing. We observed the evolution of phase fractions, grain size and crystallographic texture during deformation while simultaneously recording the process parameters, like temperature, force and length change.

Here we present the hot compressive deformation behaviour of a Ti-43Al-4Nb-1Mo-0.1B (in at.%) alloy. Several specimens were deformed at three temperatures each with two compression rates. During the experiments the Debye-Scherrer diffraction rings were continuously recorded.

INTRODUCTION

The reduction of fuel consumption and greenhouse gas emissions requires the development of novel engineering materials, which combine light weight with high strength at elevated temperatures. Among these, intermetallic γ-TiAl based alloys are the most promising candidates for replacing the twice as dense Ni base super-alloys in high temperature applications such as turbine blades in jet engines and industrial gas turbines [1,2]. In recent years, research activities are focused on TiAl alloys containing additional amounts of ductile bcc high-temperature β phase in order to improve their hot-workability [3,4].

Thermo-mechanical treatments (TMT) as hot-rolling or forging are well established processes to improve mechanical properties and to homogenize the microstructure as well as for near net-shape production. On the other hand TMT can produce unwanted mechanical anisotropy due to the formation of crystallographic textures. This makes it necessary to find suitable process parameters specific for material and application due to the fact that several microstructure parameters, e.g. phase fractions, grain size or crystallographic texture, change during hot-forming. While conventional analysis methods can only infer the changing mechanisms from *post process* metallographic studies, in our study we monitor the evolution of these parameters *in situ*, i.e. during processing, using synchrotron radiation.

EXPERIMENT

The hot-deformation experiments were performed on a quenching and deformation dilatometer DIL805A/D from BÄHR-Thermoanalyse GmbH, Germany, which was modified to work in the HZG beamline HARWI II at DESY [5]. As starting material an as-cast and HIPed TNM alloy with a nominal composition of Ti-43Al-4Nb-1Mo-0.1B (in at.%) was used for this study.

Cylindrical samples with 5 mm in diameter and 10 mm in length were used for the experiments. In order to avoid oxidation the experiments were performed in Ar atmosphere. The temperature was controlled by a spot welded type B thermocouple. A scheme with the process parameters is shown in figure 1a. Three temperatures were selected for the experiments: 1100 °C (below the eutectoid temperature T_e), 1200 °C (above T_e) and 1230 °C (near the γ-solvus temperature $T\gamma_{solv}$) [4]. At each temperature specimens were deformed with a compression rate of $1 \cdot 10^{-2}$ s^{-1} (slow) and $3 \cdot 10^{-2}$ s^{-1} (fast), respectively. The specimens were heated up to process temperature with 400 °C·min^{-1}, followed by an isothermal holding at temperature for 10 min. Subsequently, they were deformed up to a total deformation of $\varphi = -0.6$ corresponding to about 45 % height reduction. Immediately after deformation, within 1 s, the samples were quenched to room temperature (RT) with a cooling rate of more than -50 °C·s^{-1} by blowing with gas in order to keep the deformed microstructure.

During the experiments high-energy X-ray diffraction (HEXRD) was performed in transmission geometry (see fig. 1b). In order to transmit the samples a high-energy X-ray beam with an energy of 100 keV (corresponding to a wavelength of 0.124 Å) and a beam size of $1 \cdot 1$ mm^2 was used. The resulting diffraction rings were continuously recorded during deformation using a mar555 flat panel detector (Marresearch GmbH, Germany) with an exposure rate of 0.25 Hz and an exposure time of 1 s. The diffraction rings were azimuthally integrated using the fit2d software [6]. Phase fractions and crystallographic textures were calculated with the MAUD program [7].

Subsequently, the deformed samples were cut and electrolytically polished for microstructure analysis. Scanning electron microscopy (SEM) studies were done with a Zeiss EVO 50 (Zeiss, Germany).

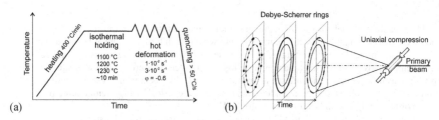

Figure 1. Scheme of the compression experiments. (a) Processing parameters. (b) Diffraction geometry.

RESULTS AND DISCUSSION

Microstructure

The as-cast and HIPed alloy mainly consists of lamellar ($\alpha_2+\gamma$) colonies and smaller globular γ and β_o grains at colony boundaries and triple points (fig. 2a). After heating up to process temperature primarily the ratio between α_2/α and γ phase in the lamellar colonies changes and the amount of globular β_o/β increases (fig. 2b). After slow deformation at 1100 °C the lamellar colonies are strongly elongated perpendicular to the load direction and the β phase is squeezed around these colonies. Additionally, small areas of fine-grained dynamically recrystallized (DRX) γ phase can be observed at several colony boundaries (fig. 2c). After slow deformation at 1200 °C the DRX areas between the colonies are larger, forming a typical necklace structure (fig. 2d). Additionally, the β phase is less elongated. After fast deformation at 1230 °C the amount of DRX α phase is significantly increased. The DRX α grains show a globular shape and grain sizes in the range of 2 to 10 µm (fig. 2e). After slow deformation at 1230 °C the microstructure is almost exclusively formed by globular DRX α grains with grain sizes up to 30 µm and a few β grains at triple points (fig.2f). From the SEM micrographs it appears that the phase fractions of the both specimens deformed at 1230 °C differ significantly, however, as shown in the next paragraph, they were almost equal during deformation.

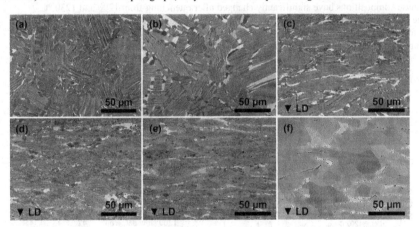

Figure 2. SEM micrographs taken in BSE mode. (a) HIPed starting material and (b) heated up to 1200 °C and quenched after 10 min holding time. Slow deformation at (c) 1100 °C, (d) 1200 °C and (f) 1230 °C. (e) Fast deformation at 1230 °C The γ phase appears in dark grey, α_2/α in medium grey and β_o/β in light grey. LD: Load direction.

Phase fractions

The Debye-Scherrer diffraction rings were continuously recorded during the experiments. Because the direction of the uniaxial deformation is oriented parallel to the detector plane (fig.

1b) the azimuthally integrated XRD patterns are not influenced by texture effects. Thus the phase fractions can be calculated by Rietveld analysis with the MAUD program.

Figure 3 shows the XRD patterns of the starting material and of the slow deformation experiments each with one representative pattern during deformation and one after quenching at room temperature. Additionally, the calculated phase fractions of all experiments are listed in table I. In the as-cast and HIPed condition the alloy consists of 23 vol.% α_2, 7 vol.% β_o and 70 vol.% γ. At 1100 °C the superlattice reflections of the ordered phases α_2 and β_o are clearly observed indicating a temperature below T_e. At 1200 °C the alloy still consists of three phases, however, now with the disordered high temperature variants α and β indicated by the extinction of the superlattice reflections $(100)\alpha_2$, $(101)\alpha_2$, $(110)\alpha_2$, and $(100)\beta_o$. Additionally, the amount of α significantly increases on cost of γ with increasing temperature. The temperature of 1230 °C is close to $T\gamma_{solv}$. During slow deformation no γ was detected in the diffraction pattern, whereas in the experiment with fast deformation a small amount of about 4 vol.% γ is still observable. This little discrepancy might be attributed either to a low temperature overshot during heating up, small chemical inhomogeneities or little differences of the spot welded thermocouple junctions. However, compared to an experimentally received phase diagram of a similar alloy [4] the measured phase fractions during slow and fast deformation lie within a narrow temperature range of about ±10 °C.

After quenching all deformed specimens consist of α_2, β_o and γ. It is interesting to note that the phase compositions have significantly changed after quenching from 1200 and 1230 °C. These quenched samples show about 15 to 20 vol.% more γ and less α. This high amount of transformed α could be attributed to a high driving force due to the chemical disequilibrium of the high temperature phase compositions at lower temperatures. Additionally, the transformation

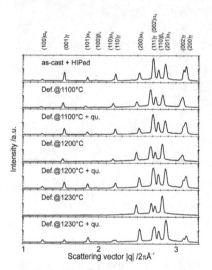

Figure 3. XRD pattern of the starting material and the slow ($1 \cdot 10^{-2}$ s^{-1}) deformation experiments each during deformation and after quenching at RT.

Table I. Phase fractions in vol.% during deformation and after quenching at RT.

Condition	α_2/α	β_o/β	γ
as-cast + HIPed	23	7	70
1100 slow def.	26	12	62
1100 slow def. + qu	24	13	63
1100 fast def.	25	12	63
1100 fast def. + qu	24	12	64
1200 slow def.	66	10	24
1200 slow def. + qu	42	13	45
1200 fast def.	68	9	23
1200 fast def. + qu	46	10	44
1230 slow def.	89	11	–
1230 slow def. + qu	76	8	16
1230 fast def.	89	7	4
1230 fast def. + qu	54	11	35

rate could be increased because the high dislocation density and defect concentration after deformation facilitates diffusion and also acts as nucleants. Nevertheless, this high amount of transformed α in spite of a quenching rate comparable to oil-quenching demonstrates the difficulties in evaluating correct high temperature compositions by conventional *post process* analysis.

In situ observation of texture evolution during hot compression

In figure 4 the texture evolution is exemplarily shown for the α phase at slow deformation ($1 \cdot 10^{-2}$ s^{-1}) at 1230 °C. In order to study the evolution of grain size and formation of crystallographic texture during deformation the diffraction ring of a specific reflection, in the present example the 002 reflection of the hexagonal α phase, was unrolled. The unrolled rings of consecutive images were combined to an azimuth angle vs. time diagram [8]. Single reflection spots on the diffraction ring occur as 'time-lines' in the diagram. Before deformation starts, sharp spots indicate a relatively coarse grained microstructure with almost perfect crystals. No intensity agglomeration can be observed at preferred azimuth angles, i.e. the crystals show an almost random orientation distribution. During elastic deformation the intensity distribution does not change significantly. As plastic flow starts, the spots become more and more diffuse indicating the increase of dislocation density, crystal defects and tilting between crystallite blocks. During further deformation, starting from about 15 % deformation, almost symmetric intensity maxima are formed at about 20°-30° tilt to the load direction. The texture calculated for the α phase shows a tilted basal fibre texture which is a typical deformation texture of the hexagonal α phase [9]. After about 30 % deformation, again weak sharp spots occur in between as well as on the diffuse intensity maxima, indicating the increasing influence of DRX, e.g. the nucleation and growth of new defect-free crystallites. This process weakens the pure deformation texture, however, the texture does not vanish during further deformation.

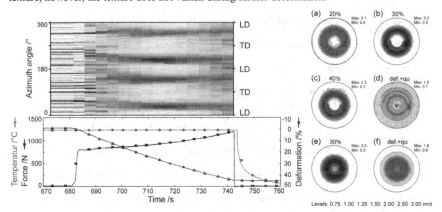

Figure 4. Azimuth angle vs. time diagram of the α 002 reflection during compression at 1230 °C with a compression rate of $1 \cdot 10^{-2}$ s^{-1}. The intensity is coded in greyscale. The process parameters are displayed below. LD: Load direction. TD: Transverse direction

Figure 5. Recalculated α 002 pole figures in load direction of (a-d) the slow and (e,f) the fast deformation experiment at 1230 °C.

75

The texture evolution can be quantitatively described by means of recalculated α 002 pole figures, as shown in figure 5, for several stages of the deformation and after quenching. Up to 30 % compression the maximum intensity raises from 2.1 to 3.3 multiples of random distribution (mrd) and the sharpness increases (fig. 5a,b). Subsequently the texture diminishes due to DRX to 2.3 mrd (fig. 5c) and after quenching the texture additionally becomes weaker (1.6 mrd) and significantly diffuser (fig. 5d). The texture evolution during fast deformation shows a similar tendency, however, the reduction is less pronounced (fig. 5e,f). Probably the higher diffusiveness of figure 5d after quenching is caused by the large DRX grains which were more disturbed by internal stresses increasing the mosaicity.

Future studies should focus also on the influence of static recrystallization and specific heat treatments after deformation on texture formation and/or texture degradation.

CONCLUSIONS

Several hot compression tests were performed on an advanced TNM alloy. The X-ray diffraction rings were continuously recorded during the deformation experiments using high-energy synchrotron radiation. This enables an *in situ* observation of the evolution of phase fractions and the formation of crystallographic textures during deformation, which is not possible by conventional metallographic methods. So, it was possible to separate the effect of defor-mation and dynamic recrystallization on the texture evolution during the *in situ* experiment, which is often difficult to determine from *post mortem* investigations.

The results demonstrate that *in situ* synchrotron experiments are a powerful tool in order to develop suitable process parameters especially for hot-forming of multiphase alloys, such as novel γ-TiAl based alloys.

REFERENCES

1. F. Appel, J.D.H. Paul and M. Oehring, *Gamma Titanium Aluminide Alloys - Science and Technology* (Wiley-VCH, Weinheim, Germany, 2011).
2. H. Clemens and S. Mayer, *Advanced Engineering Materials* in print, DOI: 10.1002/adem.201200231.
3. H. Clemens, W. Wallgram, S. Kremmer, V. Güther, A. Otto and A. Bartels, *Advanced Engineering Materials* 10, 707–713 (2008).
4. M. Schloffer, F. Iqbal, H. Gabrisch, E. Schwaighofer, F.-P. Schimansky, S. Mayer, A. Stark, T. Lippmann, M. Göken, F. Pyczak and H. Clemens, *Intermetallics* 22, 231-240 (2012).
5. P. Staron, T. Fischer, T. Lippmann, A. Stark, S. Daneshpour, D. Schnubel, E. Uhlmann, R. Gerstenberger, B. Camin, W. Reimers, E. Eidenberger, H. Clemens, N. Huber and A. Schreyer, *Advanced Engineering Materials* 13, 658-663 (2011).
6. A.P. Hammersley, S.O. Svensson, M. Hanfland, A.N. Fitch and D. Häsermann, *High Pressure Research* 14, 235–248 (1996).
7. L. Lutterotti, M. Bortolotti, G. Ischia, I. Lonardelli and H.-R. Wenk, *Z. Kristallogr. Suppl.* 26, 125-130 (2007).
8. K.-D. Liss, T. Schmoelzer, K. Yan, M. Reid, R. Dippenaar and H. Clemens, *Journal of applied physics* 106, 113526 (2009).
9. A. Stark, F.-P. Schimansky and H. Clemens, *Solid State Phenomena* 160, 301-306 (2010).

Mater. Res. Soc. Symp. Proc. Vol. 1516 © 2012 Materials Research Society
DOI: 10.1557/opl.2012.1666

Oxidation Behavior of Intermetallic Titanium Aluminide Alloys

Michael Schütze[1], Simone Friedle[1]
[1] Dechema-Forschungsinstitut, Theodor-Heuss-Allee 25, 60486 Frankfurt am Main, Germany

ABSTRACT

Above 750-800°C oxidation becomes a serious life time issue for the new group of intermetallic light-weight high temperature alloys based on titanium aluminides (TiAl). Fast growing titanium oxide competes with protective alumina as a surface scale in the oxidation reaction by which the formation of a slow-growing protective oxide scale is prevented. The key to the development of alloys with sufficient oxidation resistance is the understanding of the thermodynamic and kinetic situation during the oxidation process. The latter is influenced by the type of alloying elements, the Al- and Ti-activities in the alloy, the oxidation temperature and the environment (e.g. dry or humid air, etc.). This paper provides a comprehensive summary of the oxidation mechanisms and the parameters influencing oxide scale formation. Besides the role of metallic alloying elements, the halogen effect will also be discussed. The paper finishes with recent results concerning the prevention of oxidation-induced room temperature embrittlement of TiAl alloys.

INTRODUCTION

The oxidation behavior of intermetallic γ-based titanium aluminides alloys as a new group of high temperature lightweight materials has been an issue since the first investigations of their properties [1-3]. Over the years the development of technical versions of these alloys has been focused on base compositions with aluminum contents in the range of 40-50 at.%, i.e. the microstructure is defined by γ-TiAl as the major phase and α_2-Ti$_3$Al as a second phase whose amount depends on the aluminum content. While there are also alloys with aluminum contents lower than 40 at.%, such as Ti$_3$Al and orthorhombic titanium aluminides, this paper focuses exclusively on those with an aluminum content of at least 40 at.%, which are simply called TiAl alloys throughout the text.

Concluding from oxidation results of commercial nickel- and iron-based alloys, one would expect that aluminum contents of 40 at.% and more should be sufficient to form a protective slow-growing alumina scale in air during high temperature exposure [4]. Such a scale suppresses rapid oxidation and therefore reduces metal consumption. However, due to thermodynamic and kinetic reasons (for details see below) in most cases a mixed scale containing fast growing titania in addition to alumina is formed. At least for temperatures of up to 650°C, virtually all TiAl alloys reveal a sufficient and reliable oxidation resistance. By adding further alloying elements to the bulk material, a reasonable oxidation resistance can be achieved for temperatures of up to 850°C [5,6]. Future operation conditions of TiAl alloys, however, envisage maximum temperatures of up to 900°C and 1050°C in jet engines and for automotive turbocharger rotors, respectively, at which special non-conventional surface treatments are required in order to maintain an adequate oxidation protection [7]. Examples of such coatings will be shown later in this paper.

THE EFFECT OF OXYGEN AND NITROGEN ON THE OXIDATION BEHAVIOR OF TiAl ALLOYS

One major reason why the oxidation resistance decreases significantly at temperatures above 650°C, despite a relatively high aluminum content of 40 at.%, is the simultaneous formation of aluminum and titanium oxides on the alloy surface at these temperatures due to the nearly identical thermodynamic stabilities of Al_2O_3 and TiO_2 over a large temperature range. As a result, oxide scale growth can be five times higher (or even more) than that for mere Al_2O_3 formers. Oxide scale growth occurs by a diffusion-controlled mechanism and TiO_2 shows a significantly higher degree of lattice disorder than Al_2O_3 [8], which leads to an increased oxidation rate. Furthermore, the formation and growth of alumina leads to depletion of aluminum in the alloy subsurface and thus, to an increase in titanium activity in the alloy subsurface promoting the formation of TiO_2.

When comparing the oxidation behavior of TiAl alloys in air and in pure oxygen, significant differences are observed [9,10]. It was found that the temperature, at which a significant acceleration of the oxidation rate occurs, increases clearly with decreasing nitrogen content in the atmosphere (see figure 1). Meier and Pettit showed that pure alumina scales formed up to temperatures of about 1000°C on pure γ-TiAl in an atmosphere of pure oxygen, whereas scales of mixed TiAl-oxides were present at higher temperatures [9]. Traces of nitrogen lead to a decrease in transition temperature by about 50°C while in air this temperature was as low as ~850°C. It has to be noted that an oxidation time of 60 hours is rather short and breakaway effects that lead to a massive acceleration of the oxidation rates will eventually occur after longer exposure times of about 300 to 600 hours [3], which can be explained by the destruction of the continuous partial layer of alumina, functioning as a diffusion barrier, due to thermodynamic instability or oxidation-induced growth stresses [3,11]. Although the presence of nitrogen seems to promote the breakaway effect, accelerated oxidation was found even in pure oxygen at 1000°C [3].

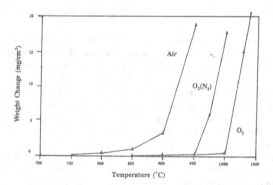

Figure 1. Mass change curves for γ-TiAl oxidized for 60h at various temperatures and atmospheres [9].

All of these findings call for a model description of the oxidation mechanisms. At first the situation for pure oxygen will be discussed. According to the Richardson-Ellingham diagram for oxygen alumina is thermodynamically slightly more stable compared to titania, but this situation may change if the metal activities are not equivalent anymore. When starting off with stoichiometric γ-TiAl, a certain dominance of alumina formation has to be expected [9,12]. This dominance will prevail after the first oxide nuclei have formed on the metallic TiAl surface consisting of both types of oxides due to their similar thermodynamic stability at the respective oxygen partial pressure in air or pure oxygen. By the formation of alumina a reduction in aluminum activity in the metal subsurface will occur and that of titanium will increase. Therefore, the alumina barrier formed at the beginning of oxidation will eventually be interspersed with newly formed titania if it cracks due to oxide growth stresses. This situation can also occur in pure oxygen, which would explain the accelerated oxidation reaction at 1000°C where growth stresses are higher and after longer exposure times at temperatures below 1000°C when these stresses had sufficient time to build up.

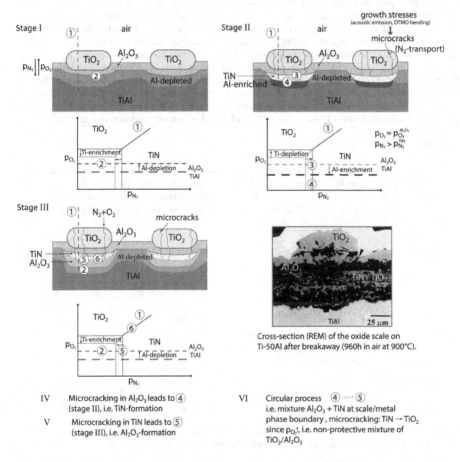

Cross-section (REM) of the oxide scale on Ti-50Al after breakaway (960h in air at 900°C).

IV   Microcracking in Al$_2$O$_3$ leads to ④ (stage II), i.e. TiN-formation

V   Microcracking in TiN leads to ⑤ (stage III), i.e. Al$_2$O$_3$-formation

VI   Circular process ④ ---- ⑤ i.e. mixture Al$_2$O$_3$ + TiN at scale/metal phase boundary, microcracking: TiN → TiO$_2$ since p_{O_2}↑, i.e. non-protective mixture of TiO$_2$/Al$_2$O$_3$

Figure 2. The oxide-nitride cycle during high temperature exposure of γ-TiAl in air.

In nitrogen-containing oxidizing environments, such as air, the situation is more complex and is described schematically in figure 2. The oxygen partial pressure gradient increases from the oxide/metal interface (position 2) towards the environment (position 1) while in the pores and microcrack network of the oxide scale the nitrogen partial pressure increases in the opposite direction [13]. Stage I describes the initial situation where titania and alumina form side by side (nucleation step) whereby titania is undergrown by alumina. As illustrated in stage II, oxide growth stresses in the scales can lead to the formation of microcracks at 900°C in air after 5-10 hours followed by repeated cracking and healing of the scale [11]. Therefore, the oxide scale can be considered as a "living system". The microcracks allow access of nitrogen to the scale/metal interface and titanium nitride (TiN) formation occurs as the oxygen partial pressure underneath the oxide is below the equilibrium pressure of alumina for the aluminum depleted phase and, thus, insufficient to prevent nitride formation (position 3, stage II). The formation of TiN in the underlying metal subsurface leads to Al-enrichment so that again alumina formation becomes favored (stage III, position 4). Once an alumina scale has been established underneath the TiN, a situation comparable to position 2 in stage I occurs again at the scale/metal interface (position 5 in stage III) and the cycle restarts. During the ongoing oxidation process the TiN is shifted to positions with higher oxygen partial pressures by thickness growth (including microcracking) of the oxide, i.e. it will no longer be stable and will be converted into titania. As a consequence, the presence of nitrogen that drives this cycle leads to the formation of a mixed titania/alumina scale resulting in increased oxidation growth rates eventually leading to breakaway behavior. This process will ultimately apply to the entire surface of the alloy (see cross-section in figure 2).

THE WATER VAPOR EFFECT

The presence of water vapor in air and oxygen can accelerate the oxidation rates of TiAl alloys significantly compared to the corresponding dry atmospheres [14]. This effect can occur at temperatures as low as 700°C [15] (figure 3), whereby it also affects the creep properties of the material [16]. At higher temperatures, at 900°C, in moist atmospheres containing oxygen, a significant increase in oxidation rate is found [17].

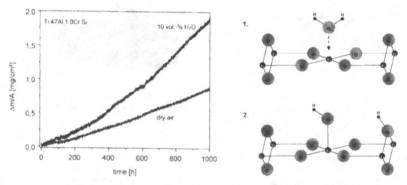

Figure 3. Left: Mass gain as a function of oxidation time for dry air and air with 10 vol.% H_2O, 700 °C. Right: Schematic diagram of the two-step adsorption process of H_2O on the rutile (110) surface[15].

Interestingly, it was observed in the 700°C study that when switching back from moist to dry atmosphere the oxidation rates immediately decreased to their original level while changing the atmosphere back to water vapor was followed by an incubation period of about 20 hours before accelerated oxidation reoccurred [15]. It is assumed that the rutile phase catalyzes water dissociation into free hydrogen atoms and hydroxide (OH⁻) groups, which alter its defect chemistry and lead to preferential growth of this phase [15]. Therefore it can be concluded that, similar to the nitrogen effect, water vapor and its decomposition products are unfavorable from a standpoint of oxidation resistance.

EFFECTS OF ALLOYING

The aim of all alloying approaches of TiAl is the stimulation of a slow-growing continuous alumina scale and the suppression of titania formation. Alloying can also lead to the formation of other protective scale phases, but so far this approach has not been successfully applied and will not be discussed here. The most successful "conventional" alloying approach for oxidation protection has been the addition of niobium and so all currently developed technical TiAl alloys contain at least 2 at.% of this metal. It was speculated that niobium increases the Al-activity in the subsurface zone and stabilizes the γ-phase [3] which was, however, questioned by later investigations [18,19]. Furthermore, the formation of an inner nitride barrier underneath the main oxide scale was assumed to be responsible for the increased protective effect [20,21]. Schmitz-Niederau et al. discovered, however, that the nitride layer was discontinuous and that the protective effect arises from a continuous alumina partial layer (about 1 μm) due to a decreased aluminum solubility in titania by doping with Nb^{5+} [22]. A reduction of the growth rate of titania by this doping effect leading to a decrease of the number of oxygen vacancies and titanium interstitials seems to be less efficient than the formation of an alumina barrier[8].

A comprehensive summary of the role of additional alloying elements for the oxidation resistance of TiAl was reported by Fergus (figure 4) [23]. It has to be considered, that the amount of alloying element can alter the oxidation properties significantly. The addition of chromium, for example, only has a beneficial effect when the contents are higher than 8wt.%. In general, every newly developed alloy has to be tested for its oxidation properties since the interaction between the different alloying elements is difficult to predict.

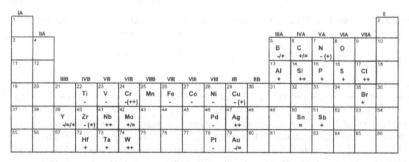

IA	IIA	IIIB	IVB	VB	VIB	VIIB	VIII	VIII	VIII	IB	IIB	IIIA	IVA	VA	VIA	VIIA	0
1																	2
3	4											5 B -/+	6 C +/=	7 N - (+)	8 O	9	10
11	12											13 Al +	14 Si ++	15 P +	16 S +	17 Cl ++	18
19	20	21	22 Ti	23 V -	24 Cr -(++)	25 Mn -	26 Fe -	27 Co -	28 Ni -	29 Cu - (+)	30	31	32	33	34	35 Br +	36
37	38	39 Y -/=/+	40 Zr - (+)	41 Nb ++	42 Mo +/=	43	44	45	46 Pd -	47 Ag ++	48	49	50 Sn =	51 Sb +	52	53	54
55	56	57	72 Hf +	73 Ta +	74 W ++	75	76	77	78 Pt -	79 Au -/=	80	81		83	84	85	86

Figure 4. Periodic table of the elements indicating the effect of alloying additions on the oxidation behavior of TiAl alloys [23].

Besides metallic alloying additions also small amounts of non-metallic elements were investigated with regard to their effect on decreasing the oxidation rates. Among these elements were boron, carbon, phosphorus and the halogens. The effect of halogens on oxidation will be addressed in more detail in the next section as they showed the highest efficiency at temperatures above 850°C. Boron and carbon showed a beneficial effect on the oxidation behavior of TiAl in 100 hours exposure tests in air at 900°C [6]. From these investigations it can be concluded that boron and carbon additions of around 35 ppm and 600 ppm, respectively, have a certain beneficial effect at least in these relatively short tests, which, however, does not exceed that of alloying with metallic elements. It can be suspected that the addition of carbon leads to a grain refinement of the microstructure with a more homogenous aluminum supply to the surface and therefore to a different oxidation behavior. No beneficial effect of boron and carbon were observed for specimens micro-alloyed by ion implantation, whereby boron even seemed to be detrimental in these investigations [24]. Furthermore, phosphorus implantation lead to a significant improvement of oxidation resistance at 900°C in air for about 200 hours. The kinetics were comparable to mere alumina kinetics and the same as for chlorine implanted samples [24,25]. While the latter, however, showed alumina kinetics even after 1000 hours of exposure there was a clear breakaway effect for phosphorus after 200 hours, figure 5.

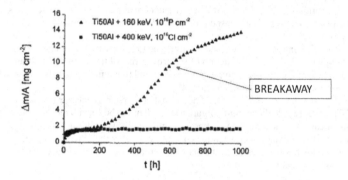

Figure 5. Comparison of the oxidation kinetics at 900°C in air of TiAl samples implanted with P and Cl, respectively [25].

THE HALOGEN EFFECT

The addition of halogens to the alloy surface so far appears to be the most effective way to achieve an oxidation resistance of TiAl alloys that is characterized by mere alumina kinetics. The effect has been proven to work, in particular for fluorine, at temperatures up to 1050°C for up to 1000 hours[26] and at 900°C for oxidation times of at least one year [27]. The effect is maintained in air containing water vapor and also under thermo cyclic conditions [28]. Figure 6 illustrates the halogen effect on a TiAl sample that was treated with chlorine only locally (upper left) or only on one side (lower left). Here, for an optimum halogen concentration in the surface a very thin alumina scale had formed, whereas the untreated material revealed a thick, mixed oxide scale. The comparison of oxidation rate constants at 900°C (figure 6, right) indicates that chlorine implantation on TiAl alloys has the potential to reach alumina kinetics, whereby fluorine performs even better, especially under cyclic conditions.

The mechanism of the halogen effect has been a matter of debate and different models have been suggested. One model proposes Al-enrichment aiding the formation of a pure alumina scale as a result of titanium halide volatilization at the alloy/scale interface [29]. In contrast, Kumagai et al. proposed that halogen ions in the rutile lattice decrease the number of oxygen vacancies and, thus, the diffusion of oxygen and the growth rate of the oxide scale [30]. Investigation of metallographic cross-sections, however, revealed that the oxide scale consists solely of alumina with an Ti-enriched zone underneath, not supporting any of these two models. Furthermore, ion implantation experiments were performed on pre-oxidized specimens whereby in one case the halogen ions were only implanted in the oxide scale and in the second case they were segregated in the area of the oxide/metal interface (figure 7) [31]. This experiment confirmed that the halogen effect only occurs when the halogen ions are located at the oxide/metal interface.

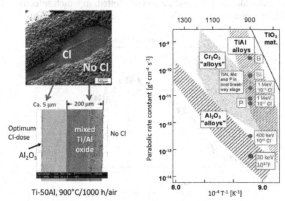

Figure 6. Left: Effect of chlorine on a γ-TiAl sample after high temperature oxidation. Right: Effect of microalloying on the oxidation behavior of TiAl alloys.

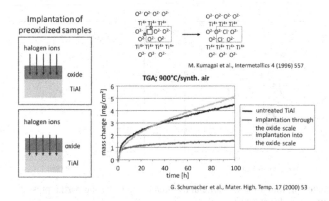

Figure 7. The lattice doping model of the halogen effect.

Concentration measurements on halogen-doped and then oxidized specimens after 100 hours at 900°C in air by proton induced gamma emission (PIGE) analysis confirmed that in the case of the halogen effect in oxidation the highest amount of halogens is found at the oxide/metal interface (figure 8) [32].

A model considering thermodynamics and kinetics was developed to determine a halogen concentration corridor in which selective aluminum transport in the gaps at the scale/metal interface via gaseous aluminum halides is possible [33]. The halogen can be brought into the alloy directly by ion implantation, or applied as a coating on the surface, for example. The activation of the halogen effect, however, can only take place at the oxidation temperature of TiAl where gaseous metal halides form. A selective aluminum transport can take place below a certain halogen concentration threshold where the vapor pressure of titanium halides is significantly lower than that of aluminum halides, indicated in figure 9 by $p_{(Fmax)}$. In contrast, a minimum amount of aluminum must be transported through the gaps to allow alumina scale growth, as indicated by $p_{(Fmin)}$. With these data a corridor vs. temperature can be plotted, in which selective transport of aluminum via the halide gas phase occurs without any participation of titanium.

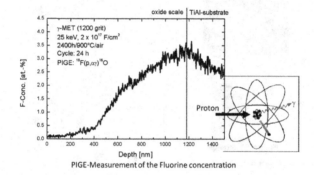

PIGE-Measurement of the Fluorine concentration

Figure 8. Halogen distribution in the surface area after oxidation for 2400h at 900°C in air.

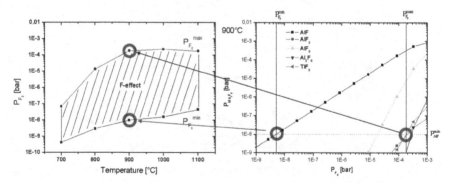

Figure 9. Calculated p_{F_2} limits for a beneficial halogen effect (left) and partial pressures of the gaseous aluminum fluorides at 900°C (right).

In figure 10, the mechanism is illustrated for chlorine as a halogen species. The reaction takes place in nano-/micro-gaps in the interface alloy/oxide scale, which arise from dislocations or dislocation pile-ups and vacancies and can potentially host gaseous compounds [34]. Furthermore, oxidation growth stresses are surprisingly high in samples containing halogens in the oxide/metal interface and microcrack networks can also provide gaps [11]. A very steep oxygen partial pressure gradient exists in these gaps due to the high thermodynamic stability of titania and alumina. The dissociation pressure of alumina (i.e. the oxygen partial pressure at the scale/metal interface) lies in the range of 10^{-40} bar at 900°C [8] and the oxygen partial pressure on top of the scale corresponds to the oxidizing environment. Chlorine reacts on the metal side of the nano-/micro-gaps with the TiAl alloy forming $TiCl_2$ and AlCl. For conditions inside the above mentioned corridor the vapor pressure of AlCl, however, is orders of magnitude higher than that of $TiCl_2$, so that the volatilization of aluminum dominates. The gaseous aluminum chloride becomes thermodynamically unstable in areas of increasing oxygen partial pressure and is converted into solid alumina that is precipitated on the more oxygen rich side of the gap. The halogen is released forming molecular chlorine and moves back to the "metal side" of the gap where again AlCl is formed and the cycle can restart. The same model applies also for other halogens [33].
Numerous experimental investigations have shown in the meantime that this effect is very stable and provides oxidation protection for very long times and under different oxidation conditions. Fluorine has turned out to be the most efficient halogen [32] while chlorine shows limitations under thermo cyclic conditions and in water vapor containing environments. According to thermodynamic considerations, bromine and iodine have an effect comparable to chlorine, but the few experiments performed so far have shown that these halogens may not reach the efficiency of fluorine.

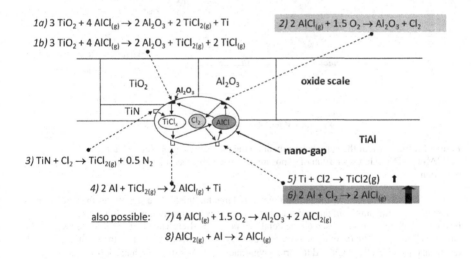

Figure 10. Possible reactions of the halogen effect during oxidation in air at elevated temperatures.

ENVIRONMENTAL EMBRITTLEMENT

Titanium and its alloys, including TiAl, show a very specific effect in that oxygen and nitrogen can become dissolved in the metal subsurface zone with a deteriorating effect on the mechanical properties of the material at room temperature. The α_2-phase consisting of Ti_3Al can dissolve up to about 20% of oxygen [3,35] and is therefore particularly susceptible to this type of environmental embrittlement. As a result, the already low rupture strains of these alloys are decreased even further [36,37]. The same applies to orthorhombic titanium aluminides as well as other Ti(Al)- phases with aluminum-contents less than about 50% [38]. Only pure γ-TiAl and the more Al-rich phases dissolve less than 1% of oxygen [39]. There has been a debate on whether environmentally induced room temperature embrittlement resulting from high temperature exposure is really due to the dissolution of nitrogen and oxygen. An interesting observation by Wu et al. was that the removal of the thin metal subsurface zone after oxidation restored the original "ductility" of the material before further high temperature exposure again lead to embrittlement of the material [37].

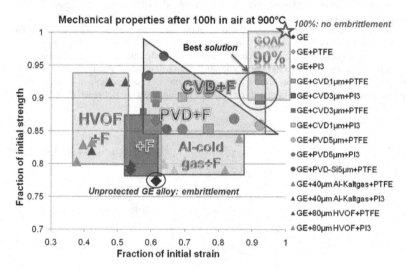

Figure 11. Influence of the surface treatment on the mechanical properties of alloy Ti-48Al-2Cr-2Nb (GE) after 100h of exposure at 900°C, given as fractions of the values before oxidation.

The halogen treatment without aluminum enrichment in the surface suppresses the formation of fast-growing titanium containing oxide, but cannot prevent embrittlement of the material. A recent research project was based on the combination of an Al-enriched metal subsurface with the halogen effect [40]. The rationale behind the coating development was to treat the surface of technical α_2/γ-TiAl alloys with Al to form a continuous γ-TiAl zone, which has low oxygen solubility, in a subsequent diffusion heat treatment. In order to facilitate the formation of a thin protective alumina scale on this coating, an additional fluorine treatment was applied before high temperature exposure. After oxidation tests at 900°C, four-point bending tests were performed.

Interestingly, for samples Al-enriched by chemical vapor deposition (CVD) and then halogen-treated, about 90% of the original strength and rupture strain were retained by this coating, while the values dropped to significantly lower values for untreated samples (figure 11). A number of other coatings and surface treatments were applied, but with less success. These preliminary results confirm that room temperature embrittlement can be a severe issue and that this coating concept can help to overcome this problem. Nevertheless significant further work is needed to shed further light on the mechanisms and to develop reliable tools for its prevention.

CONCLUSIONS

The key to improve the high-temperature oxidation resistance of TiAl alloys lies in the suppression of TiO_2 formation and the stimulation of α-Al_2O_3 growth to provide a thin and protective scale. This mechanism can be achieved by different alloying approaches or surface treatments where so far the halogen effect has proven to show the highest efficiency among all potential protection routes. More recently, environmentally-induced embrittlement of these alloys has been identified as a major issue impeding the progress of their industrial exploitation. First attempts to overcome this problem have been performed. Altogether, an optimum oxidation resistance of the technical versions of TiAl alloys requires both, a slow-growing, protective alumina scale and minimization of environmental embrittlement. In order to maintain the optimized mechanical properties of the material, a promising solution can be found in the combination of Al-enrichment of the surface and subsequent halogen treatment.

REFERENCES

1. N. S. Choudhury, H. C. Graham and J. W. Hinze, in *Properties of High Temperature Alloys*, edited by Z. A. Foroulis and F. S. Pettit (Electrochemical Society, 1976) pp. 668-680.
2. E. U. Lee and J. Waldman, *Scr. Metall.*, **22**, 1389 (1988).
3. S. Becker, A. Rahmel, M. Schorr and M. Schütze, *Oxid. Met.*, **38**, 425 (1992).
4. G. W. Goward, *Surf. Coat. Technol.*, **108-109**, 73 (1998).
5. Y. Shida and H. Anada, *Mat. Trans.*, **35**, 623 (1994).
6. Y. Shida and H. Anada, *Oxid. Met.*, **45**, 197 (1996).
7. F. Appel, J. D. H. Paul and M. Oehring, in *Gamma Titanium Aluminide Alloys*, edited by Wiley VCH, 2012) pp. 729-738.
8. P. Kofstad, in *High Temperature Corrosion* (Elsevier, 1988)
9. G. H. Meier and F. S. Pettit, *Mater. Sci. Eng. A*, **153**, 548 (1992).
10. N. Zheng, W. J. Quadakkers, A. Gil and H. Nickel, *Oxid. Met.*, **44**, 477 (1995).
11. W. Przybilla and M. Schütze, *Oxid. Met.*, **58**, 337 (2002).
12. A. Rahmel and P. J. Spencer, *Oxid. Met.*, **35**, 53 (1991).
13. M. Schütze and H. Fellmann, *Corrosion-Deformation Interactions*, edited by T. Magnin and J. M. Gras (1992) pp. 565-576.
14. S. Taniguchi, N. Hongawara and T. Shibata, *Mater. Sci. Eng. A*, **307**, 107 (2001).
15. A. Zeller, F. Dettenwanger and M. Schütze, *Intermetallics*, **10**, 59 (2002).
16. A. Zeller, F. Dettenwanger and M. Schütze, *Intermetallics*, **10**, 33 (2002).
17. R. Kremer and W. Auer, *Mater. Corros.*, **48**, 35 (1997).
18. M. P. Brady, B. A. Pint, P. F. Tortorelli, I. G. Wright and R. J. Hanrahan Jr., in *Materials Science and Technology: Corrosion and Environmental Degradation, Volumes I+II*, edited by R. W. Cahn, P. Haasen, and E. J. Kramer (Wiley-VCH, 2012) pp. 232-325.

19. W. J. Quadakkers, P. Schaaf, N. Zheng, A. Gil and E. Wallura, *Mater. Corros.*, **48**, 28 (1997).
20. H. Nickel, N. Zheng, A. Elschner and W. J. Quadakkers, *Mikrochim. Acta*, **119**, 23 (1995).
21. M. F. Stroosnijder, N. Zheng, W. J. Quadakkers, R. Hofmann, A. Gil and F. Lanza, *Oxid. Met.*, **46**, 19 (1996).
22. M. Schmitz-Niederau and M. Schütze, *Oxid. Met.*, **52**, 225 (1999).
23. J. W. Fergus, *Mater. Sci. Eng. A*, **338**, 108 (2002).
24. G. Schumacher, F. Dettenwanger, M. Schütze, U. Hornauer, E. Richter, E. Wieser and W. Möller, *Intermetallics*, **7**, 1113 (1999).
25. G. Schumacher, Dettenwanger, F., Hald, M., Lang, C. and Schütze, M., in *The microalloying effect in the oxidation of TiAl-alloys* (Utrecht, 1998)
26. A. Donchev, M. Schütze, R. Yankov, A. Kolitsch and W. Möller, in *Structural Aluminides for Elevated Temperatures: Proceedings of the TMS Annual Meeting 2008*, edited by Y. W. Kim, D. Morris, R. Yang, and C. Leyens (TMS, 2008) pp. 323-332.
27. A. Donchev, E. Richter, M. Schütze and R. Yankov, *J. Alloys Compd.*, **452**, 7 (2008).
28. A. Donchev, P. Masset and M. Schütze, *Mater. Res. Soc. Symp. Proc.*, **1128**, 159 (2009).
29. S. Taniguchi, *Mater. Corros.*, **48**, 1 (1997).
30. M. Kumagai, K. Shibue, M.-S. Kim and M. Yonemitsu, *Intermetallics*, **4**, 557 (1996).
31. G. Schumacher, F. Dettenwanger and M. Schütze, *Mater. High Temp.*, **17**, 53 (2000).
32. H.-E. Zschau, V. Gauthier, G. Schumacher, F. Dettenwanger, M. Schütze, H. Baumann, K. Bethge and M. Graham, *Oxid. Met.*, **59**, 183 (2002).
33. A. Donchev, B. Gleeson and M. Schütze, *Intermetallics*, **11**, 387 (2003).
34. V. Maurice, G. Despert, S. Zanna, M.-P. Bacos and P. Marcus, *Nat. Mater.*, **3**, 687 (2004).
35. A. Rahmel, W. J. Quadakkers and M. Schütze, *Mater. Corros.*, **46**, 271 (1995).
36. S. Draper, B. A. Lerch, I. E. Locci, M. Shazly and V. Prakash, *Intermetallics*, **13**, 1014 (2005).
37. X. Wu, A. Huang, D. Hu and M. H. Loretto, *Intermetallics*, **17**, 540 (2009).
38. A. Ralison, F. Dettenwanger and M. Schütze, in *Proceedings of the Fifth International Conference on the Microscopy of Oxidation,* edited by S. B. Newcomb and G. Tatlock (2003) pp. 361-383.
39. A. Menand, A. Huguet and A. Nérac-Partaix, *Acta Mater.*, **44**, 4729 (1996).
40. L. Bortolotto, M. Galetz, P. J. Masset and M. Schütze, *8th International Symposium on High-Temperature Corrosion and Protection of Materials* (France, 2012) *in press.*

Mater. Res. Soc. Symp. Proc. Vol. 1516 © 2012 Materials Research Society
DOI: 10.1557/opl.2012.1562

Oxidation Protection of γ-TiAl Alloys by Intermetallic Ti-Al-Cr-Zr Coatings

Reinhold Braun[1], Klemens Kelm[1], Arutiun P. Ehiasarian[2] and Papken Eh. Hovsepian[2]
[1]DLR – German Aerospace Center, Institute of Materials Research, D-51170 Cologne, Germany.
[2]Nanotechnology Centre for PVD Research, Sheffield Hallam University, Howart Street, Sheffield S1 1WB, United Kingdom.

ABSTRACT

The oxidation behavior of γ-TiAl specimens coated with an intermetallic Ti-49Al-34Cr-4Zr layer was investigated at 1000°C under cyclic conditions in laboratory air. The 11 μm thick coating was produced using a combined technique of high power impulse magnetron sputtering and unbalanced magnetron sputtering. The as-deposited coating exhibited a dense layered structure and excellent adhesion to the substrate. The Ti-Al-Cr-Zr coating possessed high oxidation resistance associated with the formation of a thin continuous alumina scale for exposure time periods exceeding 1000 cycles of 1 h dwell time at 1000°C. During the high temperature exposure, the coating being amorphous in the as-deposited condition became crystalline exhibiting different polytypes of Ti(Cr,Al)$_2$ Laves phases with Ti probably partially substituted by Zr and Nb. Due to alumina formation and interdiffusion the coating was depleted in aluminum and chromium as well as enriched in titanium. After 1000 cycles at 1000°C, the coating consisted of an outer layer of the hexagonal C14 Laves phase and an inner layer of a probably orthorhombic phase whose structure was not yet determined. In both layers, pores and fine precipitates rich in Zr and Y were found.

INTRODUCTION

Intermetallic γ-TiAl based alloys have gained considerable technical importance for rotating components in automotive and aero engines due to their low density and good mechanical properties at elevated temperatures [1,2]. At temperatures above 800°C, however, the oxidation resistance of γ-TiAl alloys even with high niobium content is poor due to the formation of fast growing non-protective mixed TiO$_2$ and Al$_2$O$_3$ scales [3]. The use of protective coatings is a suitable method to improve the oxidation resistance of γ-TiAl components [4,5]. Ternary Ti-Al-Cr alloys proved to be potential coating material for protection of titanium aluminides [4-6]. The high oxidation resistance of Ti-Al-Cr based alloys was associated with the Ti(Cr,Al)$_2$ Laves phase exhibiting low oxygen permeability and being capable of forming a protective alumina scale [7,8]. During long-term exposure at elevated temperatures, Ti-Al-Cr based coatings were found to degrade by depletion in chromium [9,10]. This Cr loss was caused by rapid diffusion of chromium into the substrate and precipitation of a quaternary phase, probably a (Ti,Nb)(Cr,Al)$_2$ Laves phase, at the coating/substrate interface, resulting in dissolution of the Laves phase in the protective layer. Similar to the Ti-Cr-Al system exhibiting a narrow Ti(Cr,Al)$_2$ C14 Laves phase field dissolving about 42 at.% Al [11], a Zr(Al,Cr)$_2$ C14 type Laves phase extends into the ternary region up to 54 at.% Cr [12]. Incorporated into the Ti(Cr,Al)$_2$ Laves phase by substituting Ti, zirconium might retard its dissolution. Therefore, in the present work, the oxidation protection capability of an intermetallic Ti-Al-Cr-Zr coating deposited on the Ti-45Al-8Nb alloy was studied performing cyclic oxidation tests and cross-sectional analyses of oxidized samples.

EXPERIMENTAL DETAILS

The materials used were extruded and annealed rods of a γ-TiAl based alloy with the nominal composition Ti-45Al-8Nb-0.2C (in at.%) provided by GfE, Nuremberg, Germany. The alloy exhibited a two-phase microstructure consisting of γ-TiAl and $α_2$-Ti_3Al phases. From these rods disc-shaped specimens with 15 mm diameter and 1 mm thickness were machined using spark erosion. The surfaces of the samples were polished and ultrasonically cleaned. The intermetallic Ti-Al-Cr-Zr coatings were manufactured by a mixed technique of high power impulse magnetron sputtering (HIPIMS) and unbalanced magnetron sputtering (UBM) using an industrial size HTC 1000-4 system (Hauzer Techno Coating, The Netherlands) equipped with four linear magnetron cathodes. Two targets (TiAlY and Cr) were operated in HIPIMS mode and two targets (Al and Zr) in UBM mode in argon atmosphere. Details of this coating process are described elsewhere [13]. Cyclic oxidation tests were carried out at 1000°C in laboratory air using automatic rigs. One cycle consisted of 1 h exposure at high temperature and 10 min at ambient temperature during which the specimens cooled down to about 70°C. Cross-sectional analyses of oxidized samples were carried out using scanning electron microscopy (SEM), energy-dispersive X-ray spectroscopy (EDS) and X-ray diffraction (XRD) measurements. The coatings were also investigated by analytical transmission electron microscopy (TEM).

RESULTS AND DISCUSSION

As revealed by cross-sectional SEM examination, the as-deposited coating exhibited a dense layered structure with a layer thickness of 700 nm. The coating was tightly adherent to the substrate. Its chemical composition was determined by EDS area analysis to Ti-48.3Al-34.2Cr-3.9Zr in at.% (table I). No yttrium was detected by EDS. XRD pattern indicated that the as-deposited coating was amorphous or microcrystalline with weak broad reflexes of TiO_2.

Figure 1a shows an SEM micrograph of a Ti-Al-Cr-Zr coated sample after 10 cycles of exposure to air at 1000°C. A thin alumina scale formed on top of the coating. Interdiffusion between coating and substrate resulted in the formation of two single-phase layers. The outer layer exhibited a high amount of fine pores, whereas rod-like precipitates were found in the inner layer. The chemical compositions of both layers determined by EDS were quite similar except for the concentrations of Nb and Zr (table I). XRD analysis confirmed the presence of corundum and indicated the formation of hexagonal C14 and cubic C15 Laves phases with lattice parameters of $a = 0.510$ nm and $c = 0.832$ nm as well as $a = 0.721$ nm, respectively (figure 1b). In the ternary Ti-Al-Cr system, the hexagonal C14 Laves phase extends over the composition range from 5 to 42 at.% Al at 1000°C [11]. For lower Al concentrations the C15 and the C36 Laves phases exist. Although the Al amount was high in both layers, the presence of the cubic Laves phase might be associated with the alloying elements Zr and Nb. The lattice constants determined for the C14 Laves phase were quite similar to values obtained for the Al stabilized C14 phase in a $Ti_{31.5}Cr_{26.7}Al_{41.8}$ alloy quenched from 1000°C ($a = 0.50442$ nm and $c = 0.82462$ nm) [14]. For the C15 $TiCr_2$ Laves phase a lattice constant $a = 0.6945$ nm was reported [15]. The outer layer exhibited a high amount of porosity probably caused by Kirkendall voids due to rapid outward and inward diffusion of aluminum and chromium, respectively, forming alumina and the inner Laves phase layer. The diffusion of zirconium was slow, being enriched in the outer layer.

Similar results of microstructural examinations by SEM were found after 100 cycles of exposure (figure 2). The two Ti-Al-Cr based layers with quite similar chemical compositions were still observed. Compared to the EDS analysis of the coated sample after 10 cycles of exposure, both layers depleted further in aluminum, and the concentrations of Nb and Zr increased in the outer and inner layers, respectively (see table I). Spectrum profiling across coating and substrate confirmed the interdiffusion of niobium and zirconium as well as the deep diffusion of chromium into the substrate (figure 2b).The XRD pattern of the coated sample after 100 cycles of exposure could be fitted with hexagonal and cubic Laves phases as well as α-Al$_2$O$_3$ grown on the outer, pores containing layer. The lattice parameters obtained from the fit were $a = 0.718$ nm for the cubic C15 polytype and $a = 0.496$ nm and $c = 1.634$ nm for the hexagonal polytype, indicating that the latter phase had probably a C36 structure.

Table I. Chemical composition of various phases found in the intermetallic Ti-Al-Cr-Zr coating and reaction zone between coating and substrate after exposure to air at 1000°C for different cycles by standardless EDS analysis (concentration in at.%). Phases are shown in figures 1 - 3.

	Ti	Al	Cr	Zr	Nb
	as deposited				
coating	13.6	48.3	34.2	3.9	
	10 cycles (figure 1a)				
coating	26.7	40.8	25.0	6.4	1.1
reaction zone	28.7	39.1	27.2	0.5	4.5
	100 cycles (figure 2a)				
coating	26.8	35.7	28.8	5.4	3.3
reaction zone	29.3	34.2	30.1	1.5	4.9
	1000 cycles (figure 3a)				
1	21.7	35.9	27.8	4.4	10.2
2	41.7	30.4	19.0	4.3	4.6
3	42.9	30.2	19.5	2.7	4.7
γ-TiAl	43.1	47.2	2.5	0.1	7.1

Figure 3a presents a SEM cross-section of the coated sample after 1000 cycles of exposure. A protective alumina scale was still present on the coating. The pores containing zone of the coating consisted of two different layers (1 and 2) with different compositions (table I). The outer, occasionally discontinuous layer 1 was depleted in titanium and enriched in niobium. The layer 2 below contained a high amount of titanium and was depleted in chromium; Zr and Nb concentrations were similar. The chemical composition measured for the pores free layer 3 was similar to that of the latter layer with a slightly reduced amount of Zr. Their microstructure probably consisted of the same phase. The adjacent γ-TiAl phase was slightly enriched in Al compared to the alloy composition. Very often, a thin α_2-Ti$_3$Al zone formed beneath layer 3. As revealed by spectrum profiling across coating and substrate, chromium deeply diffused into the γ-TiAl substrate, whereas the diffusion of zirconium was slow. XRD pattern taken from the surface of the coated sample after 1000 cycles of exposure indicated the presence of alumina, rutile and a hexagonal C14 Laves phase (figure 3b). The remaining reflexes could be quite well fitted to a body-centered orthorhombic unit cell with $a = 3.24$, $b = 0.518$ and $c = 1.333$ nm. The presence of rutile was mainly associated with severe oxidation at the rim of the sample where sharp edge effects related to the applied bias voltage during the HIPIMS deposition process led to a change in the coating composition, particularly to depletion in aluminum.

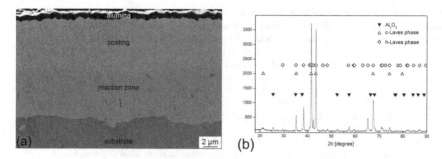

Figure 1. (a) Scanning electron micrograph and (b) XRD pattern of a γ-TiAl sample with Ti-Al-Cr-Zr coating which was exposed to air at 1000°C for 10 cycles

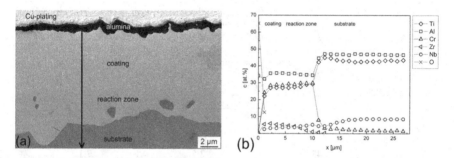

Figure 2. (a) Scanning electron micrograph and (b) quantitative EDS profile (see arrow) of a γ-TiAl sample with Ti-Al-Cr-Zr coating which was exposed to air at 1000°C for 100 cycles

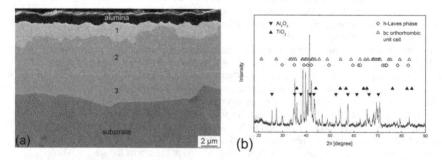

Figure 3. (a) Scanning electron micrograph and (b) XRD pattern of a γ-TiAl sample with Ti-Al-Cr-Zr coating which was exposed to air at 1000°C for 1000 cycles.

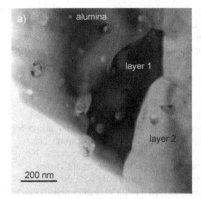
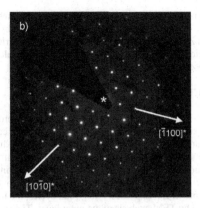

Figure 4. (a) TEM bright field image and (b) SAED pattern of the hexagonal C14 Laves phase of layer 1 (outer layer) formed in the Ti-Al-Cr-Zr coating after 1000 cycles of exposure at 1000°C. Viewing direction is [0001].

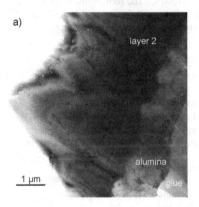
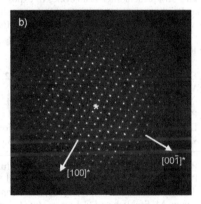

Figure 5. (a) TEM bright field image and (b) SAED pattern of an unknown phase formed in layer 2 (inner layer) of the Ti-Al-Cr-Zr coating after 1000 cycles of exposure at 1000°C. Viewing direction is [010].

TEM examinations confirmed the present of a hexagonal C14 phase in the outer layer 1 (figure 4). The selected area diffraction (SAED) pattern was assigned to a C14 type Laves phase with lattice parameters of $a = 0.53$ nm and $c = 0.83$ nm. The phase of the layer 2 could not yet be identified (figure 5). Its SAED patterns were indexable assuming a body-centered orthorhombic unit cell with lattice parameters $a = 3.24$ nm, $b = 0.518$ nm and $c = 1.333$ nm, also used for analysis of the XRD measurements (figure. 3b). The unidentified phase formed large patches with diameters of up to 10 μm and more in the subsurface region below the oxide scale. Bright field images also showed pores and fine precipitates in the outer (1) and inner (2) layers of the coating as well as in the alumina scale. As found by EDS TEM analysis, the small particles were rich in Zr, Y and O, being probably zirconium-yttrium oxides, which might improve the oxidation resistance due to the so called reactive element effect [16]. Pores and precipitates were not observed in layer 3.

CONCLUSIONS

Intermetallic Ti-49Al-34Cr-4Zr coatings produced by HIPIMS technology possessed high oxidation resistance at 1000°C, providing effective oxidation protection to γ-TiAl alloys for exposure time periods exceeding 1000 cycles of 1 h dwell time at elevated temperature. The amorphous as-deposited coating was well adherent to the Ti-45Al-8Nb substrate material. No spallation of oxide scale and coating occurred during the thermal cycling test. Interdiffusion between coating and substrate resulted in the formation of a layered structure with different polytypes of Laves phases. After 1000 cycles, the hexagonal C14 Ti(Al,Cr)$_2$ Laves phase was verified by TEM analysis of the outer layer. Ti sites might be partially occupied by Nb and Zr. The phase of the layers below was not yet identified. The SAED pattern highly correlated with an orthorhombic phase exhibiting fairly large lattice parameters.

During high temperature exposure in air, a thin continuous α-Al$_2$O$_3$ scale grew on top of the coating still present after 1000 cycles of exposure at 1000°C. The formation of alumina was associated with the presence of Ti(Cr,Al)$_2$ Laves phases. Pores and fine precipitates of Zr and Y oxides were found in the α-Al$_2$O$_3$ scale and the different layers of the original coating.

REFERENCES

1. H. Clemens and H. Kestler, *Adv. Eng. Mater.* **2**, 551 (2000).
2. X. Wu, *Intermetallics* **14**, 1114 (2006).
3. M. Yoshihara and Y.-W. Kim, *Intermetallics* **13**, 952 (2005).
4. M.P. Brady, W.J. Brindley, J.L. Smialek and I.E. Locci, *JOM* **48**, 46 (November 1996).
5. C. Leyens, R. Braun, M. Fröhlich and P.Eh. Hovsepian, *JOM* **58**, 17 (January 2006).
6. G.S. Fox-Rabinovich, G.C. Weatherly, D.S. Wilkinson, A.I. Kovalev and D.L.Wainstein, *Intermetallics* **12**, 165 (2004).
7. M.P. Brady, J.L. Smialek, J. Smith and D.L. Humphrey, *Acta Mater.* **45**, 2357 (1997).
8. M.P. Brady, J.L. Smialek, D.L. Humphrey and J. Smith, *Acta Mater.* **45**, 2371 (1997).
9. M. Fröhlich, R. Braun and C. Leyens, *Surf. Coat. Technol.* **201**, 3911 (2006).
10. R. Braun, M. Fröhlich and C. Leyens, *Int. J. Mat. Res.* **101**, 637 (2010).
11. T. J. Jewett and M. Dahms, *Z. Metallkd.* **87**, 254 (1996).
12 V.Y. Markiv and V.V. Burnashova, *Poroshk. Metall.* **12**, 53 (1970).
13. P.Eh. Hovsepian, A.P. Ehiasarian, Y.P. Purandare, R. Braun and I.M. Ross, *Plasma Process. Polym.* **6**, S118 (2009).
14. T.J. Jewett, B. Ahrens and M. Dahms, *Mater. Sci. Eng.* **A225**, 29 (1997).
15. W. Zhuang, J. Shen, Y. Liu, L. Ling, S. Shang, Y. Du, J.C. Schuster, *Z. Metallkd.* **91**, 121 (2000).
16. B.A. Pint, *Oxid. Met.* **45**, 1 (1996).

Mater. Res. Soc. Symp. Proc. Vol. 1516 © 2012 Materials Research Society
DOI: 10.1557/opl.2012.1563

High Temperature Oxidation Protection of Multi-Phase Mo-Containing TiAl-Alloys by the Fluorine Effect

Alexander Donchev[1], Raluca Pflumm[1], Svea Mayer[2], Helmut Clemens[2], Michael Schütze[1]

[1]DECHEMA-Forschungsinstitut, D-60486 Frankfurt am Main, Germany
[2]Department of Physical Metallurgy and Materials Testing, Montanuniversitaet Leoben, A-8700 Leoben, Austria

ABSTRACT

Intermetallic titanium aluminides are potential materials for application in high temperature components. In particular, alloys solidifying via the β-phase are of great interest because they possess a significant volume fraction of the disordered body-centered cubic β-phase at elevated temperatures ensuring good processing characteristics during hot-working. Nevertheless, their practical use at temperatures as high as 800°C requires improvements of the oxidation resistance. This paper reports on the fluorine effect on a multi-phase TiAl-alloy in the cast and hot-isostatically pressed condition at 800°C in air. The behavior of the so-called TNM material (Ti-43.5Al-4Nb-1Mo-0.1B, in at %) was compared with that of two other TiAl-alloys which are Nb-free and contain different amounts of Mo (3 and 7 at%, respectively). The oxidation resistance of the fluorine treated samples was significantly improved compared to the untreated samples. After fluorine treatment all alloys exhibit slow alumina kinetics indicating a positive fluorine effect. Results of isothermal and thermocyclic oxidation tests at 800°C in air are presented and discussed in the view of composition and microstructure of the TiAl-alloys investigated, along with the impact of the fluorine effect on the oxidation resistance of these materials.

INTRODUCTION

The increasing demand for more efficient automotive or jet engines triggers the replacement of heavy Ni- or Co-based superalloys currently in use by lighter TiAl-alloys. This group of inter-metallic materials offers promising properties for several high temperature applications. Hence, they are currently being used for turbocharger rotors in automotive engines [1] and have now been introduced for turbine blades in General Electric's GENx engine which powers the Boeing 787 [2]. Recent developments have led to multi-phase alloys with a significant amount of the β/$β_0$-phase next to the γ- and $α_2$-phases which are common in technical TiAl-alloys [3, 4]. Their machinability and mechanical properties would allow their use at elevated temperatures. Even though, in the envisaged temperature range above 750°C their oxidation resistance must be im-proved. At such high temperatures, TiAl-alloys do not form a protective alumina scale like other aluminides (e.g. NiAl) because of the similar thermodynamic stabilities of Ti- and Al-oxide [5]. A fast growing and non-protective mixed TiO_2/Al_2O_3/TiN scale develops on TiAl-alloys after high temperature exposure in air [6]. It has been proven that the fluorine effect changes the oxi-dation behavior of TiAl-alloys at temperatures higher than 750°C. Very small amounts of halo-gen, micro-alloyed at the metal/oxide interface, lead to the formation of a slow-growing alumina barrier, which protects the substrate against further oxidation even in wet environments and un-der thermocyclic conditions [7, 8]. This effect works for technical $α_2$/γ-alloys with an Al-content above 40 at.% [9]. In this paper β-phase containing alloys are investigated regarding the efficiency of the fluorine effect on their oxidation resistance for the first time to the knowledge of the authors. Results of isothermal and thermocyclic high temperature exposure experiments of untreated and F-treated specimens of three alloys (TNM, TiAl-3Mo, TiAl-7Mo) with different compositions are presented. The behavior of the TNM material was compared with that of two other TiAl-alloys which do not contain any Nb, but different amounts of Mo [10, 11]. The results were discussed in view of the potential use of such multi-phase alloys at temperatures higher than 750°C.

EXPERIMENTAL

Specimens of three TiAl-alloys with a composition of Ti-43.5Al-4Nb-1Mo-0.1B (TNM), Ti-44Al-3Mo-0.1B (3Mo) and Ti-44Al-7Mo-0.1B (7Mo) (in at.%) were cut from bars into coupons of 15 x 10 x 1 mm³ dimension. The coupons were ground up to 1200 grit using SiC-paper, cleaned with ethanol, rinsed with distilled water and dried in air prior to further use. The F-compound was sprayed in a homogeneous manner over the whole sample surface including the side faces. The amount of F-compound deposited per square centimeter was comparable for all treated samples.

First tests were conducted at 800°C in synthetic air under isothermal conditions with a duration of 120 h. The in-situ mass change for each investigated alloy has been measured using a thermogravimetric (TGA) device. The heating phase was conducted in an inert Ar atmosphere in order to minimize oxidation during this time. The atmosphere was changed to synthetic air when 800°C were reached. This was also the moment when the recording of the mass change due to the oxidation was started. Hence, the mass loss of the F-treated samples due to the evaporation of the organic residues of the F-compound during heating up was not considered for the mass gain.

In addition, thermocyclic experiments were performed in a box furnace in lab air as follows: 24 h hot dwell and 1 h cold dwell. The heating up took 10 min within these experiments. The furnace was operated at test temperature all the time. The total duration of one cycle was 25 hours and the total exposure time of each sample was 120 h hot. During the cold dwell, the specimens were cooled down to room temperature (RT) within 10 minutes and weighed. For the untreated samples the mass gain due to oxidation was calculated by subtracting the sample mass before exposure from the actual mass after each day (i.e. cycle). In case of the F-treated samples the calculation was slightly different because the evaporation of the organic residues of the F-compound took place simultaneously with the oxidation process at the beginning of the first cycle (i.e. day) and could not be evaluated separately in terms of mass gain. The mass gain due to oxidation did not exceed the complementary mass loss for the first 24h of exposure. For this reason the mass change after the first day was therefore set as a starting point for the oxidation during the test. Hence, the specimen mass during the test is lower than the one before exposure. The mass value prior to exposure is defined as the sum of the amount of organic F-compound and the weight of the metallic substrate.

Post experimental investigations included examination of the surface with a Philips XL 40 scanning electron microscope (SEM). All the samples were galvanic platted with Ni prior to the preparation of the metallographic cross sections. This led to a better visibility of the very thin oxide scales grown on the F-treated samples. Furthermore, the cross sections were analyzed with a Leica light-optical microscope (LOM) and SEM with an EDAX energy dispersive X-ray spectroscopy (EDX).

RESULTS AND DISCUSSION

Isothermal exposure

The TGA-mass change curves of the untreated alloys are presented in figure 1a.The two alloys with Mo-additions, but no Nb, show similar oxidation kinetics at 800°C. The mass gain curves are situated above that for the TNM-alloy, which contains both Mo and Nb. This is not surprising considering that Nb additions up to 10 at% improve the oxidation resistance of the TiAl by reducing the TiO_2 growth and promoting the development of a protective layer [12]. Nevertheless, the protection degree of such scales is of course limited by the exposure temperature. The mass gain data of the fluorine treated specimens stay below those of the untreated ones (comparing fig. 1a and b). The oxidation kinetics of all alloys is slowed down by the fluorine effect.

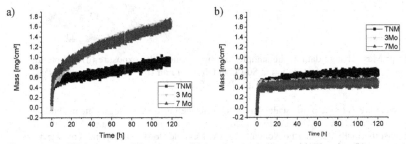

Figure 1a, b: TGA-curves of the untreated (a) and treated alloys with Fluorine (b)

For all investigated alloys the mass gain curves show a short incubation time with an enhanced mass gain followed by a remarkable diminution of the oxidation kinetics in the presence of F compared to the F-free condition.

Thermocyclic exposure

The mass gains of the untreated samples are higher for all tested alloys than those of the F-treated ones (fig. 2a). The mass gains of the untreated TNM- and 7Mo-sample show a steady evolution. The samples show no spallation at the end of the experiment and their appearance is similar (fig. 2b, f). The 3Mo-sample revealed a higher mass gain during the first two days followed by a mass loss. The reason therefore is partial spallation occurred during each cooling step in the last three days of exposure (fig. 2d). The reasons for this form of stress relief are the thermal expansion mismatch between the intermetallic substrate and the oxide scale stresses arising from oxide growth which are proportional to the scale thickness and also stresses developing at the sharp edges of the sample. These cumulated stresses can lead to spallation if a critical value is exceeded [13]. In our case the spallation is located near the edges of the sample.

All F-treated specimens exhibit a mass loss after the first day, which is, as mentioned before, due to the evaporation of the organic residues of the F-compound. Additionally, the mass of the F-treated sample of the 7Mo-alloy decreased until the third day of exposure. This was probably caused by a loss of some residual flakes during the cooling step, since this sample had the highest F-addition compared with the other two samples. The figures 2c, e, g show the F-treated samples.

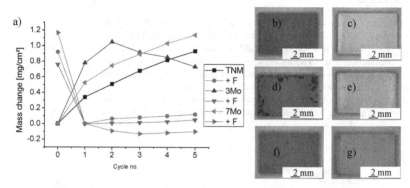

Figure 2a-g: Mass change data of the samples during thermocyclic exposure (a) and photos of the specimens after 5 days of oxidation at 800°C in laboratory air (b = TNM no F, c = TNM + F, d = 3Mo no F, e = 3Mo + F, f = 7Mo no F, g = 7Mo + F)

A detailed analysis of the oxide scales has been performed using metallographic methods. Figures 3-5 show relevant LOM- and SEM-micrographs of different oxide scales. Since the images for the isothermally and thermocyclically exposed samples look similar, only cross-sections of the thermocyclically oxidized samples are shown here for the sake of simplicity. The untreated samples have thicker oxide scales than those found on the F-treated specimens. Their morphology is typical for untreated TiAl-alloys and consists of a mixed $TiO_2/Al_2O_3/TiN$-scale. The residual flakes on the alumina layer left from the F-treatment are also visible in the micrographs.

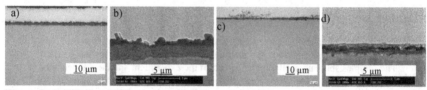

Figure 3a-d: LOM- and SEM-images of oxide scales on the untreated (a, b) and F-treated (c, d) TNM samples after 5 days (cycles) of oxidation at 800°C in air

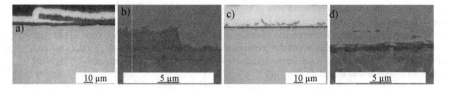

Figures 4a-d: LOM- and SEM-images of oxide scales on the untreated (a, b) and F-treated (c, d) 3Mo-samples after 5 days (cycles) of oxidation at 800°C in air

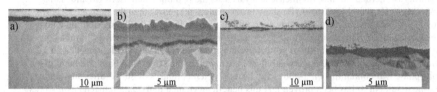

Figure 5a-d: LOM- and SEM-images of oxide scales on the untreated (a, b) and F-treated (c, d) 7Mo-samples after 5 days (cycles) of oxidation at 800°C in air

EDX-spectra are shown representatively for TNM in fig. 6a-c. The Nb and Mo signals reflect the influence of the substrate. The Ni signal originates from the Ni-plating. The investigated oxide layers are thin compared to the excitation volume of EDX, so that influences of the adjacent domains cannot be excluded. The EDX-spectra of the F-treated samples reveal an Al-rich oxide (fig. 6b). Additionally, the Ni-peak is higher than that of the untreated samples. This demonstrates that the scales are thinner than those found on the untreated materials (fig. 6a). Ti, Nb and Mo from the substrate were detected due to the larger excitation volume of EDX, too. Some fluorine was found under the oxide scale on all F-treated samples (fig. 6c), representative for TNM. This so called "fluorine pockets" [14] can act as a reservoir which can supply fluorine for crack healing processes. Cracks might be caused by different stresses (see above) that might appear in the scale prior to spallation.

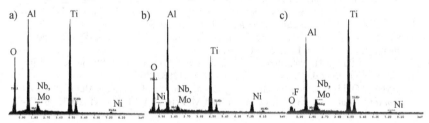

Figures 6a-c: EDX-spectra of the oxide scales on the untreated (a) and on the F-treated (b) TNM-samples for thermocyclic exposure. An EDX-spectrum of the remaining fluorine ("fluorine pocket") on the F-treated TNM sample after thermocyclic oxidation is shown in (c).

The results show that Mo-addition into TiAl-alloys does not impede the fluorine effect. We observed a positive fluorine effect on all F-treated alloys independent of their chemical composition and/or microstructure. The presence of ordered β_0-phase in the sub-surface zone of the materials does not hinder the formation of the protective alumina layer. Although the oxidation rate is still low at 800°C, the mixed scale formed on the untreated materials is not protective. It spalled on the 3Mo-alloy under thermocyclic load.

SUMMARY

During isothermal thermogravimetric analysis tests the two untreated TiAl-Mo alloys (Ti-44Al-3Mo-0.1B, Ti-44Al-7Mo-0.1B) show faster oxidation kinetics than the untreated TNM alloy (Ti-43.5Al-4Nb-1Mo-0.1B). However, after a fluorine treatment all alloys exhibit a positive fluorine effect. The fluorine effect leads to the formation of a slow-growing alumina barrier. This alumina layer has the ability to protect the substrate against further oxidation since no TiN formation has been observed underneath it and the mixed oxide scale formation was suppressed. Multi-phase Mo-containing TiAl-alloys can be used at temperatures of 800°C after an optimized fluorine treatment. It has been proven that this beneficial change in oxidation mechanism works also for this new type of alloys.

REFERENCES

1. T. Tetsui, Adv. Eng. Mat. **3**, 307-310 (2001).
2. R. R. Boyer and J. C. Williams, "Developments in Research and Applications in the Titanium Industry in the USA", *Ti 2011*, ed. L. Zhou, H. Chang, Y. Lu and D. Xu, (Science Press Beijing 2012) pp. 10-19.
3. H. Clemens, M. Schloffer, E. Schwaighofer, R. Werner, A. Geitzenauer, B. Rashkova, R. Pippan and S. Mayer, these proceedings.
4. H. Clemens and S. Mayer, *Adv. Eng. Mat.* **14**, DOI 10.1002/adem.201200231 (2012).
5. A. Rahmel and P. J. Spencer, *Oxid. Met.* **35**, 53-68 (1991).
6. J. M. Rakowski, F. S. Petit, G.H. Meier, F. Dettenwanger, E. Schumann and M. Rühle, *Scripta Met. Mat.* **33**, 997-1003 (1995).
7. A. Donchev, B. Gleeson and M. Schütze, *Intermetallics* **11**, 387-398 (2003).
8. A. Donchev, E. Richter, R. Yankov and M. Schütze, *Intermetallics* **14**, 1168-1174 (2008).
9. P. Masset and M. Schütze, *Adv. Eng. Mat* **10**, 666-674 /2008).
10. T. Schmoelzer, S. Mayer, C. Seiler, F. Haupt, V. Güther, P. Staron, K,-D. Liss and H. Clemens, *Adv. Eng. Mat.* **13**, 306-311 (2011).
11. S. Mayer, C. Sailer, H. Nakashima, T. Schmoelzer, T. Lippmann, P. Staron, K.-D. Liss, H. Clemens and M. Takayama, *Mat. Res. Soc. Proc.* **1295**, 113-118 (2011).
12. C. Leyens, "Oxidation and Protection of Titanium Alloys and Titanium Aluminides", *Titanium and Titanium Alloys*, ed. C. Leyens and M. Peters (WILEY-VCH, 2003) pp.187-230.
13. M. Schütze, „Protective Oxide Scales and their Breakdown", ed. D.R. Holmes and R. B. Waterhouse (John Wiley & Sons, 1997) pp.67-102.
14. A. Donchev and M. Schütze, *Mat. Sci. Forum* **638-642,** 1294-1298 (2010).

Poster Session

Mater. Res. Soc. Symp. Proc. Vol. 1516 © 2012 Materials Research Society
DOI: 10.1557/opl.2012.1681

An Overview of Dry Sliding Wear of Two-Phase FeNiMnAl Alloys

Xiaolan Wu[1], Fanling Meng[1], Ian Baker[1], Hong Wu[2] and Paul R. Munroe[3]
[1]Thayer School of Engineering, Dartmouth College, Hanover NH 03755, U.S.A
[2]State Key Laboratory of Powder Metallurgy, Central South University, Changsha 410083, P.R.C.
[3]Electron Microscope Unit, University of New South Wales, Sydney NSW 2052, Australia

ABSTRACT

The pin-on-disc wear behavior of nanostructured two-phase $Fe_{30}Ni_{20}Mn_{20}Al_{30}$ and eutectic lamellar-structured $Fe_{30}Ni_{20}Mn_{35}Al_{15}$ is compared emphasizing the influence of the microstructure and mechanical properties of alloys as well as the effect of test environment. Although the wear of both alloys was greater in oxygen-containing environments, eutectic $Fe_{30}Ni_{20}Mn_{35}Al_{15}$ is less sensitive to oxygen than nanostructured $Fe_{30}Ni_{20}Mn_{20}Al_{30}$. Abrasive wear dominated during the wear in all cases, while plastic deformation also occurred during the wear of eutectic $Fe_{30}Ni_{20}Mn_{35}Al_{15}$. A tribolayer of zirconia, which was embedded in the surface of the wear pin, was characterized using a scanning transmission electron microscope equipped with an energy dispersive spectrometer.

INTRODUCTION

Recently, a range of two-phase FeNiMnAl alloys with elemental compositions ranging from 15-35 at. % has been investigated for their assorted possible phase transformations and the potential complexity of microstructures and mechanical properties [1]. Detailed microstructural characterization and mechanical properties studies have been performed on two of these alloys, $Fe_{30}Ni_{20}Mn_{20}Al_{30}$ [2] and $Fe_{30}Ni_{20}Mn_{35}Al_{15}$ [3]. The former alloy, $Fe_{30}Ni_{20}Mn_{20}Al_{30}$ [2], consists of (Fe, Mn)-enriched B2 (ordered b.c.c.) and (Ni, Al)-enriched $L2_1$ (further ordered B2) phases, which exist in a cube-on-cube relationship. The phase width was ~5 nm for the as-cast alloy (Figure 1(a)), and ~25 nm for the alloy given a 72 h anneal at 823 K (Figure 1(b)) with corresponding Vickers hardness values of 514 Hv and 547 Hv. Both the as-cast and annealed alloy fractured before yield under compression with a fracture strength of ~ 1350 MPa at room temperature. However, the annealed alloy showed a yield strength of ~1450 MPa at 573 K, while at the same temperature, the as-cast alloy still showed a brittle fracture before yielding with a similar fracture strength to that found at room temperature. In contrast, the microstructure of $Fe_{30}Ni_{20}Mn_{35}Al_{15}$ [3] consists of (Fe, Mn)-enriched f.c.c lamellae with a width of ~500 nm and (Ni, Al)-enriched B2 lamellae with a width of ~200 nm (Figure 1(c)). This novel eutectic alloy showed a yield strength of ~ 600 MPa, a fracture strength of ~ 840 MPa, and an elongation of ~ 8% at room temperature. The hardness of the as-cast alloy is 303 ± 6 HV.

In two previous papers the dry sliding wear of $Fe_{30}Ni_{20}Al_{20}Al_{30}$ [4] and $Fe_{30}Ni_{20}Mn_{35}Al_{15}$ [5] was studied in different environments using pin-on-disk wear tests with an yttria stabilized zirconia disk. Unfortunately, in those papers pins of different geometry were used and so it was impossible to compare their wear rates. In this paper, we compare the wear behavior of these two alloys using the same wear pin geometry in three different environments. The wear behavior of $Fe_{30}Ni_{20}Mn_{20}Al_{30}$ was also examined both the as-cast and annealed states because of the improved mechanical properties in the latter state.

Figure 1. Bright field transmission electron micrographs of $Fe_{30}Ni_{20}Mn_{20}Al_{30}$ in (a) the as-cast state and (b) after a 72h anneal at 823 K, and (c) as-cast $Fe_{30}Ni_{20}Mn_{35}Al_{15}$.

EXPERIMENT

Ingots of $Fe_{30}Ni_{20}Mn_{20}Al_{30}$ and $Fe_{30}Ni_{20}Mn_{35}Al_{15}$ were prepared by arc-melting using the constituent elements of >99.9% purity in a water-chilled copper crucible under argon. The resulting ~50 g buttons with ~5 cm diameter were flipped over and re-melted three times to ensure a homogeneous mixture. Excess Mn was added to each melt to amend for its loss during melting.

The $Fe_{30}Ni_{20}Mn_{35}Al_{15}$ buttons were machined into cylindrical wear pins ~9.5 mm diameter and ~ 1 cm long with a hemispherical tip. However, the wear pins could not be machined from the as-cast $Fe_{30}Ni_{20}Mn_{20}Al_{30}$ buttons because of its brittle nature. Thus, the $Fe_{30}Ni_{20}Mn_{20}Al_{30}$ buttons were cut into small pieces and arc-melted again and directly cast into the wear pins with the same dimensions as the $Fe_{30}Ni_{20}Mn_{35}Al_{15}$ pins. Some $Fe_{30}Ni_{20}Mn_{20}Al_{30}$ pins were annealed at 823 K in air for 72 h, followed by air-cooling. All the pins were ground and polished to a mirror finish.

Pin-on-disc wear tests were performed against an yttria-stabilized zirconia counterface polished to a surface finish of ~ 0.1 Ra (~ 0.0254 μm), using a home-made device [6]. All tests were conducted on the new disk surface at ~ 25 °C at a constant sliding speed of 1 m s^{-1} for a total sliding distance of 1 km with a normal load of 23 N in three different environments, i.e. oxygen, air, argon. The humidity in the oxygen and argon was less than 5%, while the humidity of the air was ~45%. Three tests were performed in each environment. Wear mass loss was determined by measuring the mass difference of the pins before and after testing, using an electronic balance of ± 0.1 mg precision.

The wear surface of pins were examined using both secondary electron (SE) and backscattered electron (BSE) imaging on an FEI Field emission gun XL-30 scanning electron microscope (SEM) operated at 15 kV, equipped with an EDAX Li-drifted EDS. Cross-sectional TEM specimens, used to examine the subsurface of the wear pins, were prepared using a FEI Nova

200 Nanolab FIB using the lift-out method. The specimens were examined using a Philips CM200 TEM operating at 200 kV. Elemental X-ray maps were collected in scanning transmission electron microscope (STEM) mode using EDS.

RESULTS AND DISCUSSION

Wear loss

The mass losses of as-cast and annealed $Fe_{30}Ni_{20}Mn_{20}Al_{30}$ pins and as-cast $Fe_{30}Ni_{20}Mn_{35}Al_{15}$ pins after wear in different environments are summarized in Figure 2. It can also be seen that the humidity plays little if any role in the wear loss, since the error bars overlap between the tests for all the pins tested in air and oxygen. When tested in an inert atmosphere (argon) there is little difference in the wear behavior of the two alloys in the as-cast condition, which is surprising given the large difference in their hardness values. The annealed $Fe_{30}Ni_{20}Mn_{20}Al_{30}$ perhaps shows a slightly lower wear rate even though it is only 6% harder than the cast alloy at room temperature. However, the higher temperature mechanical properties may be more important since calculations suggest that the temperature at the tip of the pin is >750K [4, 5]. It is worth noting that the annealed $Fe_{30}Ni_{20}Mn_{20}Al_{30}$ shows a lower ductile to brittle transition temperature than the as-cast alloy [2]. It is evident that an oxygen environment is detrimental to the wear behavior of all alloys with a greater deterioration in wear behavior for $Fe_{30}Ni_{20}Mn_{20}Al_{30}$ than $Fe_{30}Ni_{20}Mn_{35}Al_{35}$.

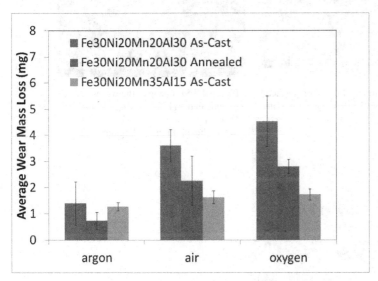

Figure 2. Average mass loss of as-cast and annealed $Fe_{30}Ni_{20}Mn_{20}Al_{30}$ pins and as-cast $Fe_{30}Ni_{20}Mn_{35}Al_{15}$ pins after 1 km sliding tests with a 23 N load in oxygen, air or argon. Three tests were performed in each environment. Error bars signify standard deviations.

Worn surface

All the worn surfaces show morphologies characteristic of abrasive wear, e.g. long parallel grooves induced by plowing and wear pits from material pullout, indicated as A and B respectively in Figure 3. Cracks perpendicular to the sliding directions (Figure 4) were observed on the worn surface of the nanostructured $Fe_{30}Ni_{20}Mn_{20}Al_{30}$ pins, which is consistent with the brittleness of this alloy. For the worn surface of the eutectic $Fe_{30}Ni_{20}Mn_{35}Al_{15}$ pins, material flow was also observed (Figure 5), and this further confirmed the good ductility and, hence plastic flow of the eutectic alloy. In all the cases, abrasive wear dominated during the wear. However, for eutectic $Fe_{30}Ni_{20}Mn_{35}Al_{15}$, plastic deformation also occurred.

The worn surfaces of the pin looked significantly different on pins tested in argon and in an oxygen-containing environment, presumably because of oxide layers on the surface of the pin in the latter case [4, 5]. Figure 6 shows an example of these differences.

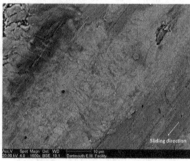

Figure 3. SE image of the worn surface of as-cast $Fe_{30}Ni_{20}Mn_{35}Al_{15}$ pin tested in air, showing A) plowing grooves and B) wear pits.

Figure 4. SE image of the worn surface of annealed an $Fe_{30}Ni_{20}Mn_{20}Al_{30}$ pin tested in air, showing cracks perpendicular to the sliding direction.

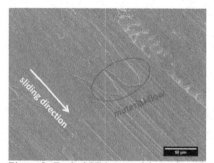

Figure 5. Typical SE image of the worn surface of an as-cast $Fe_{30}Ni_{20}Mn_{35}Al_{15}$ pin tested in argon, showing the plastic flow on the the wear pin.

106

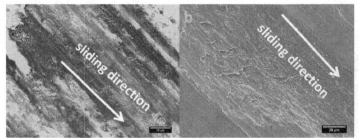

Figure 6. SE images of as-cast $Fe_{30}Ni_{20}Mn_{35}Al_{15}$ pins tested in a) oxygen and b) argon.

Tribolayer

During the wear of pins of both alloys, a similar complex tribolayer was formed on the surface of the wear pins. Figure 7 is a STEM image of the specimen produced by FIB lift-out from the tribolayer and corresponding X-ray elemental maps. The outermost layer is mostly compacted zirconia. This layer is porous, severely deformed and even contains some cracks, see Figure 8. The next layer is a mechanically mixed layer with some zirconia particles embedded in it. A similar tribology was observed when testing Cu-15wt%Ni-8%Sn against zirconia [4]. Beneath the tribolayer is mainly the original pin material.

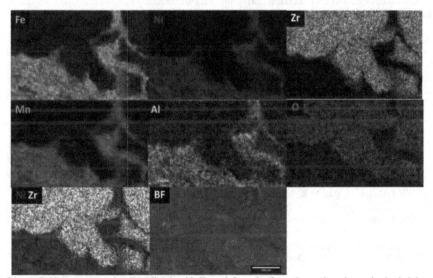

Figure 7. X-ray maps using Fe, Ni, Mn, Al, Zr and O peaks from the region shown in the bright field STEM image of an as-cast $Fe_{30}Ni_{20}Mn_{35}Al_{15}$ pin worn in argon. Note the outermost layer is mostly zirconia and the next layer is a mechanically mixed layer.

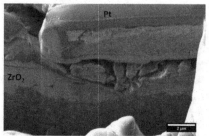

Figure 8. SE image from a pit in an as-cast $Fe_{30}Ni_{20}Mn_{35}Al_{15}$ pin wear-tested under air from which a TEM specimen was removed using FIB milling. The Pt strip was deposited on the top to protect the surface during preparation. Note the cracks in the outmost layer.

CONCLUSIONS

Even though the mechanical properties of nanostructured $Fe_{30}Ni_{20}Mn_{20}Al_{30}$ and lamellae-structured $Fe_{30}Ni_{30}Mn_{35}Al_{15}$ are substantially different, their wear behavior in an inert atmosphere was very similar. For both $Fe_{30}Ni_{20}Mn_{35}Al_{15}$ and $Fe_{30}Ni_{20}Mn_{20}Al_{30}$, wear was greater in oxygen and air than in argon, $Fe_{30}Ni_{20}Mn_{35}Al_{15}$ is less sensitive to oxygen than the nanostructured $Fe_{30}Ni_{20}Mn_{20}Al_{30}$. The dominant wear mechanism was abrasive wear in all cases, while plastic deformation also occurred during the wear of eutectic $Fe_{30}Ni_{20}Mn_{35}Al_{15}$. Zirconia particles were mechanically mixed into the tribolayers on the surface of the pins.

ACKNOWLEDGMENTS

This research was supported by U.S. Department of Energy (DOE), Office of Basic Energy Science Award #DE-FG02-10ER46392 and National Science Foundation (NSF) award #DMR-0905229. The views and conclusions contained herein are those of the authors and should not be interpreted as necessarily representing official policies, either expressed or implied of the DOE, NSF or the U.S. Government.

REFERENCES

1. M.W. Wittmann, I. Baker, J.A. Hanna and P.R. Munroe PR, MRS Proceedings 842, S5.17, 35 (2005).
2. X. Wu, I. Baker, H. Wu, M.K. Miller and P.R. Munroe, in preparation.
3. Y. Liao, F. Meng and I. Baker, Intermetallics 19, 1533 (2011).
4. X. Wu, I. Baker and P. R. Munroe, Intermetallics, 23, 116 (2012).
5. F. Meng, I. Baker and P. R. Munroe, Journal of Materials Science, 47, 4827 (2012).
6. B.J. Johnson, F.E. Kennedy, and I. Baker, Wear 192, 241(1996).
7. J.B. Singh, W. Cai and P. Bellon, Wear 263, 830 (2007).

Mater. Res. Soc. Symp. Proc. Vol. 1516 © 2013 Materials Research Society
DOI: 10.1557/opl.2013.62

Microstructure Control of Nb-Si Based Alloys with Cr, W, Ta and Zr by Using Nb₃Si Phase Stability Control

Yuting WANG[1], Seiji MIURA[2], Akira YOSHINARI[1]

[1] Materials Research Centre, Hitachi Research Laboratory,
1-1, Omika-cho 7-chome, Hitachi, Ibaraki,
319-1292, Japan
[2] Faculty of Engineering, Hokkaido University, N13 W8, Sapporo, Hokkaido,
060-8628, Japan

ABSTRACT

Recently, Nb-Si based alloys have attracted considerable attention as potential candidate materials for ultra-high temperature applications, because of their low densities and high melting points. However, it is still very difficult to obtain materials with a good balance of high-temperature strength and room-temperature toughness. To address this issue, microstructure control is considered to be a promising method. In applying microstructure control to Nb-Si based alloys with a eutectic reaction (L → Nb_{ss} + Nb_3Si) and a eutectoid reaction (Nb_3Si → Nb_{ss} + Nb_5Si_3), the key is the control of Nb_3Si phase stability. Nb_{ss} (Nb solid solution) is considered as a ductile phase. In previous reports, it was revealed that different elements had different effects on the stability of Nb_3Si. In particular, Mo and W (>3 at %) destabilize the Nb_3Si phase, while Ti and Ta stabilize it, and Zr acts as an accelerator for decomposition of Nb_3Si. On the other hand, Cr is known to enhance the formation of the ductile Nb_{ss} phase. In the present study, we investigated the effects of adding combinations of stabilizing, destabilizing, and accelerating elements with Cr, such as Cr and W, Cr and Ta, Cr and Zr. According to SEM observation, different microstructures were obtained with different combination of additives, and the fracture toughness at room temperature of these samples were also evaluated to reveal the effects of the microstructure on the mechanical properties of Nb-Si based alloys.

INTRODUCTION

In order to improve the performance and achieve higher efficiency of thermal systems, materials with higher thermal functioning than current materials, nickel (Ni)-based superalloys have been receiving considerable attention [1-3]. Recent studies have identified alloys based on Nb-Si as having great potential for application to turbine engines and land-based gas turbines because of their high melting points [1-5]. The lower density of a Nb-Si system is also considered to be a significant advantage over Ni-based superalloys. However, the lack of balance between low-temperature damage tolerance and high-temperature strength is still one of the major issues for practical application. To improve the mechanical properties of Nb-Si based alloys, microstructure control is considered to be a promising method.

Microstructure control involves two kinds of phase reactions: eutectic solidification (Eq. 1), followed by a eutectoid composition reaction (Eq. 2)

$$L \rightarrow Nb_{ss} + Nb_3Si \qquad (1)$$

$$Nb_3Si \rightarrow Nb_{ss} + Nb_5Si_3 \qquad (2)$$

The primary Nb_{ss} dendrite phase and the Nb_3Si phase form through solidification (Eq. 1). The Nb_{ss} phase is considered as a ductile phase, as the larger size of Nb_{ss} grains is helpful in suppressing the propagation of cracks and provides better toughness at low temperatures [6-8].

The fine Nb_{ss}/Nb_5Si_3 lamellar structure can be formed through the eutectoid decomposition (Eq. 2) of Nb_3Si. Therefore the Nb_3Si phase which transiently appears during the alloy preparation is very important for the microstructure control. In previous reports, it was revealed that different elements had different effects on the stability of Nb_3Si. Notably, Mo and W (>3 at %) destabilize the Nb_3Si phase [7], while Ti and Ta stabilize it [7]. On the other hand, Cr is known to enhance the formation of the ductile Nb_{ss} phase [9]. In this study, we investigated the effects of adding combinations of stabilizing and destabilizing elements, such as Cr+Ta and Cr+W. Also, it has been reported that Zr accelerates the eutectoid decomposition of the Nb_3Si phase [5]. So the effect of a Cr+Zr combination was also investigated. The microstructures of these samples obtained with different combinations of additives, and the fracture toughness at room temperature and strength at high temperatures were also evaluated to reveal the effects of the microstructure on the mechanical properties of Nb-Si based alloys.

EXPERIMENT

To compare the microstructures, the five alloys listed in Table 1 were prepared from high purity elements (>99.9%) using a high-frequency induction melting furnace in an Ar atmosphere at 500 Pa. The alloys were in re-melted several times in order to ensure their homogeneity. After that, the alloys were denoted by their atomic concentrations. Alloy 1 and Alloy 2 were prepared as base alloys to compare with Alloy 3-5 with the combination additions. Alloy ingots were cut into several pieces for heat-treatment and microstructure observation. Heat treatment was conducted under an Ar-flow atmosphere at a temperature of 1300°C for 20 hours.

Table 1: The Compositions of Experimental Alloys (at%).

Alloy	at%
1	Nb-16Si
2	Nb-16Si-7.5Cr
3	Nb-16Si-7.5Cr-3Ta
4	Nb-16Si-7.5Cr-3Zr
5	Nb-16Si-7.5Cr-3W

Specimens were carefully polished for scanning electron microscope (SEM) observation of as-cast or heat-treated microstructures using a Hitachi S3200 with back scattered electron (BSE) detector. The phases were identified by a combination of energy dispersive spectroscopy (EDS) in SEM, and powder x-ray diffraction. The room temperature toughness of Alloy 3-5 as-cast and heat-treated at 1300°C were evaluated by four-point bending tests using chevron notched specimens and analytical calculation using the Bluhm's slice model [10]. The dimensions of bending test specimen were $5 \times 5 \times 30$ mm, and the chevron notch was introduced by electro-discharge machining (EDM) to set up the initial ligament length being 0.8 against the width of the bending bar. High temperature compression tests of Alloy 3-5 heat-treated at 1300°C were carried out with a strain rate of $1.0 \times 10^{-4}\text{s}^{-1}$ in an Ar atmosphere at 1200°C. The size of compression test specimens was 3 mm × 3 mm × 6 mm.

RESULTS AND DISCUSSION

<u>Microstructure of as-cast alloys</u>

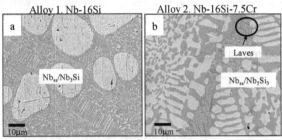

Fig 1. The microstructure of as-cast alloys.

Fig. 1 shows BSE images of the as-cast microstructures of the alloys listed in Table 1. In Alloy 1, only Nb (bright) and Nb_3Si (gray) are recognized.

In Alloy 2, mainly Nb_{ss} (bright) and Nb_5Si_3 (dark) are recognized and no Nb_3Si phase is found. Coarse primary Nb_{ss} grains larger than several tens of microns are formed. This suggests that Cr has a destabilizing effect on Nb_3Si and also helps to form coarse Nb_{ss} grains during the solidification. A small amount of laves phase is also formed, indicated by the circle in Fig. 1b.

In Alloy 3, Nb_3Si (gray) co-exists with small amount of fine Nb_{ss} (bright) within the eutectic cell structure and dark α-Nb_5Si_3 appears between cells of the primary Nb_{ss} (bright) phase. The Nb_{ss}-Nb_3Si eutectic reaction occurs during the solidification, and forms α-Nb_5Si_3 near the edge of the eutectic cell, an area rich in Cr according to the EDS analysis result. Nb_3Si remains in the center of the eutectic cells, which are found to be rich in Ta. These results imply Cr destabilizes Nb_3Si phase, while Ta stabilizes as was suggested by previous study. Furthermore, a lamellar structure can be found in the middle of the zone rich in Cr and Ta, as show in Fig. 1c. This suggests that Cr and Ta exert their own effect on the rich Nb-Si microstructure as mentioned above even in combination.

The as-cast microstructures of Alloy 4 and Alloy 5 are quite similar, but different from that of Alloy 3. In these alloys, mainly Nb_{ss} and Nb_5Si_3 are observed, but no Nb_3Si. The eutectic cell and coarse Nb_{ss} grains can also be seen in the as-cast microstructures (Fig. 1 d and e). This suggests that the combinations of Cr+Zr and Cr+W have strong tendencies to form Nb_5Si_3 during the solidification. Laves phase is observed as indicated by the circles in Fig. 1d and 1e.

Five fields of SEM view were used to determine the average size of primary Nb_{ss} grain. The mean value of primary Nb_{ss} grain size of Alloy 3, 4, 5 are 25.8 μm, 16.3 μm and 23.1 μm, respectively.

Further investigation of the microstructure over a broad composition range with different elements combinations is necessary. However, the as-cast microstructure is quite easily controlled by the addition of elements.

Microstructure of heat-treated alloys

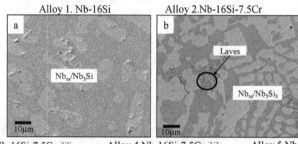

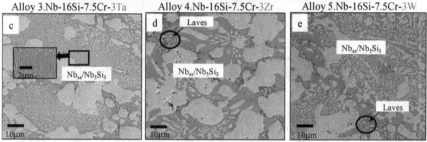

Fig 2. The microstructure of heat-treated alloys of 1300°C for 20 h.

Fig. 2 shows the microstructure of the heat-treated alloys listed in Table 1. Nb_{ss} grains in all alloys grow larger and the coarse Nb_{ss} grains in Alloy 2-5 show a tendency to connect with each other after heat treatment. The mean value of coarse Nb_{ss} grain size of Alloy 3, 4, 5 after heat-treatment are 31.8 μm, 26.3 μm and 26.1 μm, respectively.

As can be seen in Alloy 3 (Fig. 2c), a eutectoid Nb_3Si reaction occurs. Nb_3Si (at the center of the eutectic cell) decomposes into Nb_{ss}/Nb_5Si_3 with a lamellar structure as shown in Fig. 2c.

In Alloys 4 and 5, since there is no Nb_3Si phase after solidification, no lamellar structure can be formed by a eutectoid reaction. Alloy 5 with Cr + W shows some micron-sized structures in the eutectic cell even after heat-treatment.

It is known that the lamellar structure can enhance strength at elevated temperatures, and coarse Nb_{ss} grains are effective in improving the toughness at room temperature. Thus, the alloy with added Cr + Ta seems to have a very interesting microstructure, which is composed of lamellar

112

structures and coarse Nb$_{ss}$ grains. This kind of microstructure is considered to have the potential to improve the high temperature strength and lower temperature damage tolerance at the same time.

Room-temperature toughness

To compare the room temperature toughness of alloys with different microstructures, 4-point bending tests were carried out (Alloy 3-5 as-cast and heat-treated). The results are shown in Table 2. The improvement of room temperature toughness can be clearly seen after heat treatment, which proves that the control of coarse Nb$_{ss}$ grains scale is helpful for improving room temperature toughness. A larger microstructural scale of the Nb$_{ss}$ phase tends to improve the fracture toughness. However, the room temperature toughness of these alloys is still not sufficient for manufacturing, which requires more than 20 MPa·m$^{1/2}$[11]. To realize the further improvement, details of the rupture mechanism of these alloys with different microstructures should be studied.

Table 2 Fracture Toughness at Room Temperature (MPa·m½)

Alloy	As-cast	Heat-treatment
Alloy 3 (Cr+Ta)	8.3	11.2
Alloy 4 (Cr+Zr)	9.3	10.7
Alloy 5 (Cr+W)	9.5	10.4

High-temperature strength.

High temperature compression tests of Alloys 3-5 heat-treated at 1300°C were carried out at 1200°C with a strain rate of $1.0 \times 10^{-4} s^{-1}$ in an Ar atmosphere. Compressive stress–strain curves are shown in Fig. 3. All of the specimens show compression strengths over 420 MPa at 1200°C. The stresses of all the alloys decrease gradually after the beginning of plastic deformation at 1200°C. The difference of the microstructure is one of the possible reasons for different curves as shown in Fig. 3. However, the detail effect of the microstructure differences of these alloys on high temperature strength is still unclear and remains for future work.

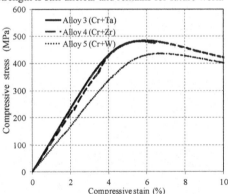

Fig.3 Compressive stress-strain curves of alloy 3,4,5 (heat-treated) with a strain rate of $1.0 \times 10^{-4} s^{-1}$ at 1200°C

CONCLUSIONS

The effect of adding combinations of stabilizing, destabilizing, and accelerating elements of Nb_3Si phase was investigated. The combinations of Cr + Ta, Cr+Zr and Cr+W provided three different microstructures after solidification and heat-treatment. In particular, the alloy with a combination of Cr and Ta shows an interesting microstructure after heat treatment, possessing both a high-temperature strength lamellar structure and a ductile coarse-Nb_{ss}-grain structure. The combination of Cr and Ta shows a strong improvement in room-temperature toughness.

REFERENCES

1. B.P. Bewlay, M.R. Jackson, J-C. Zhao, P.R. Subramanian, M.G. Mendiratta, and J.J. Lewandowski, MRS bull. 28, 646 (2003).
2. B.P. Bewlay, M.R. Jackson, and M.F.X Giliotti, *Intermetallic Compounds- Principles and Practice* . (John Wiley & Sons, 2001) P.541.
3. J. Kajuch, J. Short and J.J. Lewandowski, Acta Metal. Mater., 43, 1955 (1995).
4. B.P. Bewlay, M.R. Jackson, J-C.Zhao, P.R. Subramanian, Metall. Mater. Trans. A, 34A, 2043 (2003).
5. S. Miura, M. Aoki, Y. Saeki, K.Ohkubo, Y. Mishima, and T. Mohri, Metall. Mater. Trans. A, 36A, 489 (2005).
6. S. Miura, Y. Saeki, and T. Mohri, MRS Symp. Proc., 552, KK 6.9.1(1999).
7. S. Miura, T. Tanahashi, Y.Mishima, T.Mohri, Mater. Sci. Forum, 645-656,444 (2010).
8. S. Miura, Y. Murasato, Y. Sekito, Y. Tsutsumi, K. Ohkubo, Y. Kimura, Y. Mishima and T. Mohri, Metall. Mater. Trans. A, 510-511, 317 (2009).
9. A. Deal, W. Heward, D.Ellis, J.Cournoyer, K.Dovidenko, B.P.Bewlay , Microsc. Microanal., 13 (Suppl. 2) , 90-91(2007).
10. D. G. Munz, Int. J. Fract. 16, 137–140 (1980).
11. R. Tanaka, A. Kasama, M. Fujikura, I. Iwanaga, H. Tanaka, Y. Matsumura, J. Gas Turbine Soc. Jpn, 31(2), 81-86 (2003).

Mater. Res. Soc. Symp. Proc. Vol. 1516 © 2012 Materials Research Society
DOI: 10.1557/opl.2012.1726

Microstructure and tribological properties of gray cast iron specimens coated by aluminizing, boronizing, chromizing and siliconizing

T. Murakami[1], K. Matsuzaki[1], Y. Gomi[2], S. Sasaki[2] and H. Inui[2]

[1] National Institute of Advanced Industrial Science and Technology (AIST), Tsukuba Ibaraki 305-8564, Japan
[2] Department of Mechanical Engineering, Tokyo University of Science, Tokyo 102-0073, Japan
[3] Department of Materials Science and Engineering, Kyoto University, Kyoto 606-8501, Japan

ABSTRACT

In this study, aluminized, boronized, chromized and siliconized gray cast iron plate specimens were prepared, and their microstructures and tribological properties were investigated. The surfaces of the aluminized, boronized, chromized and siliconized specimens mainly consisted of FeAl, Fe_2B, $(Cr, Fe)_{23}C_6$ and FeSi phases, respectively. Also, the surface of the boronized specimen exhibited the highest microvickers hardness of all the specimens. The aluminized, boronized and chromized specimens exhibited friction coefficients as low as the non-coated specimens when sliding against AISI 52100 steel ball specimens in poly-alpha-olefin. In addition, the boronized and chromized specimens exhibited much higher wear resistance than the non-coated specimens.

INTRODUCTION

Gray cast iron has been used variously as a sliding material because of its low friction coefficient in various oils. However, gray cast iron has low strength and low wear resistance. Therefore, the gray cast iron surface is sometimes nitrided [1, 2] and quenched to increase the strength and wear resistance, although this does not sufficiently improve the corrosion resistance. Pack cementation [3-6] of gray cast iron may improve such problems because it is thought that adhesive and thick coating layers with high hardness, high wear resistance and high corrosive resistance can be formed using this process. However, the friction and wear properties of such coating layers have not been investigated sufficiently. In this study, gray cast iron specimens coated with Al-, B- and Cr-rich layers were prepared by aluminizing [3], boronizing [4, 5] and chromizing [5] using an electric furnace, recpectively. Also, gray cast iron specimens coated with Si-rich layers were prepared by siliconizing [6] using a hot pressing machine at a pressure of 20 MPa because the Si-rich layers obtained using the electric furnace did not adhere adequately to the gray cast iron substrates as reported by Tatemoto et al [7]. After the coating, the microstructures, microvickers hardness

Table 1 Pack cementation conditions used in this study.

	Composition of pack cementation mixtures (mass%)		Temperature (K)	Time (h)
Aluminizing	Al	10	1273	8
	NH₄Cl	5		
	Al₂O₃	85		
Boronizing	B₄C	8	1273	12
	SiC	82		
	KBF₄	10		
Chromizing	Cr	25	1323	12
	Al₂O₃	69		
	NH₄Cl	6		
Siliconizing	Si	10	1373	1
	Si₃N₄	88		
	MgF₂	2		

and tribological properties of the coated specimens were investigated.

EXPERIMENTAL PROCEDURES

The gray cast specimens used in this study have a composition of Fe-4.6C-2.0Si (mass%). These specimens were cut from an ingot bar, polished using 4000-grit SiC papers and cleaned in a mixture of 50 vol% acetone and 50 vol% petroleum benzene using an ultrasonic cleaner for 1.2 ks before the pack cementation. The aluminizing, boronizing and chromizing and siliconizing were performed in an electronic furnace and hot pressing machine in an argon gas atmosphere under the conditions shown in Table 1. After the coating, the surface of each coated specimen was polished and cleaned using an ultrasonic cleaner again.

The friction and wear properties of each plate specimen sliding against AISI 52100 steel ball specimens were investigated using a rotating ball-on-plate tribometer. The diameter of the ball specimen was 10 mm, and its microvickers hardness was 7.9 GPa at a load of 9.8 N and a loading time of 15 seconds. In the friction tests, the plate specimens were immersed in 100 ml of poly-alpha-olefin (PAO) with an ISO viscosity of VG 32. The friction testing conditions were as follows; load: 9.8N, sliding speed: 6.3 mm/s, track radius: 4 mm, testing time: 3.6 ks, and temperature: 298 K. The PAO was not replenished during each friction test. After the friction tests, all of the plate specimens and their paired ball specimens were cleaned using an ultrasonic cleaner for 1.2 ks. The specific wear rates of the plate and ball specimens were calculated using a surface profilometer and an optical microscope [8]. In addition, the worn surfaces of each plate specimen were observed using a scanning electron microscope (SEM) with an energy

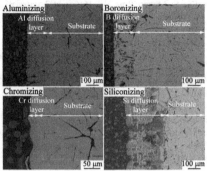

Fig. 1 SEM photographs showing the cross sections of the coated specimens.

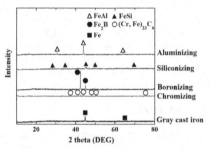

Fig. 2 X-ray diffraction patterns of the coated specimens.

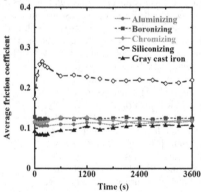

Fig. 3 Friction coefficients of the coated specimens.

116

dispersive spectroscopy (EDS) attachment, and x-ray photoelectron spectroscopy (XPS).

RESULTS

Figure 1 shows the cross sections of the coated specimens. The aluminized, boronized, chromized and siliconized specimens were coated with Al-, B-, Cr- and Si-rich layers, respectively. The thickness of the Al-, B- and Si-rich layers was more than 100 μm, while the thickness of the Cr-rich layer was approximately 20 μm. These results indicate that Cr atoms diffuse most sluggishly in the gray cast iron because the chromizing was performed at the highest temperature as shown in Table 1. On the other hand, most of the graphite phases remained in the Al-rich phases after the aluminizing processes. In addition, many pores were observed in the B-rich and Si-rich layers of the boronized and siliconized specimens, respectively. However, it

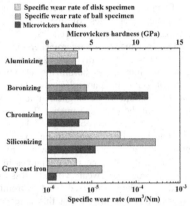

Fig. 4 Microvickers hardness and specific wear rates of the coated specimens, and specific wear rates of their paired ball specimens.

is not clear why such pores were observed in the B- and Si-rich layers. Moreover, neither graphite phases nor pores were observed in the Cr-rich phase of the chromized specimens. It is considered that the graphite phases near the surfaces of the chromized specimens became carbide phases by reacting with the Cr content.

Figure 2 shows the X-ray diffraction pattern of each coated specimen. The aluminized, boronized, chromized and siliconized specimens were coated with FeAl, Fe_2B, $(Cr, Fe)_{23}C_6$ and FeSi phases, respectively. The Cr-rich phases became carbide phases as expected in Fig. 1.

Figure 3 shows the friction coefficients of the coated specimens sliding against the AISI 52100 steel ball specimens in the PAO with no additives. The low friction coefficients of the non-coated specimens were due to the presence of the graphite phase, which is a low friction material and also become an oil sump. On the other hand, the aluminized, boronized and chromized specimens exhibited friction coefficients as low as the non-coated specimens. However, the siliconized specimens exhibited much higher friction coefficients than the other specimens. It is considered that the low friction coefficients of the aluminized, boronized and chromized specimens were due to the formations of low friction $Al(OH)_3$, H_3BO_3 and chromium oxides on the worn surfaces, respectively, while such low friction materials were not formed on the worn surfaces of the siliconized specimens. It is expected that the worn surfaces of the aluminized, boronized and chromized specimens were covered with thin iron oxide, Al_2O_3, B_2O_3 and chromium oxide films because the worn surfaces of metal and alloy substrates were generally covered with thin oxide films during oil lubrication [9, 10]. It is also known that $Al(OH)_3$ and H_3BO_3 are formed on the surfaces of Al_2O_3 and B_2O_3 substrates by reacting with water molecules in air [11, 12], and that the $Al(OH)_3$ and H_3BO_3 exhibit low friction because they have layered crystal structures [12, 13]. Furthermore, CrO_3 is thought to be a low friction material because CrO_3 has a low melting point (473 K) [14]. Erdemir reported that low melting point oxides generally exhibit low friction [15].

Figure 4 shows the microvickers hardness and specific wear rates of the coated plate specimens, and the specific wear rates of their paired ball specimens. The microvickers hardness of the surfaces of the gray cast iron specimens increased by aluminizing, boronizing, chromizing and siliconizing. In particular, the boronized specimens exhibited the highest microvickers hardness of all the plate specimens. These results were due to the formations of the FeAl, Fe_2B, (Cr, Fe)$_{23}C_6$ and FeSi layers, which were harder than the gray cast iron specimens, as shown in Figs. 1 and 2. Concerning the specific wear rates of the plate specimens, no wear volume loss was observed on the boronized and chromized specimens. It is considered that this absence of wear volume loss was due to the formations of hard Fe_2B and (Cr, Fe)$_{23}C_6$ layers and low friction materials such as H_3BO_3 and CrO_3. However, the specific wear rates of the aluminized specimens were as high as those of the non-coated specimens although the microvickers hardness of the aluminized specimens was slightly higher than that of the chromized specimens. The higher specific wear rates of the aluminized specimens might have been due to the mechanical or chemical properties of the $Al(OH)_3$. We will investigate such mechanisms in the near future. Moreover, the specific wear rates of the AISI 52100 steel ball specimens sliding against the aluminized, boronized and chromized specimens were lower than those of the ball specimens sliding against the non-coated specimens. It is considered that these lower specific wear rates were due to the formations of low friction materials such as $Al(OH)_3$, H_3BO_3 and CrO_3 on the worn surfaces. On the other hand, the siliconized specimens and their paired ball specimens exhibited the highest specific wear rates of all the plate and pin specimens, although the siliconized specimens exhibited higher microvickers hardness than the aluminized and chromized specimens. This would be due to the forming of few low friction materials on the worn surfaces of the siliconized plate and steel ball specimens.

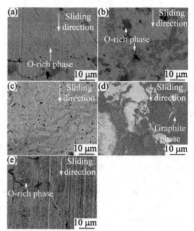

Fig. 5 SEM photographs showing worn surfaces of the (a) aluminized, (b) boronized, (c) chromized, (d) siliconized and (e) non-coated specimens.

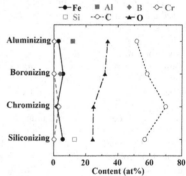

Fig. 6 Fe, Al, B, Cr, Si, C and O content on the worn surfaces of the coated plate specimens.

Figure 5 shows the worn surfaces of the plate specimens. Smooth surfaces with few scratch marks were observed on the worn surfaces of the boronized and chromized specimens, indicating that both of these specimens exhibited high wear resistance as shown in Fig. 4. In addition, shallow scratch marks were observed on the worn surfaces of the aluminized specimens while more distinct scratch marks were observed on the worn surfaces of the non-coated specimens. On the other hand, worn surfaces with cleavage along the graphite phase were observed on the siliconized specimens. This cleavage might have been due to the low bonding strength between the iron silicide and graphite phases in the Si-rich layer. Also, the high friction coefficients and high specific wear rates of the siliconized specimens (Figs. 3 and 4) would be due to the abrasive wear of the wear debris consisting of hard iron silicides. However, it is unclear why the iron silicide phases and graphite phase had low bonding strength. More study will be needed to clarify this mechanism.

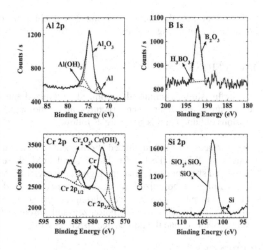

Fig. 7 XPS spectra of Al 2p, B 1s, Cr 2p and Si 2p of the worn surface areas of the aluminized, boronized, chromized and siliconized specimens, respectively. Dotted lines indicate backgrounds and fitted peaks.

Figure 6 shows the Fe, Al, B, Cr, Si, C and O content on the worn surfaces of the coated plate specimens as obtained by XPS analyses. Large amounts of carbon and oxygen were observed on the worn surfaces of all the plate specimens. Also, small amounts of Fe, Al, B, Cr and Si were observed on the worn surfaces. It is considered that most of the carbon contents were due to the carbon contamination, graphite phase, acetone and petroleum benzene [16], and that most of the oxygen contents were due to the oxide phases and acetone. Acetone and petroleum benzene were used for the ultrasonic cleaning of the plate and ball specimens before and after the friction tests.

Figure 7 shows the XPS spectra of Al 2p, B 1s, Cr 2p and Si 2p of the worn surface areas of the aluminized, boronized, chromized and siliconized specimens, respectively. Strong Al_2O_3 and weak $Al(OH)_3$ peaks were observed in the spectrum of the aluminized specimen, while strong B_2O_3 and weak H_3BO_3 peaks were observed in the spectrum of the boronized specimen. It is considered that most of $Al(OH)_3$ and H_3BO_3 became Al_2O_3 and B_2O_3, respectively, due to the evaporation of water molecules in the high vacuum XPS chamber. It is known that $Al(OH)_3$ and H_3BO_3 are formed on the surfaces of Al_2O_3 and B_2O_3 substrates by absorbing water molecules [12, 13]. Therefore, it is considered that large amounts of low friction $Al(OH)_3$ and H_3BO_3 were present on the worn surfaces of the aluminized and boronized specimens during their friction tests, respectively. On the other hand, Cr_2O_3 peaks were observed in the spectrum of the chromized specimen, while SiO_2 peak was observed in the spectrum of the siliconized specimen. It is considered that the worn surfaces of the Cr_2O_3 became higher valence oxides such as low friction CrO_3 during the friction tests, while the SiO_2 did not work as a low friction material. Panas et al. reported that the surface of Cr_2O_3 is covered with a monolayer of Cr(VI) oxide [17].

We think that the aluminized, boronized and chromized specimens exhibited low friction coefficients for foregoing reasons.

CONCLUSIONS

In this study, aluminized, boronized, chromized and siliconized gray cast iron specimens were prepared, and their microstructures and tribological properties were investigated. The conclusions obtained in this study are as follows.

1. The surfaces of the aluminized, boronized, chromized and siliconized specimens mainly consisted of FeAl, Fe_2B, $(Cr, Fe)_{23}C_6$ and FeSi phases, respectively. The surface of the boronized specimen exhibited the highest microvickers hardness of all the specimens.
2. The aluminized, boronized and chromized specimens exhibited friction coefficients as low as the non-coated specimens when sliding against AISI 52100 steel ball specimens in poly-alpha-olefin. In addition, the boronized and chromized specimens exhibited much higher wear resistance than the non-coated specimens.

REFERENCES

1. B.B. Nayak, O.P.N. Kar, D. Behera, B.K. Mishra, *Surf. Eng.*, **27**, 99-107 (2011).
2. O.D. Sokolov, O.V. Mannapova, A.I. Kostrzhyts'kyi, A.P. Olik, *Mater. Sci.*, **42**, 849-852 (2006).
3. R.S. Dutta, S. Majumdar, A. Laik, K. Singh, U.D. Kulkarni, I.G. Sharma, G.K. Dey, *Surf. Coat. Technol.*, **205**, 4720-4725 (2011).
4. C. Li, M.S. Li, Y.C. Zhou, *Surf. Coat. Technol.*, **201**, 6005-6011 (2007).
5. S.Y. Lee, G.S. Kim, B.S. Kim, *Surf. Coat. Technol.*, **177-178**, 178-184 (2004).
6. S.W. Choi, Y.C. Kim, S.H. Chang, I.H. Oh, J.S. Park, C.S. Kang, *Trans. Nonferrous Met. Soc. China*, **19**, 875-878 (2009).
7. K. Tatemoto, Y. Ono, R.O. Sizuki, *J. Phys. Chem. Solids*, **66**, 526-529 (2005).
8. T. Murakami, Y. Hibi, H. Mano, K. Matsuzaki, H. Inui, *Intermetallics*, **20**, 68-75 (2012).
9. T. Murakami, K. Kaneda, M. Nakano, H. Mano, A. Korenaga, S. Sasaki, *Intermetallics*, **15**, 1573-1581 (2007).
10. T. Murakami, K. Kaneda, M. Nakano, Y. Xia, S. Sasaki, *Tribol. Int.*, **43**, 312-319 (2010).
11. R. Schoen, C.E. Roberson, *American Mineralogist*, **55**, 43-77 (1970).
12. B.R. Burroughs, J.H. Kim, T.A. Blanchet, *Tribol. Trans.*, **42**, 592-600 (1999).
13. R. Demichelis, B. Civalleri, Y. Noel, A. Meyer, R. Dovesi, *Chem. Phys. Lett.*, **465**, 220-225 (2008).
14. W.K. Jozwiak, W. Ignaczak, D. Dominiak, T.P. Maniecki, *Appl. Catal. A: Gen.*, **258**, 33-45 (2004).
15. A. Erdemir, *Tribol. Lett.*, **8**, 97-102 (2000).
16. T. Murakami, H. Mano, Y. Hibi, K. Matsuzaki, H. Inui, *Tribol. Int.*, **56**, 1-8 (2012).
17. I. Panas, J.E Svensson, H. Asteman, T.J.R. Johnson, L.G. Johansson, *Chem. Phys. Lett.*, **383**, 549-554 (2004).

Mater. Res. Soc. Symp. Proc. Vol. 1516 © 2013 Materials Research Society
DOI: 10.1557/opl.2013.109

Characterization of Ni₃(Si,Ti) intermetalic alloys synthesized by powder metallurgical method

Yuki Miura[1], Yasuyuki Kaneno[1], Takayuki Takasugi[1] and Atsushi Kakituji[2]

[1]Department of Materials Science, Graduate School of Engineering, Osaka Prefecture University, 1-1Gakuen-cho Naka-ku, Sakai, Osaka 599-8531, Japan
[2]Technology Research Institute of Osaka Prefecture, Izumi 594-1157, Japan

ABSTRACT

A Ni₃(Si,Ti) intermetalic alloy was synthesized by the powder metallurgy method using elemental powders. The raw powder mixtures with various compositions were sintered by a spark plasma sintering apparatus and then homogenized at high temperatures. Microstructure, hardness, tensile properties and density of the sintered alloys were investigated as functions of the chemical composition and sintering temperature. It was found that a highly-densified Ni₃(Si,Ti) sintered alloy was obtained by choosing proper chemical composition and sintering temperature. Also, the Ni₃(Si,Ti) sintered alloy with an $L1_2$ single-phase microstructure exhibited high hardness and tensile strength.

INTRODUCTION

Many intermetallic compounds possess a superior high temperature characteristic such as high phase stability and high strength at elevated temperature, and therefore they have been considered to be a candidate for high temperature structural materials. However, polycrystalline intermetallic compounds with an $L1_2$ crystal structure have been known to mostly suffer from propensity for intergranular fracture. A Ni₃(Si,Ti) intermetallic alloy with an $L1_2$ crystal structure, which was developed by Ti addition to Ni₃Si [1,2], shows high strength and fairly good tensile ductility even at low temperature. Also, the Ni₃(Si,Ti) intermetallic alloy have a unique high temperature hardness property; the reduction of hardness with increasing temperture for this intermetallic alloy is relatively small compared with that for conventional metallic materials. However, the hardness of the Ni₃(Si,Ti) alloy at room temperature is inferior to that of existing wear resistant materials. It is well known that dispersion strengthening by hard particles such as carbide, niteride and boride is an effective method enhancing the hardness of metallic materials. It has been reported that the Ni₃Al base composite material dispersion-strenghened with hard particles showed a superior characteristic [3,4]. These composite materials can not be made by the ingot metallurgical (melting and casting) method. Alternatively, they could be fabricated by the powder metallurgical method.

In this study, the Ni₃(Si,Ti) intermetallic alloy was synthesized by the powder metallurgy method using elemental powders. The raw powder mixtures with various compositions were sintered and densificated by a spark plasma sintering apparatus and then homogenized at high temperatures. The effects of the chemical composition of powder mixtures, sintering condition such as temperature and heating rate, and homogenization temperature on the microstructure and the mechanical properties were examined.

EXPERIMENTAL

In this study, a $Ni_3(Si,Ti)$ intermetalic alloy was synthesized by the powder metallurgy method using elemental powders. Raw elemental powders used in this study were Ni (99.8% purity, particle size of 4–7 μm), Si (99.8% purity, particle size of 100 μm under), Ti (99% purity, particle size of 100 μm under) and B (99.9%purity, particle size of 1 μm under). The composition of each sample is shown in Table 1. Boron doping of the $Ni_3(Si,Ti)$ alloy is necessary to suppress the intergranular fracture due to environmental embrittlement at room temperature [5]. The powder mixtures were milled in a mortar for 30 min. The mixed powders were formed by means of cold isostatical pressing (CIP), and then sintered by spark plasma sintering (SPS) method in a vacuum at 1073 K and 1173 K, respectively. The sintered alloys were homogenized at 1223 K for 48 h in a dynamic vacuum.

Microstructural observations were carried out by using a scanning electron microscope (SEM). X-ray diffraction (XRD) analysis was performed by using $CuK\alpha$ radiation at an accelerated voltage of 40 kV mostly with a scan speed of 1°/min to determine constituent phases and second-phase dispersions. Bulk densities of the homogenized specimens were measured by the Archimedes method, and the relative densities were calculated by comparing the density of the homogenized specimens made from the cast alloys. Vickers hardness tests of the specimens after the homogenization were conducted in the temperature range from room temperature to 1073 K under the conditions of holding time of 20 s and a load of 1 kg. High temperature hardness tests were conducted in atmosphere of argon gas 90 % + H_2 10 %. Hardness data points more than five were collected and averaged for each experimental condition. Room-temperature tensile tests were carried out in air by using the homogenized specimens.

Table 1 Chemical compositions, sintering temperature and relative density of the alloys used in this study.

Sample	Composition	Sintering temperature	Relative density
#1	$Ni_{79.5}Si_{11.0}Ti_{9.5}$+50wt.ppmB	1073 K	96.4%
#2	$Ni_{79.5}Si_{11.0}Ti_{9.5}$+50wt.ppmB	1173 K	99.8%
#3	$Ni_{79.0}Si_{11.3}Ti_{9.7}$+50wt.ppmB	1173 K	99.8%
#4	$Ni_{78.0}Si_{11.8}Ti_{10.2}$+50wt.ppmB	1173 K	99.7%

RESULTS and DISCUSSION

Microstructures

The microstructures of the as-sintered samples are shown in Fig.1. Intermediate phases in addition to an $L1_2$ phase (or Ni solid solution (A1)) were detected in all the as-sintered samples irrespective sintering temperature: it was remarkable in the samples sintered at lower temperature or in the samples with lower Ni content. From the results of SEM observations and XRD analysis, these intermediate phases were identified as Ni_3Ti and Ni_5Si_2. It was considered that the minor elements of Ti and Si reacted to the major element of Ni in the early stage of the reaction (sintering), resulting in formation of binary metastable Ni_3Ti and Ni_5Si_2 phases.

122

The intermediate phases such as Ni₃Ti and Ni₅Si₂ (Fig.1) disappeared in all the samples after homogenization at 1123 K for 48 h. The homogenized samples #1~#3 exhibited the microstructure consisting of an $L1_2$ single-phase region and ($L1_2$+A1) two-phase region, as typically shown in Fig.2. In the later region, A1 (fcc Ni solid solution) phase existed in the channels between the cuboidal shaped $L1_2$ precipitates, as has been observed in the microstructure of the conventional γ/γ' type superalloys (Fig.2(b)). A similar two-phase microstructure has been observed in the Ni-rich Ni₃(Si,Ti) alloys [4]. The volume fraction of the ($L1_2$+A1) two-phase region decreased with decreasing Ni content. For sample #4 with the lowest Ni content, the ($L1_2$+A1) two-phase region completely disappeared and $L1_2$ single-phase phase dominated the entire microstructure. As shown in Fig.3, fine oxide particles with bright contrast were frequently observed in all the samples. These fine particles are assumed to be oxides of the raw Ti powders. In fact, the oxide particles were preferentially formed in the raw Ti powders.

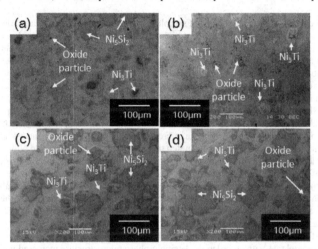

Fig.1 SEM backscattered electron (BE)-images of as-sintered samples; (a) #1, (b) #2, (c) #3 and (d) #4.

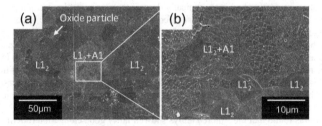

Fig.2 SEM secondary electron (SE)-images of sample#2 which was sintered and subsequently homogenized; (a) typical low magnification image and (b) high-magnification image in ($L1_2$+A1) two-phase region.

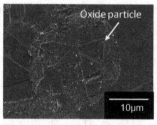

Fig.3 Oxide particles formed in raw Ti powder in sample #4.

Properties

Relative density of the homogenized samples is listed in Table 1. The relative density of the sample sintered at 1173 K (sample #2) was higher than that of the sample sintered at 1073 K (sample #1). High relative density more than 99% (almost 100%) was consequently attained in the samples sintered at 1173 K.

The room temperature (RT) Vickers hardness of the homogenized alloys is shown in Fig.4. Hardness datum of the cast material whose composition is the same as samples #1 and #2 is also included in this figure. The hardness of sample #2 was almost identical to that of the cast material, while the hardness of sample #1 was slightly lower than other two alloys (sample #2 and cast material). The lower hardness for sample #1 may be primarily due to inhomogeneous microstructure (i.e., ($L1_2$+A1 (fcc)) two-phase microstructure containing pores) caused by the low (insufficient) sintering temperature. It is noted that the hardness of the sintered alloy increased with decreasing Ni content. Possibly, this is due to the fact that the volume fraction of a softer phase of disordered fcc Ni solid solution than ordered $L1_2$ phase decreased with decreasing Ni content. Furthermore, it has been reported that the yield and tensile strength properties of the $Ni_3(Si,Ti)$ alloys increased with decreasing Ni content [2]. Therefore, the higher hardness observed in the sample with a lower Ni content may be attributed to the higher hardness of the $L1_2$ phase with lower Ni content in addition to disappearance of the soft phase of a disordered A1 (fcc) Ni solid solution. Fig.5 shows changes in hardness with test temperature. The hardness of sample #4 with an ordered $L1_2$ single-phase microstructure increased with increasing test temperature and showed a peak at around 573 K, then gradually decreased with further increasing test temperature. On the other hand, the hardness of the sample #2 with an ($L1_2$+A1) two-phase microstructure monotonically decreased with increasing test temperature. It is obvious from Fig.5 that the $L1_2$ single-phase microstructure is favorable in the wide range of temperature for enhancing the hardness or attaining high hardness.

Fig.6 shows the nominal stress versus nominal strain curves for the homogenized samples #2, #3 and #4. The datum for the cast-alloy with a base composition [4] is also included in this figure as reference. The yield stress of the sintered alloys was higher than that of the cast alloy because grain sizes of the sintered alloys were smaller (~12 μm) than that of the cast alloy (several mm). The yield stress of the sintered alloys increased with decreasing Ni content because the volume fraction of soft phase of A1 (fcc) phase decreased with decreasing Ni content. Also, it was found that the sintered alloys showed fairly good elongation (ductility) possibly due to high densities (almost pore-free) and fine grain sizes (homogenous deformation).

SEM fractographies of the sintering alloys after tensile test at RT are shown in Fig.7. A brittle intergranular fracture surface was partially observed for sample #4 while a ductile dimple pattern was dominated in sample #3. Therefore, the low tensile ductility of sample #4 is assumed to be caused by the intergranular fracture. On the other hand, the chemical composition analysis by EPMA revealed that the Ti content in the $L1_2$ phase for sample #2 was lower than that for sample #3. It has been reported that tensile ductility of the $Ni_3(Si,Ti)$ alloy was affected by the content ratio of Ti to Si, that is, a large content ratio of Ti to Si led to large tensile ductility [2]. Therefore, the low elongation value observed for sample #2 may be due to low Ti content in the $L1_2$ matrix.

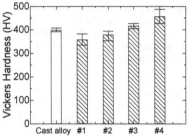

Fig.4 Vickers hardness of the homogenized alloys. Composition of the cast alloy is the same as that of samples #1 and #2.

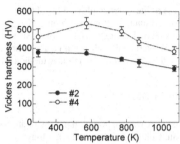

Fig.5 Changes in Vickers hardness with test temperature for the homogenized samples #2 and #4.

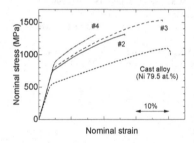

Fig.6 Nominal stress versus nominal strain curves for the homogenized alloys.

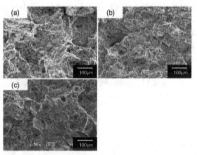

Fig.7 SEM fractography of samples (a) #2, (b) #3 and (c) #4 tensile-deformed at RT in a vacuum.

CONCLUSIONS

A $Ni_3(Si,Ti)$ intermetallic alloy was synthesized by the pulse current sintering from the elemental powder mixture and subsequent annealing (homogenization) process. The following results were obtained from the present study:

1. The $Ni_3(Si,Ti)$ alloy sintered at 1073 K showed lower density than that at 1173 K, indicating that the densification of the sintering body finished at 1173 K.
2. The as-sintered alloys showed a microstructure containing intermediate phases of Ni_3Ti and Ni_5Si_2 in the $L1_2$ matrix irrespective of sintering temperature and alloy composition. These intermediate phases completely disappeared by the homogenization at 1223 K for 48 h. After the homogenization, the alloys with Ni content of 79.5 and 79.0 at.% showed an ($L1_2$+A1) two-phase microstructure. The A1 (fcc) phase decreased with decreasing Ni content, resulting in an $L1_2$ single-phase microstructure for the alloy with Ni content of 78.0 at.%.
3. Hardness of the sintered alloys showed the same level as that of the cast alloy. Also, yield stress of the sintered alloys was higher than that of the cast alloy due to fine grained microstructure, keeping a relative high tensile elongation. It was concluded that a sintered $Ni_3(Si,Ti)$ intermetallic alloy with a good hardness and tensile property was successfully fabricated by the powder metallurgy method using elemental powders.

ACKNOWLEDGMENTS

This work was supported in part by Grant-in-aid for Scientific Research for the Ministry of Education, Culture, Sports and Technology.

REFERENCES

1. T. Takasugi, D. Shindo, O. Izumi and M. Hirabayashi: Acta metal. mater. **38**, 739 (1990).
2. T. Takasugi, M. Nagashima and O. Izumi: Acta metal. mater. **38**, 745 (1990).
3. S. Buchholz, Z.N. Farhat, G.J. Kipouros and K.P. Plucknett: Int. Journal of Refractory Metals and Hard Materials **33**, 44 (2012).
4. T. Nakamura, Y. Kaneno, H. Inoue and T. Takasugi: Mater. Sci. Eng. A **383**, 259 (2004).
5. D. Imajo, Y. Kaneno and T. Takasugi: MRS Proceedings Publication **1295**, 219 (2011).
6. T. Takasugi and M. Yoshida, J. Mater. Sci. **26**, 3032 (1991).

Mater. Res. Soc. Symp. Proc. Vol. 1516 © 2012 Materials Research Society
DOI: 10.1557/opl.2012.1727

Alloy design for reducing V content of dual two-phase Ni3Al-Ni3V intermetallic alloys

Takahiro Hashimoto [1], Yasuyuki Kaneno [1] and Takayuki Takasugi [1,2]
[1]Department of Materials Science, Graduate School of Engineering, Osaka Prefecture
University, 1-1 Gakuen-cho Naka-ku, Sakai, Osaka 599-8531, JAPAN
[2]Kansai Center for Industrial Materials Research, Institute for Materials Research,
Tohoku University, 1-1 Gakuen-cho Naka-ku, Sakai, Osaka 599-8531, JAPAN

ABSTRACT

The objective of this study is to establish alloy designing which can reduce the amount of
V for a Ni-base dual two-phase intermetallic alloy, without degenerating the dual two-phase
microstructure. It was demonstrated that the favorable dual two-phase microstructure will be
maintained as far as the valence electron concentration (e/a) of the alloys added by Cr is not so
much different from that of the base alloy (i.e. the alloy without additive elements).
Consequently, it was found that the dual two-phase microstructure was maintained even though
the amounts of V were reduced by 7 at.%, 7 at.%, and 10at.% by substituting of Cr for V, Cr for
both of Ni and V, and Cr for Ni, respectively. The hardness of the alloys with reduced V content
was higher than that of the base alloy.

INTRODUCTION

A dual two-phase Ni_3Al-Ni_3V intermetallic alloy is a new type of high temperature
structural materials which have been developed in our laboratory. A feature of this alloy is its
unique microstructure: the dual two-phase microstructure is comprised of upper two-phase
microstructure with a micron scale and lower two-phase microstructure with a sub-micron scale.
The upper two-phase microstructure is composed of primary cuboidal precipitates of Ni_3Al (L1$_2$)
and Ni solid solution (A1) at high temperature. Then, the prior A1 phase is decomposed into
Ni_3Al (L1$_2$) and Ni_3V (D0$_{22}$) by a eutectoid reaction at low temperature, resulting in the lower
two-phase microstructure in the channel region. It is found that the dual two-phase
microstructures are highly coherent among primary cuboidal precipitates of Ni_3Al (L1$_2$) phase
and channel regions of Ni_3V (D0$_{22}$) and Ni_3Al (L1$_2$) phases, and display high microstructural
stability at high temperature [1-4]. In addition, the dual two-phase intermetallic alloys are found
to show excellent mechanical properties, e.g., higher high-temperature tensile and creep strength
properties than many conventional superalloys, therefore promising for the development of a
next generation-type high temperature structural material [4-6].
The dual two-phase intermetallic alloys however contain relatively high amount of
vanadium which are inferior in the oxidization characteristic, density and cost. Therefore, it is a
critical issue to reduce the content of constituent element V. The objective of this study is to
establish alloy designing which can reduce the amount of V, without degenerating the dual
two-phase microstructure, i.e., maintaining their superior mechanical and chemical properties.
From the thermodynamic point of view to achieve this purpose, additive elements are
needed to have a large partition coefficient to the channel region containing high amount of V.
Figure 1 [7] shows partition coefficients of alloying elements between L1$_2$ and A1 phases, and
between D0$_{22}$ and A1 phases. The elements that the partition coefficient of L1$_2$/A1 is small can
become candidates as additive because those elements preferentially enter into the channel

region (A1) rather than Ni_3Al (L1$_2$) phase above eutectoid temperature. At the same time, those elements should have large partition coefficient of $D0_{22}$/A1 because those elements prefer to enter into Ni_3V ($D0_{22}$) phase containing high amount of V during eutectoid reaction at low temperature. Therefore, Nb and Cr that have the same group number as, and the adjoining group number to V, respectively in a periodic table fit to this purpose.

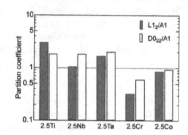

Figure 1   Partition coefficient of alloying elements between L1$_2$ and A1 phases and between $D0_{22}$ and A1 phases [7].

On the other hand, it has been reported that the phase stability of the constituent phases in the pseudo-ternary Ni_3Al-Ni_3V-Ni_3X alloy systems is dependent on valence electron concentration (e/a). The existing phase field of each GCP Ni_3X intermetallic phase observed at 1273 K more likely fit to the equi-contour lines for the valence electron concentration (e/a) than those for the atomic radius ratio (R_X/R_{Ni}) as reported in Ref. [8]. In other words, the crystal structure and phase field of each GCP Ni_3X are maintained if the e/a is kept the same value even though additive elements are added. Thus, Nb (and Cr) satisfy the condition as additive elements because their additions have the same e/a as that of V or the similar e/a to that of V, thereby maintaining the dual two-phase microstructure comprised of Ni_3Al (L1$_2$) and Ni_3V ($D0_{22}$), via not reducing the phase stability of $D0_{22}$.

EXPERIMENT

Alloy compositions used in this study are shown in Table 1. A nominal composition of $Ni_{75}Al_{10}V_{15}$ doped with 50 wt.ppm boron (expressed by at.% except for boron) was used as a base alloy composition. Nb was substituted for V while Cr was substituted for V, Ni, and both of Ni and V. Alloy buttons with a diameter of 50 mm were prepared by arc melting in argon gas atmosphere using a non-consumable tungsten electrode and copper hearth. The alloy buttons were homogenized at 1553 K for 3 h in a dynamic vacuum condition and then simply cooled in furnace to room temperature.

Microstructural observations were carried out by field emission scanning electron microscopy (FE-SEM) and transmission electron microscopy (TEM). Vickers hardness test was conducted at room temperature in conditions of a holding time of 10 s and a load of 1 kg, using the specimens with a thickness of about 1 mm that were sliced from the homogenized arc buttons. Hardness data of more than ten points were collected and averaged in each experimental condition.

Table 1. Alloy compositions used in this study.

	alloy No.	Ni (at%)	Al (at%)	Nb (at%)	Cr (at%)	V (at%)	B (wt.%)
	1	75	10	?	?	15	0.005
	2	75	10	2	2	11	0.005
V-site	3	75	10	3	3	9	0.005
	4	75	10	4	4	7	0.005
	5	74	10	1	2	13	0.005
	6	73	10	2	4	11	0.005
Ni,V-site	7	72	10	3	6	9	0.005
	8	71	10	4	8	7	0.005
	9	72	12	3	6	7	0.005
	10	71	12	4	8	5	0.005
	11	73	10	2	2	13	0.005
Ni-site	12	72	10	3	3	12	0.005
	13	71	10	4	4	11	0.005
	14	70	10	5	5	10	0.005

RESULTS AND DISCUSSION

Microstructure and phase identification

Figure 2 shows typical FE-SEM images of the homogenized alloys. The microstructure of the alloys with reduced V content by substitution of Cr for V (e.g., Fig.2b) was not so much changed compared with that of the base alloy (Fig.2a). On the other hand, for the alloys with reduced V content by substitution of Cr for Ni/V, the morphology of the cuboidal $L1_2$ precipitates is somehow degenerated and the volume fraction of the channel region was reduced when the Al content was increased (Fig.2e). For the alloys with reduced V content by substitution of Cr for Ni, the size of the $L1_2$ precipitates was almost the same as that for the base alloy (Fig.2a) though the shape of those slightly changed (Fig.2f).

Figure 3 shows FE-SEM images at low magnification of microstructures of alloys Nos.13 and 14. For the alloy with reduced V content by substitution of Cr for Ni, the GCP phases other than $L1_2$ and $D0_{22}$ appeared along grain boundaries when more than 4 at.% Nb was added to the base alloy. The X-ray diffraction analysis showed that $D0_a$ (Ni_3Nb) phase in addition to $L1_2$ (Ni_3Al) and $D0_{22}$ (Ni_3V) phases was identified for alloy No.14.

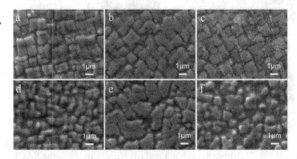

Figure 2 FE-SEM images at high magnification of microstructures of alloys (a) No.1, (b) No. 4, (c) No.7, (d) No.8, (e) No.9 and (f) No.14.

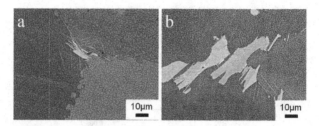

Figure 3 FE-SEM images at low magnification of microstructures of alloys (a) No.13 and (b) No.14.

Figure 4 shows TEM bright field images and selected area diffraction patterns (SADP) for the alloys with reduced V content. As understood from the existence of the extra spots from $D0_{22}$ structure in the SADPs, it was clearly shown that the dual two-phase microstructure was maintained even though reducing the amount of V by 7 at.%, 7 at.%, and 10 at.% by substituting of Cr for V, Cr for Ni/V, and Cr for Ni, respectively. For instance, alloy No.7 with 9 at.% of V exhibited the dual two-phase microstructure comprised of $L1_2$ and $D0_{22}$ phases (Fig.4a). However, for alloy No. 8 with a higher amount of Nb and Cr, the channel regions were mixed with Ni solid solution (fcc) phase region (Fig.4c) and $D0_{22}/L1_2$ phases region (Fig.4b). For alloy No. 9 which has a higher amount of Al, the whole channel regions were consisted of only Ni solid solution (fcc) phase, resulting in degeneration of the dual two-phase microstructure (Fig.4d).

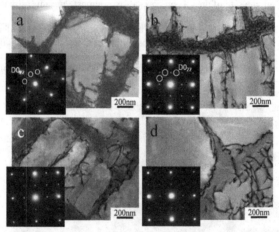

Figure 4 TEM microstructures and selected area diffraction patterns (SADP) of alloys (a) No. 7, (b) No.8, (c) No.8 and (d) No.9. Circles in the SADPs show extra spots from $D0_{22}$ structure.

Table 2 summarizes the microstructures and valence electron concentrations (e/a) of the alloys. Here, the e/a of the alloys was defined by the following relationship:

$$e/a = \Sigma e_i c_i \qquad (1)$$

where e_i is the number of the valence electron in the element i, and c_i is the atomic concentration of the element i in the alloy. The e/a was defined as the number of the electron added from the rare gas configuration. It is evident from Table 2 that the dual two-phase microstructure without containing the GCP phases other than $L1_2$ and $D0_{22}$ was maintained when the e/a values of the alloys range between 8.59 and 8.43. The latter extreme value (i.e., 8.43) for the e/a is suggested to correspond to the lowest e/a value by which the alloys added by Nb and/or Cr are permitted to exhibit the dual two-phase microstructure without containing the GCP phases other than $L1_2$ and $D0_{22}$. Thus, the alloy designing based on the valence electron concentration (e/a) may be helpful to predict whether the alloys exhibit the dual two-phase microstructure or not. This prediction is suggested to be applicable to Nb/Cr substitution, or other additive elements and their combinations.

Table 2. Microstructure and valence electron concentrations of alloys investigated.

	alloy No.	V (at%)	Microstructure of the channel region	The GCP phase other than $L1_2$ and $D0_{22}$	Valence electron concentration
V-site	1	15			8.55
	2	11			8.57
	3	9			8.58
	4	7	$D0_{22} + L1_2$	None	8.59
Ni/V-site	5	13			8.52
	6	11			8.49
	7	9			8.46
	8	7	$D0_{22} + L1_2$ fcc (A1)	None	8.43
	9	7	fcc (A1)		8.42
	10	5			8.39
Ni-site	11	13	$D0_{22} + L1_2$	None	8.47
	12	12			8.43
	13	11	$D0_{22} + L1_2$	Exist	8.39
	14	10			8.35

Mechanical properties

Figure 5 shows the hardness of all the alloys investigated in this study. The hardness of the alloys that contain less amount of V and showed the dual two-phase microstructure was higher than that of the base alloy. On the other hand, the alloys whose channel region was comprised of Ni solid solution phase showed lower hardness than the base alloy because the hardness of the ordered $D0_{22}$ phase was higher than that of the disordered Ni solid solution phase. Also, it is interesting to note that the alloys with substitution of Cr for Ni showed higher hardness than those with substitution Cr for V and Ni/V. The hardness increase of the alloys with substitution of Cr for V is supposed to be due to solid solution hardening by increasing Nb and Cr contents in the constituent phases, especially in $D0_{22}$ phase.

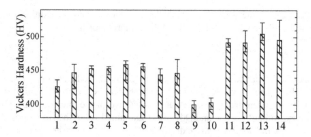

Figure 5 Hardness of the homogenized alloys.

CONCLUSIONS

To establish the designing of the dual two-phase Ni_3Al-Ni_3V intermetallic alloy which can reduce the amount of V, without degenerating the dual two-phase microstructure, Nb and Cr were added in different manners. These two elements have a large partition coefficient to the channel region and keep the valence electron concentration almost the same value as V, thereby maintaining the dual two-phase microstructure, via keeping the phase stability of especially DO_{22}. It was found from the present study that the dual two-phase microstructure was maintained even though the amounts of V were reduced by 7 at.%, 7 at.%, and 10at.% by substituting of Cr for V, Cr for Ni/V, and Cr for Ni, respectively. The alloys that contain less amount of V and have the dual two-phase microstructure showed higher levels of hardness than the base alloy.

ACKNOWLEDGMENTS

This work was supported in part by Grant-in-aid for Scientific Research for the Ministry of Education, Culture, Sports and Technology.

REFERENCES

1. Y. Nunomura, Y. Kaneno, T. Takasugi, Intermetallics 12 (2004) 389.
2. Y. Nunomura, Y. Kaneno, H. Tsuda, T. Takasugi, Acta Mater. 54 (2006) 851.
3. S. Shibuya, Y. Kaneno, H. Tsuda, T. Takasugi, Intermetallics 15 (2007) 338.
4. S. Shibuya, Y. Kaneno, M. Yoshida, T. Shishido, T. Takasugi, Intermetallics 15 (2007) 119.
5. S. Shibuya, Y. Kaneno, M. Yoshida, T. Takasugi, Acta Mater. 54 (2006) 861.
6. K. Kawahara, Y. Kaneno, T. Takasugi, Intermetallics 17 (2009) 938.
7. S. Kobayashi, K. Sato, E. Hayashi, T. Osaka, T. J. Konnno, Y. Kaneno, T. Takasugi, Intermetallics, 23 (2012) 68-75.
8. W. Soga, Y. Kaneno, T. Takasugi: Intermetallics, 14 (2006) 170-179.

Mater. Res. Soc. Symp. Proc. Vol. 1516 © 2013 Materials Research Society
DOI: 10.1557/opl.2012.1748

Effect of Re addition on microstructure and mechanical properties of Ni base dual-two phase intermetallic alloys

S. Ishii[1], T. Moronaga[2], Y. Kaneno[1], H. Tsuda[1] and T. Takasugi[1]
[1] Department of Materials Science, Graduate School of Engineering, Osaka Prefecture University, Sakai 599-8531, Japan
[2] Kobelco Research Institute. Inc, Kobe 651-0073, Japan

ABSTRACT

The effect of Re addition on microstructure and hardness of the Ni_3Al ($L1_2$) and Ni_3V ($D0_{22}$) dual two-phase intermetallic alloys was investigated as functions of alloying (substituting) method of Re and aging condition (temperature and time). Re was added to the base alloy composition by three methods: Re was substituted for Ni, Al and V, respectively. The Re-added alloys were solution-treated at 1553 K and then aged at lower temperatures of 1123 K-1248 K. Apparent age hardening occurred in the alloy where Re was substituted for Ni while no age hardening was observed in the alloys where Re was substituted for Al or V. In the case of the latter two alloys, the hardness was unchanged or reduced with a progression of aging time. These results were discussed in terms of phase separation and ordering in the channel region, and hardening due to Re-rich phase precipitation.

INTRODUCTION

Ni-based superalloys are composed of Ni solid solution (A1) and Ni_3Al ($L1_2$) whose crystal structure is geometrically close packed (GCP) structure. They are widely used as high temperature structural materials. The mechanical properties of the Ni-based superalloys are known to be improved by addition of refractory elements, such as W, Re, Ta, Mo, Nb, and Ru [1-6]. However, the strength of the Ni-base superalloys decreases significantly at high temperature due to softening of the Ni solid solution phase. Ni_3X-type intermetallic compounds categorized as the geometrically close packed (GCP) structure exhibit high phase and microstructural stability up to their melting point due to low atomic diffusivity [7]. These intermetallic compounds generally have good high-temperature properties such as high strength and good oxidization resistance. However, monolithic intermetallic polycrystals have a serious drawback, i.e. poor tensile ductility at room temperature. To overcome this drawback, the present authors have developed an Ni-based two-phase intermetallic alloys [8-10]. The Ni-based dual two-phase intermetallic alloys are composed of two intermetallic compounds of Ni_3Al ($L1_2$) and Ni_3V ($D0_{22}$). The microstructures of these alloys consist of primary Ni_3Al precipitates and Ni solid solution at high temperature (above about 1273 K) and the Ni solid solution is transformed to Ni_3Al and Ni_3V by eutectoid reaction at low temperature (below about 1273 K). These alloys display a high coherence among the constituent phases, and a high microstructural stability [11,12,14,15]. Hence, the dual two-phase intermetallic alloys have better mechanical properties (for example, higher high-temperature tensile and creep strength) than many conventional superalloys [13,15,16].

It was reported by present authors' group that the hardness of the dual-two phase intermetallic alloys where Re was added by substituting for Ni was dramatically improved by the aging heat treatment [17]. However, the effect of alloying (substituting) method for Re addition and its optimum content has not been investigated so far. In the present study, Re was added to the base alloy composition by three methods: Re was substituted for Ni, Al and V, respectively. Aging behavior and microstructure of the Re-added two-phase intermetallic alloys were investigated with Vickers hardness test, scanning electron microscopy (SEM) and transmission electron microscopy (TEM).

EXPERIMENT

Alloy compositions used in this study are shown in Table 1. An alloy composition of $Ni_{75}Al_8V_{17}$ (at.%) is used as a base alloy composition. 2 at.% Re was added to the base alloy composition by substituting for Ni, Al and V, respectively. These alloys are simply denoted as base, 2Re(Ni), 2Re(Al) and 2Re(V) in this paper. They were prepared from starting raw materials of 99.99 wt.% Ni, 99.99 wt.% Al, 99.9 wt.% V and 99.9 wt.% Re. Button ingots with a diameter of 30 mm were prepared by arc melting in argon gas atmosphere using a non-consumable tungsten electrode on a copper hearth. The button ingots were remelted more than three times to ensure chemical homogeneity. They were homogenized (solution treated) at 1553 K for 5 h in a vacuum and then furnace cooled to room temperature. The button ingots were sliced to proper sizes using an electron discharge machine (EDM). The homogenized alloys were aged in an evacuated silica tube at 1123 K and 1248 K for 0.5, 1, 10, 24 and 48 h, respectively, followed by water quenching.

Vickers hardness test was carried out with a holding time of 10 s and a load of 1 kg at room temperature. More than ten points of hardness data were collected and averaged in each experimental condition. Microstructural observations and phase identification were carried out by a scanning electron microscope (SEM) and a transmission electron microscope (TEM). Specimens used for SEM observations were abraded on SiC paper and electronically polished in a mixed solution of 15 ml H_2SO_4 + 85 ml CH_3OH at 243K. Specimens used for TEM observations with a diameter of 3 mm were mechanically thinned to approximately 0.1 mm. After that, they were jet polished in a mixed solution of 15 ml H_2SO_4 + 85 ml CH_3OH at 253 K. TEM observation was carried out using a JEM-2000FX operating at 200 kV, with the beam incident from the [100] direction of the primary Ni_3Al precipitates.

Table 1. Chemical compositions of alloys used in this study.

alloys	Ni(at.%)	Al(at.%)	V(at.%)	Re(at.%)	B(wt.ppm)
Base	75	8	17	–	50
2Re(Ni)	73	8	17	2	50
2Re(Al)	75	6	17	2	50
2Re(V)	75	8	15	2	50

RESULTS and DISCUSSION

Hardness properties

Figure 1 shows Vickers hardness of the base and Re-added alloys as a function of aging time. The base alloy showed neither age hardening nor softening by aging both at 1123 K and 1248 K. Alloy 2Re(Ni) became hardened in a short time of aging at both ageing temperatures. As well as the base alloy, alloys 2Re(Al) and 2Re(V) did not show either age hardening or softening by aging at 1123 K. However, these two alloys became softened when aged at 1248 K. The age hardening behavior observed for alloy 2Re(Ni) is consistent with our previous study [17].

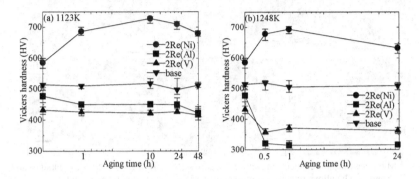

Fig.1 Changes in Vickers hardness with aging time at (a) 1123 K and (b) 1248 K.

Microstructural observation by SEM

Fig.2 shows SEM images of the alloys which were solution treated (homogenized) at 1553 K and aged 1248 K for 1 h. The solution treated (i.e., unaged) alloys showed a microstructure composed of cuboidal $L1_2$ (Ni_3Al) particles and the channel although the size of the $L1_2$ particle in alloy 2Re(Al) is somewhat smaller than that in other two alloys, i.e. alloys 2Re(Ni) and 2Re(V). After aging at 1248 K for 1 h, alloy 2Re(Ni) showed a peak hardness while alloys 2Re(Al) and 2Re(V) were softened, as shown in Fig.2b. For alloy 2Re(Ni), needle-shaped precipitates were observed in the channel regions. Therefore, the observed age hardening is due to the needle-shaped precipitates. It has been reported that age hardening for Re and Ta-added Ni_3Al/Ni_3V dual two-phase intermetallic alloy was attributed to precipitation of Re-rich particles with bcc structure [17]. For alloys 2Re(Al) and 2Re(V), no significant microstructural change was detected in the aged samples.

Volume fraction of the channel regions is also shown in Fig.2. Volume fraction of the channel regions were almost the same between the unaged and aged samples, irrespective of the substituting method. On the other hand, volume fraction of the channel regions is higher in alloy 2Re(Al) than in other two alloys. In the Ni base dual-two phase intermetallic alloys, the main constituent phase of the channel regions is Ni_3V which is harder than Ni_3Al [18,19]. If the main constituent phases of the channel regions of alloy 2Re(Al) were Ni_3V, the hardness of 2Re(Al) would be the highest in all the alloys. However, the hardness of alloy 2Re(Al) was lower than alloy 2Re(Ni) as well as the base alloy. Therefore, there is a possibility that the main constituent phase of the channel regions of alloy 2Re(Al) as well as alloy 2Re (V) is not Ni_3V. To verify it, phase identification was carried out by TEM.

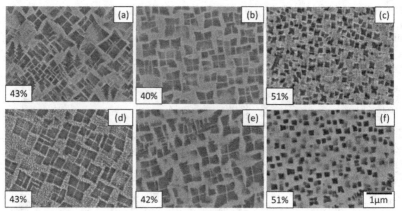

Fig.2 SEM BE (backscattered electron) images of alloys 2Re(Ni) (a and d), 2Re(V) (b and e) and 2Re(Al) (c and f). The alloys were solution treated at 1553 K for 5 h (a, b and c) and subsequently aged for 1 h at 1248 K (d, e and f). Volume fractions of channel regions are indicated in each image.

Phase identification by TEM

Fig.3 shows the simulated patterns of the superlattice reflections corresponding to $L1_2$ (Ni$_3$Al) structure and to the three variants of the $D0_{22}$ (Ni$_3$V) structure. From these simulated patterns, it is possible to identify whether Ni$_3$V exists in the channel regions or not.

Fig.4 shows TEM-bright images and selected area diffraction (SAD) patterns of 2Re(Ni) (a and d), 2Re(V) (b and e) and 2Re(Al) (c and f) that were solution treated (a, b and c) and subsequently aged for 1 h at1248 K (d, e and f), respectively. Observing selected area diffraction patterns, it is evident that the Ni$_3$Vphase is formed in as-homogenized 2Re(V) and 2Re(Al) (b and c) while superlattice reflections of Ni$_3$V are not present in as-aged 2Re(V) and 2Re(Al) (e and f). This result shows that by aging heat treat, phases of channel regions transform from Ni$_3$V to Ni solid solution (A1). This transformation is associated with the softening. In previous study [20], it was found that the eutectoid temperature of Ni base dual-two phase intermetallic alloys was decreased by Re addition. It is deduced when Re was added to base alloy composition substituted for Al and V, the eutectoid temperature would be decreased more than 150 K. In the future study, the verification about this expectation will be needed by differential scanning calorimeter (DSC) or by thermodynamic consideration.

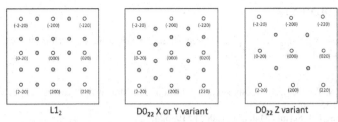

| L1$_2$ | DO$_{22}$ X or Y variant | DO$_{22}$ Z variant |

Fig.3 Simulated [0 0 1] zone axis diffraction patterns, showing positions of superlattice reflections corresponding to the Ni$_3$Al (L1$_2$) phase and to three variants of the Ni$_3$V (D0$_{22}$) phase.

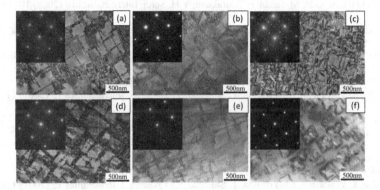

Fig.4 TEM-bright images and selected area diffraction (SAD) patterns of 2Re(Ni) (a and d), 2Re(V) (b and e) and 2Re(Al) (c and f). Specimens of images in an upper row (a, b and c) were solution treated and specimens of images in a lower row (d, e and f) were subsequently aged for 1h at1248K.

CONCLUSIONS

The effect of Re addition on microstructure and hardness of a Ni$_3$Al/Ni$_3$V dual-two phase intermetallic alloy was investigated as functions of alloying (substituting) method and aging conditions (temperature and time). The following results were obtained from the present study.
(1) Apparent age hardening was observed in the alloy where Re was substituted for Ni, and attributed to the Re-rich precipitation with bcc structure and ordering, i.e., the phase separation from disordered A1 (Ni solid solution) phase to ordered D0$_{22}$ (Ni$_3$V) and L1$_2$ (Ni$_3$Al) phases.
(2) No age hardening was observed in the-alloys where Re was substituted for Al or V. In this case, the hardness decreased with a progression of aging time. The observed softening was due to the disordering, i.e., the phase transformation from ordered D0$_{22}$ (Ni$_3$V) and L1$_2$ (Ni$_3$Al) phases to disordered A1 (Ni solid solution) phase.

ACKNOWLEDGMENTS

This work was supported in part by Grant-in-aid for Scientific Research for the Ministry of Education, Culture, Sports and Technology.

REFERENCES
1. J. Rusing, N. Wanderka, U. Czubayoko, V. Naundorf, D. Mukherji, J. Rosler, Scripta Mater. 46 (2002) 235-240
2. A.C. Yeh, S. Tin, Scripta Mater. 52 (2005) 519–524.
3. W.Z. Wang, T. Jin, J.L. Liu, X.F. Sun, H.R. Guan, Z.Q. Hu, Mater. Sci. Eng. A 479 (2008) 148–156.
4. N. Garimella, M. Ode, M. Ikeda, H. Murakami, Y.H. Sohn, Intermetallics 16 (2008) 1095–1103.
5. V. Kindrachuk, N. Wanderka, J. Banhart, D. Mukherji, D. Dwl Grnobese, J. Rosler, Acta Mater. 56 (2008) 1609–1618.
6. S. Tian, M. Wang, X. Yu, X. Yu, T. Li, B. Qian, Mater. Sci. Eng. A 527 (2010) 4458–4465.
7. D.P. Pope, S.S. Ezz, Int. Mater. Rev. 29 (1984) 136–167.
8. S. Shibuya, Y. Kaneno, M. Yoshida, T. Takasugi, Acta Mater. 54 (2006) 861–870.
9. S. Shibuya, Y. Kaneno, M. Yoshida, T. Shishido, T. Takasugi, Intermetallics 15 (2007) 119–127.
10. S. Shibuya, Y. Kaneno, H. Tsuda, T. Takasugi, Intermetallics 15 (2007) 338–348.
11. Y. Nunomura, Y. Kaneno, T. Takasugi, Intermetallics 12 (2004) 389-399.
12. Y. Nunomura, Y. Kaneno, H. Tsuda, T. Takasugi, Acta Mater. 54 (2006) 851-860.
13. S. Shibuya, Y. Kaneno, M. Yoshida, T. Takasugi, Acta Mater. 54 (2006) 861-870.
14. S. Shibuya, Y. Kaneno, H. Tsuda, T. Takasugi, Intermetallics 15 (2007) 338-348.
15. S. Shibuya, Y. Kaneno, M. Yoshida, T. Shishido, T. Takasugi, Intermetallics 15 (2007) 119-127.
16. Y. Kaneno, W. Soga, H. Tsuda, T. Takasugi, J. Mater. Sci. 43 (2008) 748-758.
17. T. Moronaga, S. Ishii, Y. Kaneno, H. Tsuda, T. Takasugi, Materials Science and Engineering A539 (2012), 30-37
18. K. Kawahara, Y. Kaneno, A. Kakitsuji, T. Takasugi, Intermetallics, 17 (2009) 938-944.
19. K. Kawahara, T. Moronaga, Y. Kaneno, A. Kakitsuji, T. Takasugi, Mater. Trans., 51 (2010) 1395-1403.
20. S. Kobayashi, K. Sato, E. Hayashi, T. Osaka, T. J. Konno, Y. Kaneno, T. Takasugi, Intermetallics23 (2012) , 68-75

Mater. Res. Soc. Symp. Proc. Vol. 1516 © 2013 Materials Research Society
DOI: 10.1557/opl.2013.141

Compressive Fracture Behavior of Bi-added $Ni_{50}Mn_{28}Ga_{22}$ Ferromagnetic Shape Memory Alloys

Hirotaka Tanimura[1*], Masaki Tahara[1], Tomonari Inamura[1] and Hideki Hosoda[1]
[1]Precision and Intelligence Laboratory, Tokyo Institute of Technology, Yokohama, Japan
*Graduate Student, Tokyo Institute of Technology

ABSTRACT

In order to develop NiMnGa/polymer composite materials, a production of single-crystal-like NiMnGa particles is important and should be developed for better quality. Although mechanical pulverization is a promising method by utilizing intrinsic intergranular brittleness of NiMnGa polycrystalline ingots, the amount of lattice defects introduced during mechanical crushing needs to be minimized. This must be achieved by enhancement of intergranular brittleness of NiMnGa particles. In this study, the effect of Bi addition on the compressive fracture behavior of polycrystalline $Ni_{50}Mn_{28}Ga_{22}$ was investigated where Bi was expected to be segregated to the grain boundaries in NiMnGa, similar to Bi segregation to the grain boundaries in Ni. It was found that only intergranular fracture was observed in $Ni_{50}Mn_{28}Ga_{22}$ polycrystals with 0.3 at.% Bi addition, although a mixture of intergranular and transgranular fracture was observed in Bi-free $Ni_{50}Mn_{28}Ga_{22}$ polycrystal. Microalloying of Bi into NiMnGa enhances intergranular embrittlement. A number of spherical particles of Bi were confirmed on the fractured surface of Bi-doped NiMnGa polycrystals. The formation of Bi particles is a proof of the grain boundary segregation of Bi in NiMnGa.

INTRODUCTION

The ferromagnetic shape memory alloy NiMnGa is a promising candidate for a new actuator material exhibiting a large magnetic field induced strain (MFIS) due to the reorientation of martensite variants. MFIS is induced not by heating/cooling but by magnetic field that enables one to control the field at a higher frequency than usual for shape memory alloys such as TiNi. That is why NiMnGa has attracted a great deal of interest during past decades. Ullako and co-workers reported 0.2% MFIS in a single crystal of Ni_2MnGa [1]. Recently, up to 6% MFIS has been reported in single crystal NiMnGa with 10M martensite [2]. The material does not, however, recover its original shape spontaneously upon removing the magnetic field. On the other hand, polycrystalline NiMnGa cannot exhibit such large MFIS. This is explained by the suppression of the twin boundary motion stemming from the constraints among grains of the martensite phase. In addition, the intrinsic intergranular brittleness of polycrystalline NiMnGa is also a drawback for practical applications. We have, therefore, proposed and developed NiMnGa/polymer composite materials which are composed of single-crystal-like NiMnGa particles and polymer matrix [3]. This composite is expected to exhibit large MFIS due to the motion of single-crystal-like NiMnGa and spontaneous shape recovery upon removing the magnetic field by elastic back stress from the polymer matrix. Based on the theory of elasticity, a calculated MFIS of around 1% in a 50vol%NiMnGa/silicone composite was expected [4]. In our previous work, single-crystal-like NiMnGa particles were fabricated by mechanical crash; utilizing the intrinsic intergranular brittleness of NiMnGa polycrystal. The actual MFIS of the composites containing mechanically-crushed NiMnGa particles was, however, around on 10^{-3}%, which was much smaller than the theoretically calculated value [4]. One of the possible reasons why the experimental MFIS was smaller than the expected value is likely that the reorientation of martensite variants is prevented by lattice defects introduced by transgranular deformation during the mechanical crushing.

Concerning the compressive fracture behavior of NiMnGa polycrystal, various studies have been made to improve the brittleness [5-7]. According to these reports, in spite of the intrinsic intergranular brittleness, the fracture of polycrystalline NiMnGa at room temperature (RT) occurs in grains as well as at grain boundaries. It is, therefore, deduced that lattice defects were introduced into NiMnGa particles during the mechanical crushing. Promoting intergranular embrittlement of NiMnGa is expected to be effective in preventing transgranular deformation during the mechanical crash and leads to minimization of introduced lattice defects.

Therefore, the microalloying of Bi into NiMnGa polycrystal is proposed to promote the intergranular brittleness of NiMnGa polycrystal. It is well known that Bi addition causes the intergranular embrittlement of polycrystalline Ni [8-9]. According to the Ni-Bi phase diagram in Figure 1(a) [10], Bi atoms in Ni segregate at grain boundaries as seen in Figure 1(b). Then, due to the weak bonding between Bi and Bi to the grain boundaries, and the segregation of Bi induces strong intergranular brittleness of Ni polycrystal [11]. Similarly, Bi additions are also expected to enhance the intergranular brittleness of NiMnGa polycrystal through the grain boundary segregation. If Bi atoms are segregated to the grain boundaries and intergranular brittleness is promoted as expected, the Bi addition will be a useful way to fabricate good quality NiMnGa particles by the mechanical pulverization. The objective of this study is to clarify the effect of Bi addition on compressive fracture behavior and Bi segregation in polycrystalline NiMnGa.

(a)

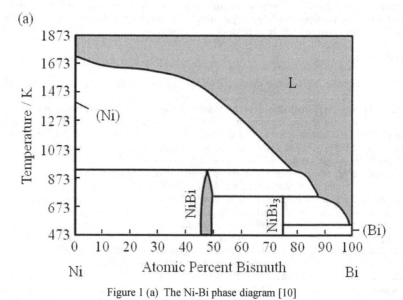

Figure 1 (a) The Ni-Bi phase diagram [10]

EXPERIMENTAL PROCEDURE

Ni-28Mn-22Ga (at.%) alloys with 0 (B-free) and 0.3 at.% Bi were fabricated by arc-melting method in Ar-1%H$_2$ from high purities of Ni (99.99%), Mn (99.9%), Ga (99.9999%) and Bi powder (<180μm, 99.999%)). The crystal structure of the martensite in Ni$_{50}$Mn$_{28}$Ga$_{22}$ is 10M at room temperature [12]. The ingots were homogenized at 1273 K for 86.4 ks in an Ar atmosphere, followed by water-quenching. Rectangular specimens with the dimension of 1.5×1.5×3mm^3 were made by electro-discharge machining from both ingots. After precise surface polishing, the specimens were compressed to fracture with a strain rate of 2×10^{-4}/s at room temperature using a Shimadzu RH-50 universal testing machine. Microstructural observations and chemical analysis of fractured surface were carried out using a scanning electron microscope (FE-SEM, Hitachi S-4500) equipped with an X-ray energy dispersive spectroscopy (EDS, Horiba EMAX ENERGY 6853H).

RESULTS AND DISCUSSIONS

Figure 2 shows the results of microstructural observation of the fracture surface of Ni$_{50}$Mn$_{28}$Ga$_{22}$ with (a) 0 at.% Bi (Bi-free) and (b) 0.3 at.% Bi. It can be seen from Fig. 2 (a) that the fracture mechanism of the Bi-free Ni$_{50}$Mn$_{28}$Ga$_{22}$ polycrystal was a mixture of intergranular and transgranular fracture. In spite of the intrinsic intergranular brittleness, the cracks in Bi-free polycrystalline Ni$_{50}$Mn$_{28}$Ga$_{22}$ propagated inside grains as well as along grain boundaries. This indicates that, when Bi was not added, a high density of dislocations must be introduced into NiMnGa particles owing to the plastic deformation during the mechanical crushing. These dislocations introduced during the compressive deformation are supposed to increase the stress required for the reorientation of martensite variants. On the other hand, only intergranular fracture was observed on the fracture surface of the Bi-added Ni$_{50}$Mn$_{28}$Ga$_{22}$ polycrystal. This means that the intergranular brittleness of Ni$_{50}$Mn$_{28}$Ga$_{22}$ is promoted by the Bi addition.

Besides, a large number of spherical particles of 2-3 μm in diameter were confirmed on the fracture surface of the Bi-added Ni$_{50}$Mn$_{28}$Ga$_{22}$ polycrystal, whereas the flat and smooth intergranular fracture surface was observed in the Bi-free Ni$_{50}$Mn$_{28}$Ga$_{22}$ polycrystal. Figure 3 shows the results of SEM-EDS measurement of the area enclosed by the solid lines in Fig. 2 (b). These spherical particles were identified as Bi by the SEM-EDS analysis. Then, Bi was confirmed to be segregated to the grain boundaries in NiMnGa. The grain boundary segregation behavior of Bi in NiMnGa was similar to that of Bi-doped Ni [13]. The overwhelming tendency of intergranular fracture in the Bi-added NiMnGa alloy is, therefore, deduced to be due to the segregation of Bi at the grain boundaries of NiMnGa.

The above mentioned results indicate that the Bi addition to polycrystalline Ni$_{50}$Mn$_{28}$Ga$_{22}$ remarkably enhances intergranular embrittlement of NiMnGa, probably due to a similar mechanism of Bi-added Ni polycrystal. This must lead to a comparably small amount of plastic deformation during the mechanical crushing. It is, however, necessary to clarify the appropriate content of Bi to NiMnGa because the Bi particles on the surface of the fabricated particle could be obstacles for the reorientation of martensite variants. Recently, the effect of surface roughness on NiMnGa single crystal was investigated by Chmielus and co-workers [14]. According to their report, the removal of the defective surface layer of NiMnGa single crystal reduces twinning stress. Similar to the defective surface layer, the motion of twin boundary could be arrested by the Bi particles on the surface of the NiMnGa particle. Therefore, 0.3 at.%

Bi addition may be excessive for the fabrication of magnetostrictive single-crystal-like NiMnGa particles by the mechanical pulverization. Optimization of the amount of Bi addition is required as the second step.

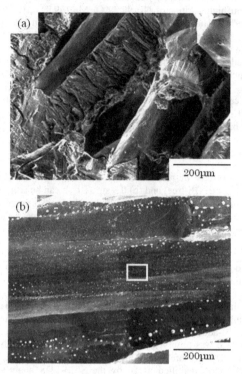

Figure 2 SEM fractographs of $Ni_{50}Mn_{28}Ga_{22}$ with (a) 0at.%Bi and (b) 0.3at.%Bi after compressive fracture at room temperature.

CONCLUSIONS

In this study, the effect of Bi addition on the compressive fracture behavior of polycrystalline $Ni_{50}Mn_{28}Ga_{22}$ was investigated. Only intergranular fracture was observed in 0.3 at.% Bi-doped $Ni_{50}Mn_{28}Ga_{22}$ polycrystals, although a mixture of intergranular and transgranular fracture was observed in the Bi-free $Ni_{50}Mn_{28}Ga_{22}$ polycrystal. These results indicate that intergranular fracture is enhanced by microalloying of Bi into NiMnGa. This must lead to a lower amount of plastic deformation during the mechanical crushing in comparison with Bi-free NiMnGa. A number of spherical particles of Bi were identified to exist on the fractured surface of the Bi-added NiMnGa polycrystal. They are the proof of Bi segregation at grain boundaries in NiMnGa. The formation of Bi spherical particles suggests that the overwhelming tendency of intergranular fracture in the Bi-added NiMnGa alloy is due to the segregation of Bi at grain boundaries in a similar manner to that of Bi-added Ni.

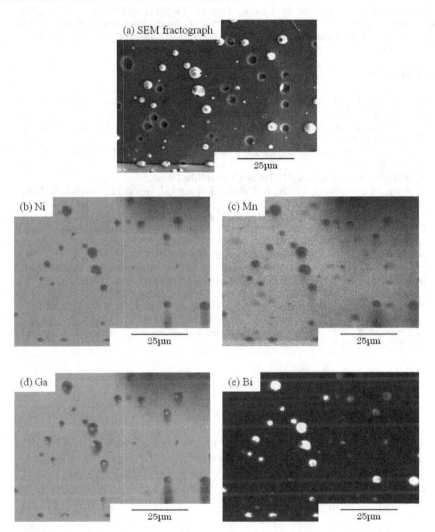

Figure 3 SEM fractograph and corresponding compositional mappings of 0.3at.%Bi-added $Ni_{50}Mn_{28}Ga_{22}$ by EDS for the area indicated by the white box in Figure 2(b): (a) SEM fractograph, (b) Ni mapping, (c) Mn mapping, (d) Ga mapping and (e) Bi mapping

ACKNOWLEDGMENTS

This work was supported by Funding Program for Next Generation World-Leading Researchers, No. LR015 (2011-2013) from Japan Society for the Promotion of Science (JSPS) and Grand-in-Aid for Young Scientists (A) (No. 24686077, 2012-2014) from the Ministry of Education, Culture, Sports, Science and Technology (MEXT), Japan.

REFERENCES

1. K. Ullakko, J. K. Huang, C. Kantner, R. C. O'Handley and V. V. Kokorin, Appl. Phys. Lett. **69**, 1966-1968 (1996).
2. S. J. Murray, M. Marioni, S. M. Allen, R. C. O'Handley and T. A. Lograsso, Appl. Phys. Lett. **77**, 886-888 (2000).
3. H. Hosoda, S. Takeuchi, T. Inamura and K. Wakashima, Sci. Technol. Adv. Mater. **5**, 503-509 (2004).
4. K. Wakashima, T. Inamura and H. Hosoda, Unpublished Work.
5. Z. Wang, M. Matsumoto, T. Abe, K. Oikawa, T. Takagi, J. Qiu and J. Tani, Mater. Trans. JIM **40**, 863-866 (1999).
6. J. Sui, X. Zhang, L. Gao and W. Cai, J. Alloy Compd. **509**, 8692-8699 (2011).
7. F. Chen, W. Cai, L. Zhao and Y. Zheng, Key Eng. Mater. **324-325**, 691-694 (2006).
8. G. H. Bishop, Trans. Metall. AIME **242**, 1343-1351 (1968).
9. N. Marie, K Wolski and M Biscondi, Scr. Mater. **43**, 943-949 (2000).
10. P. Nash, in *Binary Alloy Phase Diagrams*, edited by T. B. Massalski, J. L. Murray, L. H. Bennett, and H. Baker (American Society for Metals, Metals Park, Ohio, 1986), p.523.
11. J. Luo, H. Cheng, K. M. Asl, C. J. Kiely and M. P. Harmer, Science **333**, 1730-1733 (2011).
12. C. Jiang, Y. Muhammad, L. Deng, W. Wu and H. Xu, Acta. Mater. **52**, 2779-2785 (2004).
13. L. S. Chang and K. B. Huang, Scr. Mater. **51**, 551-555 (2004).
14. M. Chmielus, K. Rolfs, R. Wimpocy, W. Reimers, P. Müllner and R. Schneider, Acta Mater. **58**, 3952-3962 (2010).

Mater. Res. Soc. Symp. Proc. Vol. 1516 © 2013 Materials Research Society
DOI: 10.1557/opl.2013.391

Phase-Field Study on the Segregation Mechanism of Additive Elements in NbSi₂/MoSi₂ Duplex Silicide

Toshihiro Yamazaki[1,2], Yuichiro Koizumi[2], Akihiko Chiba[2], Koji Hagihara[3], Takayoshi Nakano[4], Koretaka Yuge[5], Kyosuke Kishida[5] and Haruyuki Inui[5]

[1]Department of Materials Processing, Tohoku University, 6-6-02 Aoba Aramaki, Aoba-ku, Sendai, 980-8579, Japan
[2]Institute for Materials Research, Tohoku University, 2-1-1 Katahira, Aoba-ku, Sendai, Miyagi 980-0011, Japan
[3]Department of Adaptive Machine Systems, Osaka University, 2-1 Yamada-oka, Suita, Osaka 565-0871, Japan
[4]Division of Materials and Manufacturing Science, Osaka University, 2-1 Yamada-oka, Suita, Osaka 565-0871, Japan
[5]Department of Materials Science and Engineering, Kyoto University, Sakyo-ku, Kyoto 606-8501, Japan

ABSTRACT

We have examined segregation behavior of various alloying elements at lamellar interfaces of C40-NbSi₂/C11ₐ-MoSi₂ duplex silicide by a phase-field simulation, which takes into account not only bulk chemical free energy but also segregation energy evaluated by the first principles calculation to reflect interaction between solutes and interface. The simulation suggests that segregation behaviors greatly depend on additive elements. In the case of Cr-addition, the C40-phase becomes enriched with Nb and Cr, while the C11ₐ-phase becomes enriched with Mo, which agrees with the equilibrium phase diagram. Slight segregation of Cr atoms is observed at the interface, whereas Nb and Mo concentrations monotonically change across the diffuse interface between C11ₐ and C40 phases. Significant segregations of Zr and Hf are formed at static interfaces, which are attributed to the chemical interaction between solute atoms and the static interface.

INTRODUCTION

Duplex silicide composed of NbSi₂ with C40-structure and MoSi₂ with C11ₐ-structure (Figure 1) is one of the candidates for ultrahigh-temperature structural material for future gas turbine power generation systems. In this material, an oriented lamellar structure composed of C11ₐ-type Mo-rich phase and C40-type Nb-rich phase is formed by unidirectional solidification that is followed by appropriate annealing. This lamellar structure improves properties of low-temperature toughness and high-temperature creep resistance [1, 2]. However, the lamellar structure collapses under prolonged annealing at high-temperature. Therefore, we need to improve the thermal stability of C40/C11ₐ lamellar structure for practical use. One possible approach to achieve a better thermal stability of the lamellar structure is adding another element. In a previous study [3], it was found that a small amount of Cr-addition can improve thermal stability of the lamellar structure, and Cr atoms are found to segregate at the lamellar interface. Cr-segregation at the interface is suggested to alter the lattice misfit effectively in the vicinity of the lamellar interface, which significantly improves planarity and thermal stability of the lamellar structures. But the mechanism of Cr-segregation at the C40/C11ₐ interface has not been clarified yet. Furthermore, it is unclear whether Cr is the optimum additive element that stabilizes the lamellar structure of NbSi$_2$/MoSi$_2$ duplex silicide, and the segregation mechanism is difficult to be clarified only by experiment. On the other hand, the phase field method, which is a thermodynamics-based continuum model, is a versatile technique for simulating microstructural evolution and interfacial reaction in various materials including quaternary alloy. We have been applying the phase-field method to the studies of segregation at interfaces such as stacking faults, twin boundary, anti-phase boundary and so on [4, 5]. The objective of this study is to clarify the mechanism of Cr-segregation in the lamellar silicide by the phase-field method.

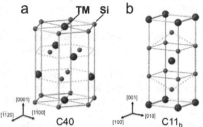

Figure 1. Unit cells of (a) C40 and (b) C11$_b$ crystal structures.

METHOD

As seen in Figure 1, C40 hexagonal structure and C11$_b$ tetragonal structure appear quite different from each other, but the atomic arrangement within an atomic layer on (0 0 0 1) plane of C40-phase and that on (1 1 0) plane of C11$_b$-phase are nearly identical (Figures 2a and b), and the only difference between the two crystal structures are the sequence of stacking of the layers. Namely, as shown in Figure 2c, C40-structure has ABDABD-type stacking and C11$_b$-structure has ACAC-type stacking, and the coherent lamellar interfaces parallel to these planes are formed. We define an order parameter φ so that $\varphi = 0$ corresponds to C40-phase and $\varphi = 1$ corresponds to C11$_b$-phase. The interface is defined as a region where the order parameter changes from 0 to 1

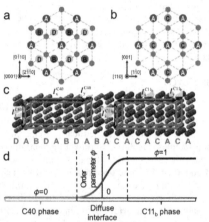

Figure 2. Atomic arrangements on planes parallel to the lamellar interface in (a) C40 structure and (b) C11$_b$ structure. (c) model of the lamellar interface. Green boxes show supercells equivalent in the two phases. (d) Definition of the order parameter.

continuously (Figure 2d) with the thickness of 6 atomic layers. In the current study of phase-field simulation, we take into account the elastic strain energy in addition to the chemical free energy and the gradient energy which are considered in most of phase-field simulation. The elastic strain energy is derived from the lattice misfit between C40- and C11$_b$-structures. The detail of chemical free energy, gradient energy and elastic strain energy are described below.

We evaluated the chemical free energy of C40-phase and C11$_b$-phase by the CALPHAD method and the thermodynamic data in the literature [6,7]. But there is no thermodynamic data of Mo-Si-Cr system, and therefore they are estimated from MoSi$_2$-CrSi$_2$ pseudo binary phase diagram [8] by the Thermo-Calc software. First, the energies of C40-phase (f_{C40}) and C11$_b$ -

phase (f_{C11b}) are calculated as functions of temperature and composition by the CALPHAD method. Free energy of intermediate state between C40-phase and C11$_b$-phase is evaluated so that each phase is connected smoothly using the order parameter as in the following equations,

$$f_{\text{chem}}(c_{\text{Nb}}, c_{\text{Si}}, c_{\text{Cr}}, \phi) = f_{\text{C40}}(c_{\text{Nb}}, c_{\text{Si}}, c_{\text{Cr}}, \phi)h(\phi) + f_{\text{C11}_b}(c_{\text{Nb}}, c_{\text{Si}}, c_{\text{Cr}}, \phi)(1 - h(\phi)) + W \cdot g(\phi) \tag{1}$$

$$h(\phi) = \phi - \frac{\sin(2\pi\phi)}{2\pi}, \quad g(\phi) = \frac{1 - \cos(2\pi\phi)}{2} \tag{2}, (3)$$

where $h(\varphi)$ is an interpolation function which increases monotonically from 0 to 1 as φ increases from 0 to 1, $g(\varphi)$ is a double well potential which penalizes the intermediate state between C40-phase and C11$_b$-phase. W is the interface energy density, which is derived from the energy of the most stable coherent interface between (0 0 0 1) plane of C40-MoSi$_2$ and (1 1 0) plane of C11$_b$-MoSi$_2$ evaluated by the first-principles calculation and the interfacial thickness of 1.3 nm (6 atomic layers). The actual value of W used in this simulation is 171 J/mol.

The gradient energy is the excess energy which is originated from the inhomogeneities of the concentrations and the order parameter. This energy is given by

$$f_{\text{grad}}(\nabla c_{\text{Nb}}, \nabla c_{\text{Si}}, \nabla c_{\text{Cr}}, \nabla\phi) = \kappa_{\text{Nb}}(\nabla c_{\text{Nb}})^2 + \kappa_{\text{Si}}(\nabla c_{\text{Si}})^2 + \kappa_{\text{Cr}}(\nabla c_{\text{Cr}})^2 + \kappa_{\phi}(\nabla\phi)^2$$
$$+ \kappa_{\text{NbSi}}\nabla c_{\text{Nb}}\nabla c_{\text{Si}} + \kappa_{\text{NbCr}}\nabla c_{\text{Nb}}\nabla c_{\text{Cr}} + \kappa_{\text{SiCr}}\nabla c_{\text{Si}}\nabla c_{\text{Cr}} \tag{4}$$

where κ_M and $\kappa_{MM'}$ (M = Mo, Nb, Si, Cr, $M \neq M'$) are the gradient energy coefficients for concentration gradients and κ_{φ} is that for order parameter gradients. The value of κ_{φ} is 2.66×10^{-11} J/m, which is estimated from the interfacial energy and the interfacial thickness on the basis of equilibrium-interface condition [9]. The gradient energy coefficients for the concentrations of constituent elements and cross terms are assumed to be 3×10^{-10} J/m, which is a typical value.

The elastic strain energy is derived from Hooke's law as in the following equations

$$f_{\text{str}}(c_{\text{Nb}}, c_{\text{Si}}, c_{\text{Cr}}, \phi) = \frac{1}{2}\int C_{ijkl}\varepsilon_{ij}^{\text{el}}\varepsilon_{kl}^{\text{el}}dV, \quad \varepsilon_{ij}^{\text{el}} = \varepsilon_{ij}^{\text{total}} - \varepsilon_{ij}^{\text{eigen}}(c_{\text{Nb}}, c_{\text{Cr}}, \phi) \tag{5}, (6)$$

$$\varepsilon_{ij}^{\text{eigen}}(c_{\text{Nb}}, c_{\text{Cr}}, \phi) = \eta_{ij,\text{Mo}}c_{\text{Mo}}\phi + \eta_{ij,\text{Nb}}c_{\text{Nb}}\phi + \eta_{ij,\text{Cr}}c_{\text{Cr}}\phi \tag{7}$$

where C_{ijkl} is elastic constant of C11$_b$-MoSi$_2$ [10], $\varepsilon_{ij}^{\text{el}}$ is the elastic strain, and $\varepsilon_{ij}^{\text{total}}$ is the total strain and $\varepsilon_{ij}^{\text{eigen}}$ is the eigen strain. The $\varepsilon_{ij}^{\text{eigen}}$ is defined as linear function of each concentration and the order parameters φ assuming Vegard's low.

Lattice misfit $\eta_{ij,\text{M}}$ of each element, required for calculating eigen strain, are defined as follows

$$\eta_{ij,\text{M}} = \frac{(l_{ij,\text{MSi}_2}^{\text{C11}_b} - l_{ij,\text{MSi}_2}^{\text{C40}})}{l_{ij,\text{MSi}_2}^{\text{C40}}} \quad (\text{M} = \text{Mo, Nb, Cr}), \quad l_{11} = l_x, l_{22} = l_y, l_{33} = l_z, l_{ij} = 0 \ (i \neq j) \tag{8}$$

Lattice sizes l_x, l_y and l_z of each disiliside are derived from the lattice parameters of C40- and C11$_b$-structures, including fictitious disilicides (Table I) evaluated by the first-principles calculation.

The total energy of the system, G_{sys}, is given by integrating the summation of the chemical free energy, the gradient energy and the elastic strain energy.

$$G_{\text{sys}} = \int [f_{\text{chem}}(c_{\text{Nb}}, c_{\text{Si}}, c_{\text{Cr}}, \phi) + f_{\text{grad}}(c_{\text{Nb}}, c_{\text{Si}}, c_{\text{Cr}}, \phi) + f_{\text{str}}(c_{\text{Nb}}, c_{\text{Si}}, c_{\text{Cr}}, \phi)]dV \tag{9}$$

Distributions of c_{Nb}, c_{Si}, c_{Cr} and φ are evolved so as to decrease the total energy of the system by solving Cahn-Hilliard equation [9] and Allen-Cahn equation [11], which are expressed as

Table I. Lattice parameters of disilicides for both C11$_b$-and C40-structure evaluated by the first-principles calculation.

Element	Structure	Lattice parameters		Lattice misfit		
		a (nm)	c (nm)	η_{11}	η_{22}	η_{33}
MoSi$_2$	C11$_b$	0.3217	0.7865	0.0310	-0.0138	-0.0156
	C40	0.4613	0.6619			
NbSi$_2$	C11$_b$	0.3290	0.7865	0.0531	-0.0345	-0.0156
	C40	0.4819	0.6627			
CrSi$_2$	C11$_b$	0.3082	0.7534	0.0265	-0.0103	-0.0123
	C40	0.4404	0.6369			

$$\frac{\partial c_M}{\partial t} = \left[\frac{D_M c_M}{kT} \left\{ (1-c_M)\nabla\left(\frac{\delta G_{sys}}{\delta c_M}\right) - \sum_{M'\neq M} c_{M'}\nabla\left(\frac{\delta G_{sys}}{\delta c_{M'}}\right) \right\} \right] \quad (M, M' = Nb, Si, Cr) \tag{10}$$

$$\frac{\partial \phi}{\partial t} = -\alpha\left(\frac{\delta G_{sys}}{\delta \phi}\right) \tag{11}$$

where D_M is the self-diffusion coefficient of element M and α is mobility for transformation between C11$_b$-phase and C40-phase. Average composition of the system is set to be (Mo$_{0.862}$Nb$_{0.138}$)Si$_2$, which gives nearly equal volumes of C40 and C11$_b$ phases in equilibrium at 1673 K. This simulation is conducted for a 12 nm long one-dimensional space using 128 grid points. In the initial condition, equal volumes of C40-phase and C11$_b$-phase having equilibrium concentrations for Cr-free system are connected, and then 1 at% of Cr atoms are uniformly distributed by replacing 0.5 at% of Mo and 0.5 at% of Nb. Here it is assumed that only transition metal elements (i.e. Mo, Nb, Cr) are able to diffuse (i.e. the mobility of Si atom is assumed to be 0). The distributions of solute concentrations and the order parameter are evolved until equilibrium is reached at 1673 K.

RESULTS & DISCUSSION

Figure 3 shows the calculated distribution of the order parameter φ and solute concentrations near the interface after equilibration for the case where Cr atoms are added as an additive element. In order to examine the effect of lattice-misfit relaxation, the result of another simulation for the case without elastic strain energy is also shown together for comparison (dotted line). In the interface region, the order parameter φ changes 0 to 1 continuously and solute elements are partitioned to each phase, but Cr-segregation is not formed contrary to the experiment. There is no significant difference between the results for the cases with and without elastic strain energy. The most significant difference is that the Nb concentration in C11$_b$-phase is lowered by the elastic strain energy by only 0.3 at%. Thus, it is difficult to attribute the experimentally observed Cr-segregation to only the relaxation of lattice misfit, and this implies that there are other mechanisms for the Cr-segregation. Segregation energies are evaluated on the basis of the first-principles calculation in order to take into account the effect of chemical interaction between additive element and the interface. The segregation energies of solute atoms located at 6 different atomic layers in the vicinity of interface are evaluated by the first-principles calculation, and the segregation energy function is defined so as to use in the phase-field calculation as below

$$E_{Cr}^{Seg}(\phi) = \Delta e_{\phi=0.2} \exp\left(\frac{-(\phi-0.2)^2}{b}\right) + \Delta e_{\phi=0.4} \exp\left(\frac{-(\phi-0.4)^2}{b}\right)$$

$$+ \Delta e_{\phi=0.6} \exp\left(\frac{-(\phi-0.6)^2}{b}\right) + \Delta e_{\phi=0.8} \exp\left(\frac{-(\phi-0.8)^2}{b}\right)$$

$$+ \Delta e_{\phi=1.0}\left(\frac{1-\cos(\pi\phi)}{2}\right). \tag{12}$$

Here the segregation energies of each layer are connected smoothly by Gaussian functions and the interpolation function. The $\Delta e_{\phi=p}$'s (p=0.2, 0.4, 0.6, 0.8) are the segregation energies of solute atoms located at the 2nd, 3rd, 4th, 5th and 6th layer counted from the C40-side edge of the interface region. A further simulation is conducted by adding the product of the segregation energy and local solute concentration to the free energy density assuming that the excess energy due to the segregation is proportional to the concentration of the additive element. The results with the segregation energies are shown in Figure 4. Unlike the result without segregation energy, a slight segregation of Cr atoms can be seen at the equilibrated interface as indicated by an arrow, although it is not so significant compared to that in the experiment. The result for the

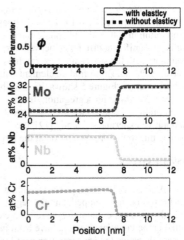

Figure 3. Profiles of the order parameter, Mo, Nb and Cr concentrations across the equilibrated interface. Dotted line shows the result for the case without elastic energy for comparison.

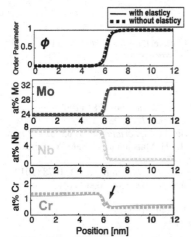

Figure 4. Profiles of the order parameter and concentrations in the case where segregation energy is taken into account.

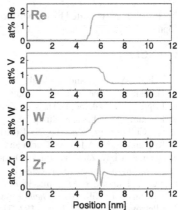

Figure 5. Concentration profiles of each additive element of Re, V, W and Zr.

case without elastic strain energy is superimposed as well. Owing to the elastic strain energy, Nb concentration is raised by 0.1at% in C40 phase and lowered by 0.2 at % in C11$_b$ phase, while Cr concentration in C40 and C11$_b$ phases are lowered and raised by 0.1% respectively. However, there is no significant difference between the segregation profiles of the two results. The simulations with the segregation energy have been conducted for the systems added individually with Re, V, W and Zr, to find an element which can segregate at the interface and to stabilize the lamellar structure. Figure 5 shows the equilibrated profiles of each element near the interface. Re, V and W exhibit no segregation. In contrast, Zr atom exhibits strong segregation and Zr concentration at the interface increases nearly to the double of that in the matrix. In our collaborative experimental study [12], the Zr-addition is found to stabilize the lamellar structure. This may due to the Zr segregation at the interface as shown in the present simulation results.

CONCLUSIONS

A phase-field model of segregation at the C40-phase/C11$_b$-phase interface in duplex disiliside has been developed, and the segregation mechanisms of Cr, whose segregation is clearly detected experimentally, have been examined. Also the effective additive elements for stabilizing the lamellar structure have been investigated. In the Cr added system, there is little difference between the concentration profiles simulated with and without elastic energy, and the segregation due to the relaxation of misfit strain has not been confirmed. In the case where the segregation energy is taken into account, Cr atoms segregate slightly at the equilibrated interface. In the cases with Re, V or W addition, solute atoms are distributed just to connect the equilibrium concentrations in each phase monotonically, and no segregation is formed. In contrast, Zr exhibits strong segregation at the interface. This result well accords with the recent experimental fact that the lamellar structure is stabilized by Zr-addition [12].

ACKNOWLEDGMENTS

This study is supported by the advanced low carbon technology Research and Development Program of Japan Science and Technology Agency.

REFERENCES

1. T. Nakano, Y. Nakai and S. Maeda, Y. Umakoshi, *Acta Mater.* **50**, 1781 (2002).
2. K. Hagihara, S. Maeda, T. Nakano and Y. Umakoshi, *Sci. Tech. Adv. Mater.* **5**, 11 (2004).
3. K. Hagihara, T. Nakano, S. Hata, O. Zhu and Y. Umakoshi, *Scripta Mater.* **62**, 613 (2010).
4. Y. Koizumi, T. Nukaya, S. Suzuki, S. Kurosu, Y. Li, H. Matsumoto, K. Sato, Y. Tanaka and A. Chiba, *Acta Mater.* **60**, 2901 (2012).
5. Y. Koizumi, S. M. Allen, M. Ouchi and Y. Minamino, *ISIJ International* **52**, 1678 (2012).
6. T. Geng, C. Li, X. Zhao, H. Xu, Z. Du and C. Guo, *CALPHAD* **34**, 363 (2010).
7. G. Shao, *Intermetallics* **13**, 69 (2005).
8. F. Wei, Y. Kimura, Y. Mishima, *Mater. Trans.* **42**, 1349 (2001).
9. J. W. Cahn, J. E. Hilliard, *J. Chem. Phys.* **28**, 258 (1958).
10. J. J. Petrovic, *Mater. Sci. Eng.* **A192**, 31 (1995).
11. S. M. Allen, J. W. Cahn, *Acta Mater.* **27**, 1085 (1979).
12. K. Hagihara, Y. Hama, M. Todai, T. Nakano, MRS Fall Meeting. 2012, Boston.

Mater. Res. Soc. Symp. Proc. Vol. 1516 © 2013 Materials Research Society
DOI: 10.1557/opl.2012.1749

Compression of Micro-pillars of a Long Period Stacking Ordered Phase in the Mg-Zn-Y system

Atsushi Inoue[1], Kyosuke Kishida[1], Haruyuki Inui[1], and Koji Hagihara[2]
[1]Department of Materials Science and Engineering, Kyoto University,
Sakyo-ku, Kyoto, 606-8501 JAPAN
[2]Department of Adaptive Machine Systems, Osaka University,
2-1, Yamada-Oka, Suita, Osaka, 565-0871, JAPAN

ABSTRACT

Deformation behavior of an 18R-type long period stacking ordered (LPSO) phase in the Mg-Zn-Y system was studied by micro-pillar compressions of single crystalline specimens prepared by focused ion beam (FIB) technique as a function of loading axis orientation and specimen dimensions. When the loading axis is inclined to the basal plane of the LPSO phase by 42°, basal slip of $(0001)<11\bar{2}0>$-type is activated irrespective of the specimen dimensions. When the loading axis is parallel to the basal plane, the formation of thick deformation bands are observed for all specimens tested. Strong size-dependence of yield stress values is observed for both types of micro-pillar specimens with different loading axis orientations.

INTRODUCTION

Recently, a new type of intermetallic phases with long-period stacking-ordered (LPSO) structures has been found in Mg-TM (transition metal)-RE (rare earth) ternary alloys and studied extensively since it has been considered as one of the key factors for endowing the Mg-TM-RE ternary alloys with both high strength and high ductility simultaneously [1-3]. However, inherent characteristics of the LPSO phases including crystal structure and deformation mechanism are largely unsolved [1-3]. We have recently studied the crystal structure of some Mg-TM-RE alloys by atomic resolution high-angle annular dark-field scanning transmission electron microscopy (HAADF-STEM) and have revealed that the crystal structure of the LPSO phase in the Mg-Zn-Y ternary system is characterized by a periodic arrangement of stacking faults within the hcp stacking of parent Mg and also by enrichment of TM and RE atoms in four atomic layers adjacent to the stacking fault as illustrated in Figure 1 [4-8]. These characteristics of the crystal structure imply that inherent deformation behavior of the LPSO phase is strongly anisotropic. As for the deformation behavior of single-phase LPSO phases, only limited studies using directionally solidified (DS) ingots composed of plate-shape grains with their basal planes being nearly perpendicular to the growth direction are available so far and they suggested that the basal slip is the dominant deformation mode in the LPSO phase, whereas abundant deformation kinking occurs when a stress is loaded parallel to the growth direction [9,10]. Studies using single crystalline ingots have never been carried out mostly because it is quite difficult to fabricate single crystals of the Mg-Zn-Y LPSO phases. In the present study, micro-pillars of single crystalline Mg-Zn-Y LPSO phase with two loading axis orientations being parallel or inclined to the basal plane were prepared from the DS ingot of the 18R-type LPSO phase by focused ion beam (FIB) technique and deformed in compression using a micro hardness testing machine equipped with a flat diamond tip under a constant loading rate at room temperature to clarify the inherent deformation behavior of the Mg-Zn-Y LPSO phase.

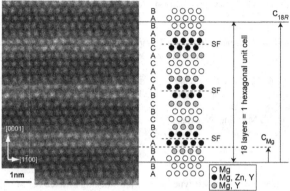

Figure 1. An HAADF-STEM image and a schematic illustration of the stacking sequence of the 18*R*-type Mg-Zn-Y LPSO phase in the [2$\overline{1}$$\overline{1}$0] projection.

EXPERIMENTAL PROCEDURE

Ingots with a nominal composition of Mg - 5at.% Zn - 7at.% Y were prepared by high-frequency induction melting in carbon crucibles. Direction solidified (DS) rods were obtained from the as-prepared ingots by the Bridgman method at a growth rate of 5 mm/h under an Ar atmosphere. Figure 2a and 2b show typical microstructures of the longitudinal and transverse sections of a DS rod. Plate-shaped grains of the LPSO phase with an average thickness of 100 μm are developed so that their basal planes tend to be parallel to the growth direction of the DS rod. Preferential growth directions of most grains of the LPSO phase were confirmed to be nearly parallel to <2$\overline{1}$$\overline{1}$0> in the DS rod [9].

Micro-pillar specimens with square cross sections and aspect ratio of 2 ~ 4 were cut from relatively large grains of the LPSO phase in the DS rods by focused ion beam (FIB) technique. Side lengths of the square cross sections were in a range from 0.5 to 7 μm in the present study. Two different loading axis orientations of [2$\overline{1}$$\overline{1}$0] and [24 $\overline{1}$2 $\overline{1}$2 1], which are inclined to the basal plane of the LPSO phase by 0° and about 42°, respectively, were selected in the present study as indicated in Figure 2c. The crystal orientations of the 18*R*-type LPSO phase are indexed with respect to a hexagonal unit cell with dimensions of $a_{Mg} \times a_{Mg} \times 9c_{Mg}$, where a_{Mg} and c_{Mg} correspond to the lattice constant of the parent hcp Mg unless otherwise noted (Figure 1). The micro-pillars were deformed in compression using a micro hardness testing machine equipped with a flat-punch diamond tip with a diameter of 20 μm at a constant stress rate of about 2 MPa/s, which corresponds to a nominal strain rate of $10^{-3} \sim 10^{-4}$ /s. Microstructures of micro-pillar specimens before and after compression tests were examined using a scanning electron microscope (SEM) equipped with a field emission gun.

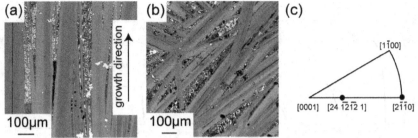

Figure 2. SEM backscattered electron images of (a) longitudinal and (b) transverse sections of a DS ingot of Mg - 5at.% Zn - 7at.% Y alloy. (c) Loading axis orientations of micro-pillar specimens.

RESULTS AND DISCUSSION

Figure 3a and 3b show typical stress - strain curves obtained for specimens compressed along [24 1̄2 1̄2 1] and [2̄1̄10], respectively. All stress - strain curves obtained for the [24 1̄2 1̄2 1]-oriented micro-pillar specimens exhibit smooth transition from elastic to plastic flow. Strain-burst behavior after about 2 % plastic deformation was observed for smaller specimens with side lengths below 1.5 μm. Such strain burst behavior is not observed for larger specimens at least at the plastic strain about 3%. In contrast, stress - strain curves for the [2̄1̄10]-oriented specimens commonly exhibit linear elastic region directly followed by sudden strain-burst behavior, i.e. very rapid plastic flow more than 10% plastic strain upon yielding. Then, 0.2% flow stress and critical stress for the sudden strain-burst are assigned as yield stress for the [24 1̄2 1̄2 1]- and [2̄1̄10]-oriented specimens, respectively (marked by arrows in Figure 3). Yield stress increases drastically with decreasing the edge length of the cross section for both the [24 1̄2 1̄2 1]- and [2̄1̄10]-oriented specimens, which is similar to that observed in micro-pillar compression tests for various metals [11-13].

Figure 4a and 4b show deformation microstructures of [24 1̄2 1̄2 1]-oriented specimens with side lengths of 2.9 and 0.72 μm, respectively. Fine slip traces corresponding to the basal plane are clearly observed for both specimens. Since the slip direction is confirmed to be on the side surface of (011̄0) by careful inspection of the shape of the micro-pillar specimen after compression test, it can be concluded that the deformation occurs by the activation of the (0001)[2̄1̄10] basal slip in compression of the [24 1̄2 1̄2 1]-oriented specimens irrespective of the specimen dimensions. The preferential activation of the basal slip in the LPSO phase is consistent with the previous study using bulk DS ingots [9,10].

Figure 5 shows deformation microstructures of [2̄1̄10]-oriented specimens with a side length of 3.5 μm. In the case of the [2̄1̄10]-oriented micro-pillar specimens, for which no shear strain is acted on the basal plane, macroscopic buckling or bending of the micro-pillars is frequently occurred and therefore determination of operative deformation modes by trace analysis is quite difficult for most specimens tested. As indicated by the sudden strain-burst behavior observed in the stress-strain curves (Figure 3b), the initiation of macroscopic buckling or bending is considered to occur much faster than the nominal strain rate of $10^{-4} - 10^{-3}$ /s

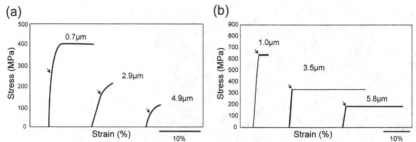

Figure 3. Typical stress-strain curves obtained for various specimen sizes with two different loading axis orientations of (a) [24 1̄2 1̄2 1] and (b) [2̄11̄0].

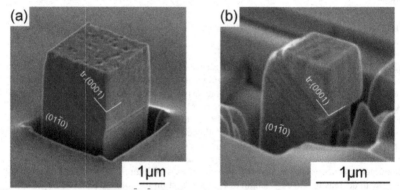

Figure 4. SEM images of [24 1̄2 1̄2 1]-oriented micro-pillars of the Mg-Zn-Y LPSO phase with side lengths of (a) 2.9 and (b) 0.7μm.

resulting in a substantial increase of the strain rate so as to keep a constant stress rate, which may accelerate the macroscopic buckling.

Figure 6 plots the CRSS values for the basal slip calculated based on the yield stresses for [24 1̄2 1̄2 1]-oriented micro-pillars of Mg-Zn-Y LPSO phase as a function of the side length of the square cross section. The CRSS of the basal slip exhibits an approximate power law relationship, i.e. $\tau_y \propto d^n$, where n has the value of about -0.63, the value of which is similar to that reported for Ni micro-pillars oriented for single slip activation [11,12]. Based on the previous reports on compression of Ni micro-pillars, CRSS value for micro-pillars with 20 ~ 30 μm size coincides with that obtained for macroscopic single crystals [11,12]. Assuming the similar trend holds true for the basal slip of the present LPSO phase, the CRSS value for macroscopic single crystals is extrapolated to be about 7 MPa, which is much lower than 10 ~ 30 MPa estimated previously from the compression tests of bulk DS ingots of the Mg - 5at.% Zn - 7at.% Y [9,10]. The estimated CRSS value for the basal slip in macroscopic single crystal is reasonable considering difference of constraint imposed from the surrounding grains in the case of the DS alloys previously reported.

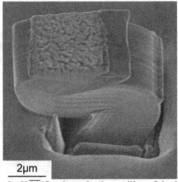

Figure 5. An SEM image of a [2$\bar{1}$10]-oriented micro-pillar of the Mg-Zn-Y LPSO phase with a side length of 3.5 μm.

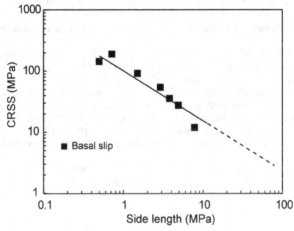

Figure 6. Specimen size dependence of the CRSS values for the basal slip in the Mg-Zn-Y LPSO phase.

CONCLUSIONS

Deformation behavior of single crystals of the Mg-Zn-Y LPSO phase was investigated as a function of the loading axis orientation by micro-pillar compression tests. When the loading axis is inclined to (0001), the basal slip is the dominant deformation mode irrespective of the specimen dimensions. When the loading axis is parallel to (0001), abundant deformation bands were observed to be formed. The yield stress for these two loading axis orientations increases with decreasing the specimen dimensions.

ACKNOWLEDGMENTS

This work was supported by Grant-in-Aid for Scientific Research from the Ministry of Education, Culture, Sports, Science and Technology (MEXT), Japan (No. 23360306, and No. 23109002) and in part by the Elements Strategy Initiative for Structural Materials (ESISM) from the MEXT, Japan.

REFERENCES

1. Y. Kawamura, K. Hayashi, and A. Inoue, Mater. Trans., **42**, 1171 (2001).
2. Y. Kawamura, T. Kasahara, S. Izumi, and M, Yamasaki, Scripta Mater., **55**, 453 (2006).
3. Y. Kawamura and M. Yamasaki, Mater. Trans., **48**, 2986 (2007).
4. H. Yokobayashi, K. Kishida, H. Inui, M. Yamasaki, and Y. Kawamura, Acta Materialia., **59**, 7287 (2011).
5. H. Yokobayashi, K. Kishida, H. Inui, M. Yamasaki, and Y. Kawamura in *Intermetallics-Based Alloys for Structural and Functional Applications*, edited by B. Bewlay, M. Palm, S. Kumar, and K. Yoshimi (Mater. Res. Soc. Symp. Proc., **1295**, Warrendale, PA, 2011), pp. 267-272.
6. K. Kishida, H. Yokobayashi, H. Inui, M. Yamasaki, and Y. Kawamura, Intermetallics., **31**, 55 (2012).
7. K. Kishida, H. Yokobayashi, H. Inui, M. Yamasaki, and Y. Kawamura in *Mg2012: 9th International Conference on Magnesium Alloys and their Applications*, edited by W.J. Poole and K.U. Kainer (ICMAA2012, Vancouver, BC, 2012), pp. 429-434.
8. K. Kishida, H. Yokobayashi, A. Inoue, and H. Inui, in in *Intermetallics-Based Alloys – Science and Technology and Applications*, edited by I. Baker, M. Heilmaier, S. Kumar, and K. Yoshimi (Mater. Res. Soc. Symp. Proc., **1516**, Warrendale, PA, 2012), in this volume.
9. K. Hagihara, N. Yokotani, and Y. Umakoshi, Intermetallics., **18**, 267 (2010).
10. K. Hagihara, Y. Sugino, Y. Fukusumi, Y. Umakoshi, and T. Nakano, Mater. Trans., **52**, 1096 (2011).
11. D.M. Dimiduk, M.D. Uchic, and T.A. Parthasarathy, Acta, Materialia., **53**, 4065 (2005).
12. M.D. Uchic and D.M. Dimiduk, Mater. Sci. Eng. A., **400**, 268 (2005).
13. J.R. Greer, W.C. Oliver, and W.D. Nix, Acta Mater., **53**, 1821 (2005) .

Mater. Res. Soc. Symp. Proc. Vol. 1516 © 2012 Materials Research Society
DOI: 10.1557/opl.2012.1667

Compression of Single-Crystal Micropillars of the ζ Intermetallic Phase in the Fe-Zn System

Masahiro Inomoto, Norihiko L. Okamoto, Haruyuki Inui

Department of Materials Science and Engineering, Kyoto University, Sakyo-ku, Kyoto 606-8501, Japan

ABSTRACT

The deformation behaviour of the ζ (zeta) phase in the Fe-Zn system has been investigated via room-temperature compression tests of single-crystal micropillar specimens prepared by the focused ion beam method. Trace analysis of slip lines indicates that {110} slip occurs for the specimens investigated in the present study. Although the slip direction has not been uniquely determined, comparison of Schmid factors and yield stress values suggests that the slip direction might be <1$\bar{1}$2>, which is inconsistent with the easiest slip system {110}[001] predicted on the basis of the primitive Peierls-Nabarro model.

INTRODUCTION

Hot-dipped galvannealed (GA) steels are widely used for the chassis of automobiles and building materials because of their high corrosion resistance, weldability, and paintability. The coating layer usually consists of a series of intermetallic phases in the Fe-Zn system, stacked on the steel substrate in the order of $\Gamma(Fe_3Zn_{10})$, $\Gamma_1(Fe_{11}Zn_{40})$, $\delta_{1k}(FeZn_7)$, $\delta_{1p}(FeZn_{10})$ and $\zeta(FeZn_{13})$ [1-3]. When the GA steels are deformed under severe conditions such as press forming operation, the coating layer occasionally fails (powdering and/or flaking), resulting in reduced corrosion resistance and paintability. The coating failure has been understood only phenomenologically, and almost nothing is known about the mechanical properties of each of the intermetallic phases. This may stem from the fact that preparing single-phase specimens in bulk form is difficult [4] because of the complicated Fe-Zn binary phase diagram [1-3]. In our preliminary compression experiments of polycrystalline micropillars of each phase prepared from the thin coating layer (~10 μm) of the GA steels, we found that the ζ phase exhibits plastic deformation to some extent followed by fracture. This fracture is considered to occur either because (1) the number of independent slip systems in the ζ phase is insufficient for the von Mises criterion, according to which five independent slip systems are required for a crystal grain to undergo an arbitrary imposed plastic deformation [5], or because (2) possible presence of hcp-Zn phase between the columnar ζ grains induces cracks to initiate from the Zn/ζ interface. In the present study, we investigated the deformation behaviour of the ζ phase via compression tests of micropillar specimens prepared from small strip-shaped single crystals by using the focused ion beam (FIB) method.

CRYSTAL STRUCTURE AND POSSIBLE SLIP SYSTEMS

The crystal structure of the ζ phase in the Fe-Zn system was first determined by Brown [6]. He assigned a base-centered monoclinic unit cell to the ζ phase as shown in the left-hand side of figure 1a. However, Gellings et al. [7] pointed out that this unit cell does not conform to the international convention to choose the axes such that the obtuse angle β is as close as 90°. Instead, Gellings et al. assigned a unit cell with a smaller β angle as shown in the right-hand side of figure 1a, although they did not carry out crystal structure refinement. Brown failed to determine the Fe atomic positions in the unit cell probably because it was difficult to obtain single crystals of good quality and also to differentiate Fe and Zn atoms by x-ray diffraction due to the small difference in their atomic scattering factors [6]. However, Belin et al. [8] later re-examined the crystal structure of the ζ phase with a single crystal of good quality and determined the Fe atoms locating exclusively at the $2c$ site, which is the midpoint of edge lines along the c axis. As shown in figure 1b, the Fe atom is the center of the Zn_{12} icosahedra, which are linked to one another [8].

Within the Peierls-Nabarro (PN) model [9, 10], shear stress required for the motion of edge dislocations (PN stress), τ_p, is given by

$$\tau_p = \frac{2\mu}{1-v}\exp\left(-\frac{2\pi d}{(1-v)b}\right),$$ (1)

where μ, v, d and b stand for the shear modulus, Poisson's ratio, atomic plane distance and magnitude of Burgers vector, respectively. Although one should consider not only Poisson's ratio but also normal/shear coupling terms as the single-crystal compliance constants, s_{15}, s_{25}, and s_{35} possess no-zero values in monoclinic (non-orthotropic) systems, we compare only the d/b ratios in the exponential which largely dominates the magnitude of the PN stress in order to estimate possible operative slip systems in the ζ phase to a first approximation. The three largest atomic plane distances and the four shortest translational vectors in the ζ phase are shown in figure 2. It should be noted that $[1\bar{1}2]$ and $[1\bar{1}\bar{2}]$ are not equivalent in monoclinic systems. Table I summarizes some d, b and d/b values for the possible operative slip systems. The {110}[001] slip system is expected to be the easiest due to the largest d/b value. (010)[001], {110}1/2<1$\bar{1}$0>, {110}1/2<1$\bar{1}$2>, (100)[001], {110}1/2<1$\bar{1}\bar{2}$> slip systems are considered to be subsequently the easiest in the order.

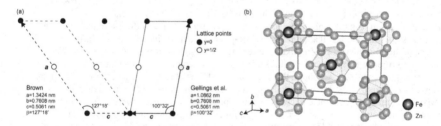

Figure 1. (a) Unit cell dimensions [6, 7] and (b) crystal structure of the ζ phase in the Fe-Zn system [7, 8].

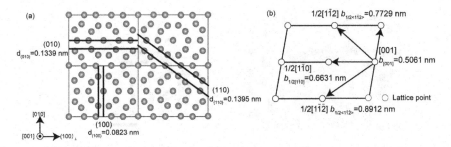

Figure 2. (a) [001] projection of the atomic structure and (b) ($1\overline{1}0$) cross section of the lattice points for the ζ phase.

Table I. Possible operative slip systems with the largest values of d/b, and Schmid factors for two loading axis orientations.

Slip system	d (nm)	b (nm)	d/b	Schmid factor	
				Orientation A [3 20 27]	Orientation B [552]
{110}[001]	0.1395	0.5061	0.2756	0.45	0.01
(010)[001]	0.1339	0.5061	0.2646	0.48	0.00
{110}1/2<1$\overline{1}$0>	0.1395	0.6631	0.2104	0.28	0.30
{110}1/2<1$\overline{1}$2>	0.1395	0.7729	0.1805	0.43	0.26
(100)[001]	0.0823	0.5061	0.1626	0.10	0.00
{110}1/2<1$\overline{1}\overline{2}$>	0.1395	0.8912	0.1565	0.31	0.22

EXPERIMENTAL PROCEDURES

We tried to grow large single crystals of the ζ phase by the flux method. Elements (4N purity) with a mol ratio of Zn:Fe = 99.5:0.5 were sealed in a quartz ampoule under vacuum. The ampoule was heated to 550°C and, then slowly cooled down to 445°C over 300 h, followed by water quenching. However, we obtained only small strip-shaped single crystals (~100×300×5000 μm³) embedded in zinc as shown in figure 3. Quenched specimens were mechanically and electrochemically polished. Square-shaped micropillars with a side length of approximately 4 μm and with an aspect ratio of approximately 1:4 were prepared from the grown single crystals

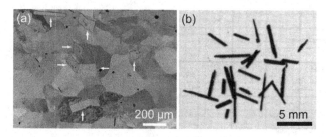

Figure 3. (a) SEM backscattered electron image of ζ crystals embedded in zinc. (b) Small strip-shaped single crystals of the ζ phase which are obtained from the quenched ingot by chemically dissolving zinc in concentrated hydrochloric acid.

of the ζ phase by the focused ion beam (FIB) technique. Two micropillars were prepared from different crystal grains to possess different loading axis orientations, which were measured by electron backscatter diffraction (EBSD) in field-emission scanning electron microscope (SEM). Compression tests for the micropillars were carried out by using a micro hardness testing machine equipped with a flat diamond tip with a diameter of 20 μm at a constant stress rate which corresponds to an initial strain rate of 8×10^{-5}- 1×10^{-4} s^{-1} in the elastic region.

RESULTS AND DISCUSSION

The stress-strain curve obtained for a micropillar with orientation A [3 20 27] is shown in figure 4(a). It exhibits a clear yield point at around 110 MPa. The micropillar displays large strain bursts soon after yielding. There seems to be almost no work hardening because the compression tests were conducted at a constant stress rate. These strain bursts are often observed for compression test of micrometer-sized single-crystal specimens [11] and considered to arise from the collective, avalanche-like motion of dislocations [12]. The other micopillar with orientation B [552] exhibits a similar deformation behaviour with the yield stress value of around 260 MPa. Thus, the ζ phase exhibits plastic deformation to a certain extent in a single-crystal form, unlike the compression deformation tests in a polycrystalline form (see INTRODUCTION). Figure 4(b) shows a SEM secondary electron image of the deformed micropillar with orientation A, observed along a direction tilted by 30° from the loading axis. Trace analysis on the slip lines observed on two orthogonal faces of the micropillar revealed that slip on ($1\bar{1}0$) plane operates. Similarly, slip on {110} plane is confirmed to operate for the micropillar with orientation B.

Among the six possible operative slip systems tabulated in Table I, (010)[001] and (100)[001] slip systems are eliminated because slip on {110} plane is confirmed to operate for two micropillars investigated. Thus, there are four possible slip systems with {110} slip plane. Schmid factors for the two loading axis orientations A and B are also tabulated in Table I. Since shear strain is produced not only by resolved shear stress (RSS) but also by normal stress (normal/shear coupling) in monoclinic systems, the Schmid law will not exactly hold true for the ζ phase. However, we deduce critical RSS (CRSS) values without regard to the normal/shear coupling effect in the present study. Calculation of the CRSS values to a first approximation by

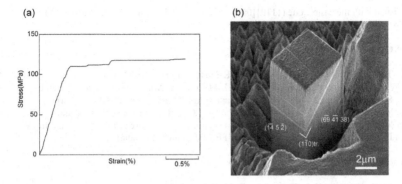

Figure 4. (a) Stress-strain curve and (b) SEM secondary electron image after deformation for a single-crystal micropillar with the loading axis orientation of A [3 20 27]. The micropillar was observed along a direction tilted by 30° from the loading axis.

using the Schmid factors and the obtained yield stress values for the four possible slip systems suggests that <1$\bar{1}$2> slip direction is the most reasonable because the two CRSS values for the slip direction for the two loading axis orientations are comparable to each other while those for the other slip directions are not. It is somewhat surprising if the operative slip system is indeed {110}<1$\bar{1}$2>, which is not the easiest one predicted on the basis of the primitive Peierls-Nabarro model. However, this discrepancy can be explained in terms of the characteristic atomic arrangement on the {110} plane; the Zn atoms are aligned along the <1$\bar{1}$2> direction, which makes it energetically unfavorable to slip along directions other than <1$\bar{1}$2>. Nevertheless, it is definitely necessary to confirm the slip direction for the {110} slip through the observation of dislocation structures as well as the characterization of the Burgers vector of operative dislocations by transmission electron microscopy.

The CRSS value averaged for the two micropillars is 57 MPa on the assumption that the operative slip system is indeed {110}<1$\bar{1}$2>. This CRSS value is relatively small in spite of the rather complex crystal structure of the ζ phase (figures 1 and 2). Moreover, a power law scaling of CRSS values against the sample size of micrometer-sized single crystals is often observed for many kinds of metals [13]. This means that the CRSS value for the ζ phase in bulk form is much smaller. The investigation of the size effect of CRSS values is under way.

CONCLUSIONS

The deformation behaviour of the ζ phase has been investigated by compression tests of single-crystal micropillars with two loading axis orientations. The ζ phase exhibits plastic deformation to some extent in a single crystal form. Slip on {110} plane is confirmed to operate for the two orientations. The slip direction is estimated to be <1$\bar{1}$2>, though the slip system is inconsistent with the easiest one ($\{110\}[001]$) predicted on the basis of the primitive Peierls-Nabarro model.

ACKNOWLEDGMENTS

This work was supported by JSPS KAKENHI grant number 24246113 and the Elements Strategy Initiative for Structural Materials (ESISM) from the Ministry of Education, Culture, Sports, Science and Technology (MEXT) of Japan, and in part by Advanced Low Carbon Technology Research and Development Program (ALCA) from the Japan Science and Technology Agency (JST). This work was also supported by Research Promotion Grant from ISIJ and Grants for Technical Research from JFE 21st Century Foundation.

REFERENCES

1. M. A. Ghoniem and K. Lohberg, Metall. 26, 1026 (1972).
2. T. B. Massalski, *Binary Alloy Phase Diagrams*, Vol.2 (ASM, Metals Park, OH, 1986) p.1128.
3. O. Kubaschewski, *Iron - Binary Phase Diagrams* (Springer-Verlag, New York, 1982) p.172.
4. M. H. Hong and H. Saka, Philos. Mag. 74, 509 (1996).
5. R. V. Mises, R. Z. Angew. Math. Mech. 8, 161 (1928).
6. P. J. Brown, Acta Crystallogr. 15, 608 (1962).
7. P. J. Gellings, E. W. Bree, G. Gierman, Z. Metallkd. 70, 315 (1979).
8. R. Belin, M. Tillard and L. Monconduit, Acta Crystallogr. C56, 267 (2000).
9. R. Peierls, Proc. Phys. Soc. 52, 34 (1940).
10. F. R. N. Nabarro, Proc. Phys. Soc. 52, 90 (1940).
11. M. D. Uchic, D. M. Dimiduk, J. N. Florando and W. D. Nix, Science 305, 986 (2004).
12. F. F. Csikor, C. Motz, D. Weygand, M. Zaiser and S. Zapperi, Science 318, 251 (2007).
13. D. M. Dimiduk, M. D. Uchic and T. A. Parthasarathy, Acta Mater. 53, 4065 (2005).

Mater. Res. Soc. Symp. Proc. Vol. 1516 © 2012 Materials Research Society
DOI: 10.1557/opl.2012.1668

Effects of the microstructure and minor elements
on the fracture toughness of Nb-Si alloy

Takuya Okawa*[1]; Seiji Miura[1]; Tetsuo Mohri[1];

[1]Division of Materials Science and Engineering, Faculty of Engineering, Hokkaido University;
Kita-13, Nishi-8, Kita-ku, Sapporo 060-8628, Hokkaido, Japan
*Graduate Student, Graduate School of Engineering, Hokkaido University

ABSTRACT

The development of a new high temperature structural material is recently required in various fields. As one of the potential materials, Nb-Si alloys have attracted attention due to their high melting point and low density. A microstructure composed of ductile Nb matrix containing finely dispersed spherical Nb_5Si_3 phase is obtained by the addition of ternary elements such as Au and it is found that such microstructure is effective in improving room temperature toughness. The main purpose of the present study is evaluating fracture toughness of Nb-Si-Au alloys using small specimens and investigating the effects of the microstructure and other minor elements on the fracture toughness. Alloy ingots of Nb-15at.%Si-3at.%Au and Nb-3at.%Au are prepared by arc-melting under Ar atmosphere, followed by heat-treatments at up to 1500°C for 100 hours. Chevron notched specimens with a size of 1.0x2.0x10mm are subjected to four-point bending tests under a laser confocal microscope for in-situ observation of crack propagation, and the effect of the microstructure and minor elements such as oxygen on the evaluated fracture toughness is investigated on both the Nb/Nb_5Si_3 alloys and the Nb solid solution (Nb_{ss}) alloys.

INTRODUCTION

A Nb-Si alloy is one of the potential materials of high temperature materials for replacing Ni based superalloys, because they exhibit superior high temperature strength to the surperalloys [1,2]. In previous studies, Miura et al. [3-5] aimed to strike a balance between high temperature strength and fracture toughness at room temperature, and it was revealed that they could obtain a suitable microstructure composed of ductile Nb matrix containing finely dispersed spherical Nb_5Si_3 phase by the addition of ternary elements such as Au. Moreover, introducing Nb dendrite is effective for improving fracture toughness. Thus, the main purpose of the present study is evaluating fracture toughness of Nb-Si-Au alloys using small specimens and investigating effects of the microstructure and minor elements on the fracture toughness.

The microstructure of Nb-15Si-3Au alloy after heat-treatment is composed of Nb_{ss} matrix and spherical Nb_5Si_3 phases, which is ideal for improving room temperature toughness. However, it is unavoidable for the Nb-Si alloy that oxygen and nitrogen contaminate from the atmosphere at high temperature. It is known that Nb_{ss} becomes worse in ductility by the contamination of oxygen [6]. Therefore, the effect of light elements such as oxygen and nitrogen on the fracture of Nb-15Si-3Au alloy is investigated.

EXPERIMENT

Alloys of about 20g were arc-melted in Ar atmosphere on a water-cooled copper hearth. In order to ensure the homogeneity, alloy ingots were turned and remelted several times. The composition of alloys was Nb-15at.%Si-3at.%Au. Heat-treatments were conducted at 1300 °C for 100 hours and at 1500 °C for 100 hours in high purity Ar-flow atmosphere with Ta foil wrapping. In order to investigate the effect of oxygen and nitrogen on the mechanical properties of Nb solid solution phase as the Nb-Si-Au alloy, Nb-3at.%Au alloy prepared by arc melting was heat-treated at 1300°C for 100 hours in high purity Ar-flow atmosphere with Ta foil wrapping. All samples were polished with Al_2O_3 powder (0.1μm in diameter) and the microstructure was observed using a scanning electron microscope (SEM, JEOL JXA-8900M), and the composition of each constituent phase was analyzed by Electron probe microanalysis (EPMA) using wavelength dispersive X-ray spectroscopy (EPMA-WDS). For the chemical analysis of light elements such as oxygen and nitrogen in present alloys, SIMS(Secondary Ion-microprobe Mass Spectrometer, stigmatic-SIMS, ims-1270) was employed. Its high spatial resolution of 0.2μm is suitable to investigate fine structures.

The in-situ bending tests were performed on all two-phases samples under a confocal laser scanning microscope (CLSM, type 1LM21, LaserTech, Co.Ltd) with a four-point bending apparatus composed of a Pieao actuator (N-214, Physik Insturumente GmbH & Co.) and a force sensor (type 9193, Kistler Instrumemte AG). The specimen size was a dimension of about 1.0x2.0x10mm and the tests were conducted at room temperature [3]. A Chevron notch with an angle of 90° was cut using an abrasive blade of as thin as 0.2mm. For four-point bending tests the spans between the outer support pins and the inner pins are 9mm and 3mm, respectively. Used for support pins were a drill rod made of cemented carbide. The speed of load point displacement is 1μm/sec. Deformation and crack propagation was recorded by a VCR with a video rate of 30 frames/sec attached to the CLSM. The applied load for arc-melted and heat-treated samples was recorded every 0.01 sec. by a data acquisition system (Yokogawa Electric Co.). Four-point bending tests were conducted twice for each alloy. After the tests, fracture surface were examined by using SEM.

In a conventional toughness measurement by using bending test, straight-notch specimens are used. Chevron notched specimens are beneficial for measuring the plane-strain fracture toughness of brittle materials, because an adequately sharp micro-crack is naturally produced during the early stage of loading so that no pre-crack is required.

Fracture toughness is described by stress intensity factor K_{1c} estimated by a formula for the Chevron notched specimens [7-9]. In the present study, considering the effect of the size of specimen with ductile Nb_{ss} phase, Nb_{ss}/Nb_5Si_3 two-phase alloys may not satisfy the small scale yield condition [10]. Therefore, the fracture toughness in the present study is dealt with as K_q.

Vickers hardness tests (0.3kg load, at room temperature) are conducted on the Nb-3Au as-cast alloy and the Nb-3Au heat-treated at 1300°C for 100 hours for the evaluation of solid solution hardening by oxygen and nitrogen contamination.

RESULT AND DISCUSSION

Figure 1 shows secondary electron images of Nb-15Si-3Au alloys. In each alloy, bright and dark phases are Nb_{ss} and Nb_5Si_3, respectively. Nb dendrites are found in the as-cast alloy as a primary phase and eutectic lamellar structures composed of Nb and Nb_5Si_3 phase are observed in Figure 1(a). Figures 1(b) and 1(c) show the microstructure of the heat-treated alloy. The Nb₅Si₃ lamellae in eutectic Nb/Nb₅Si₃ region in Nb-15Si-3Au alloy spheroidized during heat-treatment, resulting in fine Nb₅Si₃ particles embedded in Nb matrix. Nb thickness and Nb₅Si₃ particle size of Figure 1(c) are larger than those of Figure 1(b), because of the heat-treatment at higher temperature. Although silicide particles are found at narrow areas between secondary dendrite arms, precipitates are scarcely observed inside of the dendrites.

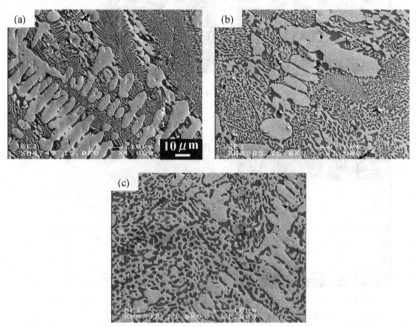

Figure 1. The microstructure of Nb-15Si-3Au alloys (a) as-cast, (b) 1300 °C x100 hr HT, (c) 1500 °C x100 hr HT.

Figure 2 and Figure 3 show SIMS observation results of O⁻ distribution and CN⁻ distribution in Nb$_{ss}$/Nb₅Si₃ two-phase alloys and the Nb-3Au alloy, respectively, after the heat-treatment at 1300 °C for 100 hours. In each alloy, bright areas are the region where SIMS detects the high intensity of O⁻ and CN⁻. Although almost no contamination was detected in the as-cast alloys, bright areas corresponding to Nb$_{ss}$ are found in the heat-treated alloys. It is concluded that the solubility of oxygen and nitrogen in Nb₅Si₃ is negligibly small, but is large in Nb$_{ss}$.

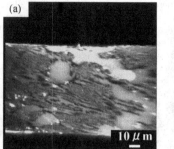
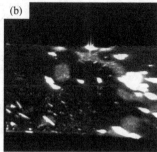

Figure 2. SIMS observation result of (a) O⁻ distribution and (b) CN⁻distribution in Nb-15Si-3Au heat-treated at 1300 °C for 100hr.

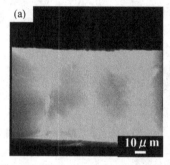
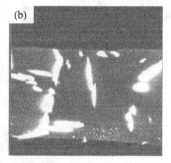

Figure 3. SIMS observation result of (a) O⁻ distribution and (b) CN⁻ distribution in Nb-3Au heat-treated at 1300 °C for 100hr.

Vickers hardness tests are conducted on the Nb-3Au alloys. Vickers hardness values of the as-cast alloy and the 1300 °Cx100hr HT alloy with a load of 0.3kg at room temperature are 172HV and 452HV, respectively. The increase in Vickers hardness values is comparable to the difference of the values of Nb with and without contamination of oxygen [11]. It strongly suggests the ductility of the Nb-3Au alloy decreases after the heat-treatment.

The values of K_q obtained for Nb_{ss}/Nb_5Si_3 two-phase alloys in the present study are summarized in Table 1. HT alloys have the lower fracture toughness than as-cast alloys. In conducting heat-treatment, the embrittlement of Nb_{ss} occurs by oxygen and nitrogen contamination, which results in the transgranular fracture in Nb_{ss}. However, the fracture toughness of the 1500 °Cx100hr HT alloys is higher than that of 1300 °Cx100hr HT alloys. Nb_{ss} in the 1500 °Cx100hr HT alloy is thought to contain oxygen comparable to or higher than Nb_{ss} in the 1300 °Cx100hr HT alloy. The higher K_q values for the 1500 °Cx100hr HT alloy strongly indicates that the progress of Nb_5Si_3 phase spheroidization is effective in improving room temperature toughness.

Table.1 Summary of the Chevron notch fracture toughness.

Nb-15Si-3Au	as-cast	1300 °Cx100hr	1500 °Cx100hr
K_q, MPa$\sqrt{m}$	7.4, 6.8	3.3, 2.9	7.3, 4.5

CONCLUSIONS

The effects of the microstructure and the solubility of light elements such as oxygen and nitrogen on the fracture of Nb-15Si-3Au alloy are investigated. By SIMS observation results of Nb_{ss}/Nb_5Si_3 two-phase alloys, it is found that the solubility of oxygen and nitrogen in Nb_5Si_3 is negligibly small, but is high in Nb_{ss}. Although the increment of Vickers hardness of the Nb_{ss} alloy by contamination of oxygen and nitrogen is consistent with the decreases of the fracture toughness of Nb_{ss}/Nb_5Si_3 two-phase alloys by heat-treatment, the progress of Nb_5Si_3 phase spheroidization is found to be effective in improving room temperature toughness.

REFERENCES

[1] B.P. Bewlay, M.R. Jackson, J.-C. Zhao, and P.R. Subramanian, *Metall. Mater. Trans.* A, **34A**, 2043-2053 (2003).
[2] S. Miura, M. Aoki, Y. Saeki, K. Ohkubo, Y. Mishima, and T. Mohri, *Metall. Mater. Trans.* A, **36A**, 489-496(2005).
[3] S. Miura, T. Hatabata, and T. Mohri, Plasticity 2012 Proceedings, p118-120(2012).
[4] S. Miura, T. Hatabata, and T. Mohri, Creep 2012 Proceedings(2012).
[5] S. Miura et al, *Metall. Mater. Trans.* A, submitted.
[6] M.J. Leadbetter, B.B. Argent, *J. Less-Common. Metals.*, **3**, 19-28(1961).
[7] J.I. Bluhm, *Eng. Fract. Mech.*, **7**, 593-604 (1975).
[8] D. Munz, R.T. Bubsey, and J.L. Shannon, Jr., *J. Am. Ceram. Soc.*, **63**, 300-305(1980).
[9] D. Munz, G. Himsolt, and J. Eschweiler, *J. Am. Ceram. Soc.*, **63**, 341-42(1980).
[10] K. Takashima, Y. Higo, *Fatigue. Fract. Engng. Mater. Struct.*, **28**, 703–710(2005).
[11] J.R. Desifano, L.D. Chitwood, *J. Nucl. Mater.*, **295**, 42-48 (2001).

Mater. Res. Soc. Symp. Proc. Vol. 1516 © 2012 Materials Research Society
DOI: 10.1557/opl.2012.1682

Role of Thermal Vacancies on Temperature Dependence of Lattice Parameter and Elastic Moduli in B2-type FeAl

Mi Zhao[1], Kyosuke Yoshimi[1], Kouichi Maruyama[1] and Kunio Yubuta[2]
[1]Graduate School of Engineering, Tohoku University, Sendai, 980-8579, Japan.
[2]Institute for Materials Research, Tohoku University, Sendai, 980-8577, Japan.

ABSTRACT

Temperature dependence of the lattice parameter and elastic moduli in Fe-40 and -43Al (at.%) was investigated by high temperature X-ray diffractometry (XRD) and the Electro-Magnetic Acoustic Resonance (EMAR) method. The thermal vacancy concentration was estimated from the activation enthalpy and entropy data of vacancy formation previously reported for FeAl. It was found that both the lattice parameter and the elastic moduli of FeAl have a linear relationship with temperature even in the temperature range where thermal vacancy concentration rapidly increases (above 400 °C), thus suggesting that newly generated thermal vacancies at elevated temperature do not make significant influence on the lattice parameter and the elastic properties of B2-type FeAl.

INTRODUCTION

Iron aluminides has been widely studied as structural materials because of their interesting performance at high temperature. For B2-type FeAl, thermal vacancy behavior is one of the major topics. B2-type FeAl contains a high concentration of thermal vacancies at high temperature and they are easy to be frozen into the material after cooling down due to the high vacancy concentration and low vacancy migration speed [1]. Excess vacancy hardening is well proved by the increasing sample hardness with increasing quenching temperature [2-3]. Chang et al. proved that in a large composition range with 40~51 at.% Al, the hardness increased with increasing aluminum content and the change in hardness with temperature is a good indication of relative vacancy concentration in B2 FeAl [4]. Moreover, the supersaturated vacancies can also result in an increase in the critical resolved shear stress (CRSS) and a decrease in the ductility of single crystals [5]. Xiao and Baker suggested that vacancies control the room temperature mechanical properties of FeAl by comparing the vacancy concentration with the yield strength as a function of Al content [6]. George and Baker proposed a vacancy hardening model to explain the positive temperature dependence of yield stress (strength anomaly) in FeAl at elevated temperature [7]. Supersaturated vacancies in FeAl can be eliminated by intermediate temperature annealing [2]. Nagpal and Baker reported that the hardness of furnace-cooled FeAl samples was apparently reduced after annealed at 400 °C for 118 hours [3]. Yoshimi et al. showed that after annealing at 425 °C for 120 hours, the hardness of Fe-48Al was lower than that of Fe-39Al (at.%) [8]. Since supersaturated vacancies influence some mechanical properties of FeAl as described above, this study focuses on the role of thermal vacancies on the lattice parameter and elastic moduli. The temperature dependence of the lattice parameter and elastic moduli was investigated by high temperature in-situ measurements.

EXPERIMENTAL PROCEDURE

Ingots with the composition of Fe-40 and -43Al (at.%) were prepared by conventional Ar arc-melting using 99.99 wt.% aluminum and 99.99 wt.% iron. Homogenization heat treatment at 1100°C for 24h was preformed on the ingots, and then they were slowly cooled at the rate of 18 K/h and aged at 400 °C for 100h to eliminate supersaturated vacancies. Powder for X-ray diffractometry (XRD) with the size less than 46μm in diameter was made by crushing the ingots in a tungsten mortar. The recovery temperature of as-crushed powder was determined by the change in the Full Width at Half Maximum (FWHM) of the <011> peak. Both residual strain and excess vacancy elimination heat treatments were performed before lattice parameter measurements. The temperature dependences of the lattice parameter and elastic moduli were obtained from high-temperature XRD and the Electro-Magnetic Acoustic Resonance (EMAR) method.

RESULTS

Temperature dependence of the lattice parameter of FeAl

The Co-Kα X-rays were produced by D8 ADVANCE (Bruker AXS) operated at 35kV, 40mA. High temperature measurements were done in the HTK1200N high temperature chamber (Anton Parr). The measurements were done in the vacuum condition in the order of 10^{-5} mbar. Residual strain was stored in the powder during the crushing process, which was reflected in the width of the diffraction peaks. Figure 1 is the high temperature XRD profiles of as-crushed powder of FeAl. At room temperature, where the powder contained the highest level of residual strain, the XRD profile showed the broadest peaks. With increasing measurement temperature, all the peaks became narrower and the <001> B2 superlattice diffraction peak became evident. This behavior indicates the removal of residual strain. Figure 2 illustrates the FWHM values of the strongest peak (the <011> peak) from 400°C to 800°C. The FWHM value decreased drastically and then reached a stable level of about 0.08° at the recovery temperatures, which are marked by arrows. It is 500°C for Fe-40Al and 600°C for Fe-43Al.

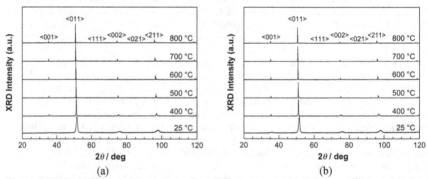

(a)   (b)

Figure 1. XRD profiles of as-crushed powder of (a) Fe-40 and (b) -43Al as a function of testing temperature.

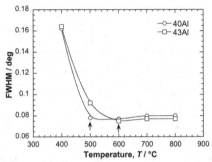

Figure 2. FWHM values of the <011> peak of Fe-40, and -43Al from 400°C to 800°C.

The powders for the lattice parameter measurements were first annealed at their recovery temperature for 1 hour to eliminate the residual strain and then further annealed at 400°C for 100 hours to eliminate supersaturated vacancies. The measurements were carried out from 25°C to 1000 °C. Before each measurement, the sample was kept at the corresponding temperature for at least 30min. Figure 3 shows room temperature and high temperature XRD profiles of FeAl measured together with a standard sample (Al_2O_3). Precise lattice parameter values were calculated by TOPAS4 software (Bruker AXS) with the Whole Powder Pattern Decomposition (WPPD) method. Figure 4 is the estimated lattice parameter from 25°C to 1000°C. The Good of Fitness (GOF) value for curve-fitting was below 1.1 and the experimental error was around ±0.001Å. A good linear relationship between the lattice parameter and temperature is found in the figure. This linear relationship is consistent with a classical theory [9], in which the mean interatomic separation increases linearly with increasing temperature (equation 1).

$$\bar{x} = \frac{3gk_BT}{4c^2} \tag{1}$$

Where g, c and k_B are positive constants and $\bar{x}$ is the average displacement of atoms at a temperature T in the two-atom model.

The linear relationship was also found by K. Ho et al. [10] and M. Kogachi et al. [11], in which the lattice parameter of FeAl was measured in-situ as a function of temperature with a high temperature powder camera or a neutron diffraction measurement, respectively. On the other hand, Y. Yang et al. [12] and M. Kupka [13] observed a slight decrease in lattice parameter of FeAl after quenched from above 700 °C. The different tendencies of lattice parameter change between the in-situ high temperature measurements and the quenched sample measurements should be paid attention to during the vacancy concentration estimation.

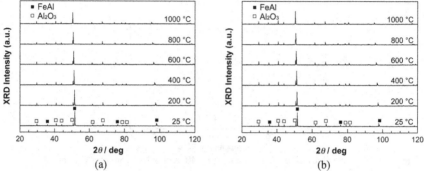

Figure 3. XRD profiles of Fe-40 (a) and -43Al (b) as a function of testing temperature.

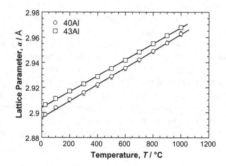

Figure 4. Lattice parameter of Fe-40 and -43Al as a function of testing temperature.

Elastic moduli by EMAR method

The EMAR method is one of the methods using the ultrasonic wave induced by an electric burst, in which the elastic constants are measured in a non-contacting way. Equation 2 is the fundamental equation for the elastic moduli analysis in EMAR [14, 15].

$$L = \frac{1}{2}\int_{\Omega}(\rho\omega^2\delta_{ij}u_iu_j - c_{ij}\varepsilon_i\varepsilon_j)dV \qquad (2)$$

Where L is the Lagrangian Function, ρ is the density of sample, ω is the resonance frequency, u_i and u_j are the displacement vector, c_{ij} is the elastic constants, and ε_i and ε_j are the elastic strain. The value of the Lagrengian function should be zero at resonance vibrations, so that c_{ij} can be calculated from the certain relationship with ω. Figure 5 shows the schematics of the EMAR measurement. The sample with the size of 5mm × 5mm × 5mm is put between two strong magnets and surrounded by a coil. The ultrasonic wave of the sample is excited when the electricity bursts through the coil and the wave in return comes to the coil to be transformed into digital signals. There are two kinds of vibration mode in this method, i.e. Ag and B1g modes.

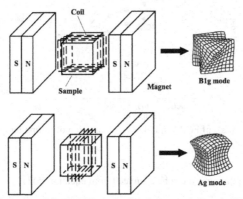

Figure 5. Schematic of the EMAR method and the two induced vibration modes.

In fact, the elastic moduli were obtained from an inverse calculation. The accuracy of the results was expressed by the root mean square (*rms*) value as shown in equation 3.

$$rms = \sqrt{\frac{\left(\frac{F_{cal} - F_{mea}}{F_{cal}}\right)^2}{N}} \tag{3}$$

Where N refers the total number of the resonance peaks, F_{mea} is the measured resonance frequency, and F_{cal} is the resonance frequency calculated from a certain c_{ij}. The c_{ij} was adopted only if *rms* <1%. The measurements were carried out from room temperature to 800°C in the frequency range between 350 and 1200kHz in a magnetic field of 0.5Tesla. Figure 6 shows the results of the measurements. Similar to the lattice parameter, the elastic moduli of FeAl and temperature also showed a good linear relationship. Early studies also showed the linear relationship [16] (equation 4).

$$c_{ij} = c_{ij}(0) - \frac{s}{t}T \tag{4}$$

Where c_{ij} is the elastic constant, $c_{ij}(0)$ is the elastic constant value at 0 K, T is the absolute temperature, s and t are constants. The experimental observations are in good agreement with the model.

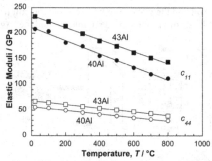

Figure 6. c_{11} and c_{44} of Fe-40 and -43Al as a function of testing temperature.

DISCUSION

A vacancy concentration, C_v, can be estimated from equation (5) [17].

$$C_V = \exp\left(\frac{S_V^F}{k_B}\right)\exp\left(-\frac{H_V^F}{k_B T}\right) \tag{5}$$

Where S_V^F and H_V^F are the activation entropy and enthalpy of vacancy formation, k_B is the Boltzmann constant and T is the absolute temperature. Using the data of S_V^F =5.7 k_B and H_V^F =0.98eV in Fe-39Al by Würchum et. al [1], thermal vacancy concentration as a function of temperature was drawn for FeAl in figure 7. The vacancy concentration exponentially increases above 400°C. However, as to the temperature dependence of the lattice parameter and elastic moduli, no visible change was found around 400°C and above. Although a large amount of thermal vacancies are generated at elevated temperature, the lattice expansion and the elastic properties in FeAl were not influenced by these vacancies.

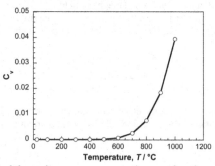

Figure 7. Calculated thermal vacancy concentration as a function of temperature [1].

CONCLUSIONS

Using high temperature in-situ measurements, both the lattice parameter and the elastic moduli of FeAl have linear relationship with temperature. The high thermal vacancy concentrations generated at high temperature do not make significant difference to the lattice parameter or elastic properties of FeAl.

REFERENCES

1. R. Würschum, C. Grupp and H. E. Schaefer, Phys. Rev. Lett., 75 (1995), 97-100.
2. J. Rieu and C. Goux, Mem. Sci. Rev. Metall., 65 (1969), 869-880.
3. P. Nagpal and I. Baker, Metall. Trans. A, 21A, (1990), 2281-2282.
4. Y. A. Chang, L. M. Pike, C. T. Liu, A. R. Bilbrey and D. S. Stone, Intermetallics, 1 (1993), 107-115.
5. K. Yoshimi, Y. Saeki, M. H. Yoo and S. Hanada, Mater. Sci. Eng. A, 258 (1998), 75-83.
6. H. Xiao and I. Baker, Acta Metall. Mater., 43 (1995), 391–396.
7. E. P. George and I. Baker, Phil. Mag. A, 77 (1998) 737-750.
8. Yoshimi, N. Matsumoto, S. Hanada and S. Watanabe, Proc. 3rd Japan Inter. SAMPE Symp., ed. by M. Yamaguchi and H. Fukutomi, (1993), 1404-1409.
9. C. Kittel, Introduction to Solid State Physics, Eighth Edition, (New York: John Wiley & Sons, Inc. Publisher, 2005), 120-121.
10. K. Ho and R. A. Dodd, Scripta Metall., 12 (1978), 1055–1058.
11. M. Kogachi, T. Haraguchi and S. M. Kim, Intermetallics, 6 (1998), 499-510.
12. Y. Yang and I. Baker, Intermetallics, 6 (1998), 167-175.
13. M. Kupka, J. Alloys Compd., 437 (2007), 373-377.
14. I. Ohno, J. Phys. Earth, 24 (1976), 355-379.
15. R. Tarumi, K. Shiraishi and M. Hirao, ISIJ International, 49 (2009), 1432-1435.
16. Y. P. Varshni, Phys. Rev. B, 2 (1970), 3952-3958.
17. A. C. Damask and G. J. Dienes, in: Point Defects in Metals, (New York: Gordon and Breach Science Publisher, 1963).

Mater. Res. Soc. Symp. Proc. Vol. 1516 © 2013 Materials Research Society
DOI: 10.1557/opl.2013.359

Compression Deformation of Single-Crystal Pt₃Al with the L1₂ Structure.

Yoshihiko Hasegawa, Norihiko L. Okamoto and Haruyuki Inui

Department of Materials Science and Engineering, Kyoto University, Sakyo-ku, Kyoto 606-8501, Japan

ABSTRACT

The plastic deformation behaviour of single crystals of Pt₃Al with a chemical composition of Pt-27 at.%Al was investigated in compression from 77K to 1,273K. The L1₂ structure is not stable below around 220 K, transforming into either D0c, D0c' or Pt₃Ga structures. {001} slip system was operative for most loading axis orientations while {111} slip system was operative for a narrow orientation region close to [001]. The CRSS for {111} slip gradually decreases in the temperature range where the L1₂ structure is stable, followed by a sharp decrease above 1,073 K, without showing positive temperature dependence. On the other hand, the CRSS for {001} slip gradually decreases below 673 K and moderately increases above 673 K. Dislocations on {111} tend to align along their screw orientation, suggesting high Peierls stress for their motion, while those on {001} are edge-oriented and fairly curved on a local scale, suggesting relatively low Peierls stress. Dislocations on both {111} and {001} with a Burgers vector $b = [\bar{1}01]$ dissociate into two collinear superpartials with $b = 1/2[\bar{1}01]$ separated by an anti-phase boundary.

INTRODUCTION

In most L1₂ compounds such as Ni₃Al, Ni₃Si and Co₃Ti, anomalous positive temperature dependence of yield stress is observed at high temperatures [1-3]. However, there are some other L1₂ compounds that do not exhibit the anomaly. Many of them are known instead to exhibit a large negative temperature dependence of yield stress at low temperatures. Pt₃Al is usually cited as the representative L1₂ compound for this second category [4,5]. The difference in the temperature dependence of yield stress of the L1₂ compounds of the two categories has been understood in terms of the dissociation scheme of the dislocation with b (Burgers vector) = $<\bar{1}01>$ and planarity of the core structure of the resultant partial dislocations [6]. For those L1₂ compounds that exhibit the yield stress anomaly at high temperatures, the dislocation with $b = <\bar{1}01>$ is believed to dissociate into two collinear superpartials with $b = 1/2<\bar{1}01>$ separated by an anti-phase boundary (APB). The core of each superpartial is believed to be planar and glissile. The yield stress anomaly is generally believed to be caused by the thermally activated cross-slip of these APB-coupled dislocations from the (111) octahedral to (010) cube planes. For those L1₂ compounds that exhibit a large negative temperature dependence of yield stress at low temperatures, on the other hand, the dislocation with $b = <\bar{1}01>$ is believed to dissociate into two superpartials with $b = 1/3<112>$ separated by a superlattice intrinsic stacking fault (SISF). The core of the SISF-coupled superpartials is believed to be non-planar and sessile [6]. Then, the Peierls stress for the motion is expected to be high and, since thermal activation is needed for the motion, the yield stress is considered to strongly depend on temperature at low temperatures. However, some first-principles calculations indicate that the SISF dissociation is unstable compared with the APB dissociation in Pt₃Al [7,8]. On top of that, since no TEM study was made on the dislocation structures for Pt$_3$Al, nothing is known about the dislocation mechanism including the dissociation schemes for the large negative temperature dependence for yield stress at low temperatures. Therefore, the purpose of this study is to reveal the mechanism of the negative temperature dependence of yields stress at low temperatures for Pt$_3$Al through the observation of the deformation microstructure using a transmission electron microscope (TEM).

EXPERIMENTAL PROCEDURES

Ingot with a nominal composition of Pt-27 at.% Al were prepared by arc-melting under an argon gas flow and were re-melted at least five times to ensure the homogeneity of the as-cast ingot. Single crystals of Pt$_3$Al were grown from these ingots by the Bridgman method in an alumina crucible at a growth rate of 8.3 mm/h. Measurements of thermal expansion were carried out by a Shimazdu TMA-60 in the temperature range from 89 to 570 K at a heating rate of 10 K/min. Compression tests were conducted using samples with dimensions of 1.5×1.5×4.5 mm^3 on an Instron-type testing machine from liquid nitrogen temperature to 1,273K at a strain rate of 10^{-4} s^{-1}. Compression tests at low temperature were performed with a sample immersed in liquid nitrogen (77K). Compression tests above room temperature were conducted in vacuum. The compression axis orientations were [$\bar{1}$ 2 12], [$\bar{1}$26], and [$\bar{2}$34] as shown in figure 1. Operative slip planes were determined by slip trace analysis on two orthogonal surfaces by optical microscopy. Dislocation structures were observed with a JEM-2000FX TEM operated at 200 kV. Selected-area electron diffraction (SAED) patterns were taken using a cryo-TEM stage (Gatan, HCHDT-3010) in the temperature range from 80 K to room temperature, to detect a possible phase transformation. Thin foils for TEM observations were cut parallel to the macroscopic slip plane (either (111) or (001)) and were subjected to ion milling with 4 keV Ar ions for perforation.

RESULTS AND DISCUSSION

Phase stability

SAED patterns of Pt$_3$Al with the composition of Pt-27 at.%Al taken along the [001], [110] and [111] incident directions are consistently indexed as those from the L1$_2$ structure. However, as the L1$_2$ phase of Pt$_3$Al has been reported to transform to tetragonal D0c and D0c' phases at low temperatures [9], the thermal stability of the L1$_2$ structure of Pt-27 at.%Al was investigated by the measurement of thermal expansion and the inspection of SAED patterns at low temperatures. The thermal expansion of Pt$_3$Al with a composition of Pt-27 at.%Al is plotted in figure 2 as relative elongation with respect to the sample length at 89 K. On heating, a discontinuous increase occurs at around 220 K in the relative elongation versus temperature curve, indicating the occurrence of phase transformation. Moreover, some additional diffraction spots appear in the SAED pattern taken along the [110] direction after cooling down to 173 K. The SAED pattern can be assigned to either D0c, D0c' or Pt$_3$Ga structures, taking it into account that orientation domains are expected to form during the cubic (L1$_2$) to tetragonal phase transformation. Another inspection of SAED patterns along other directions such as <111> is necessary to determine which tetragonal structure Pt$_3$Al with a composition of Pt-27 at.%Al possesses at low temperatures.

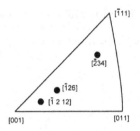

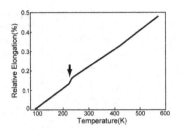

Figure 1. Stereographic projection of compression axis orientations investigated in the present study.

Figure 2. Thermal expansion of Pt-27 at.%Al plotted as relative elongation with respect to the sample length at 89 K.

Deformation behaviour

The temperature dependence of yield stress deduced from the 0.2% off-set stress are shown in figure 3. At 77 K where the L1$_2$ phase is no longer stable, failure is observed to occur soon after yielding at a plastic strain level of ~0.3% for the [$\bar{1}$ 2 12] orientation. The yield stress decreases very rapidly with increasing temperature at low temperatures for both the [$\bar{2}$ 3 4] and [$\bar{1}$ 2 12] orientations, if we extend the plot down to the low temperature region where the L1$_2$ phase is no longer stable. At room temperature and above, however, the temperature dependence of yield stress is not so significant, although there is a general tendency that the yield stress decreases with increasing temperature. Exceptionally, a slight increase in yield stress with temperature is observed above 673 K for the [$\bar{2}$ 34] orientation.

Slip lines observed for [$\bar{1}$ 2 12]-oriented samples are wavy and exclusively those corresponding to {111} slip at temperatures above 673 K as seen in figures 4(c,d), suggesting the frequent occurrence of cross slip among {111} planes. When the deformation temperature is below 673 K, slip tends to occur also on (001) plane as seen in figures 4(a,b). For the [$\bar{1}$ 2 12] orientation, the maximum Schmid factor (0.463) for (111) octahedral slip is 2.7 times larger than that (0.171) for (001) cube slip. This indicates that the CRSS for (111) slip is much higher than that for (001) slip. No strong anomalous positive temperature dependence of yield stress (CRSS) is observed in the high temperature region for (111) slip, which is quite different from that of other L1$_2$ compounds such as Ni$_3$Al [1]. On the other hand, slip lines observed for the [$\bar{2}$ 3 4] and [$\bar{1}$26] orientations are straight and exclusively those corresponding to (001) slip as seen in figures 4(e-h) and 4(i,j), respectively. Although the ratio of Schmid factors for slip on (001) to that for slip on (111) is quite different for the two orientations investigated (0.635 and 1.16, respectively, for [$\bar{1}$26] and [$\bar{2}$ 34] orientations), the values of CRSS coincide with each other within 6 % at room temperature. Therefore, the Schmid law holds true for slip on (001). The CRSS for (001) slip decreases with increasing temperature at temperatures below around 673 K and then increases to a small degree up to around 1073 K, followed by a sharp decrease at higher temperatures. Similar temperature dependence of CRSS and features of slip lines for both {111} and {001} slips for Pt$_3$Al have been reported by Wee et al. [6] and Heredia et al. [7].

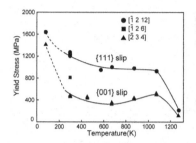

Figure 3. Temperature dependence of yield stress for Pt-27 at.%Al single crystals for three compression axis orientations: [$\bar{1}$ 2 12], [$\bar{1}$26], and [$\bar{2}$34].

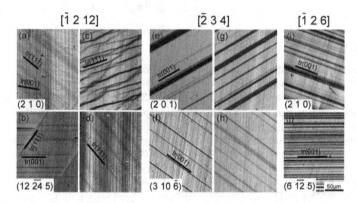

Figure 4. Slip lines observed on two orthogonal faces of [$\bar{1}$ 2 12]-oriented samples deformed (a,b) at room temperature and (c,d) at 673 K, [$\bar{2}$34]-oriented samples deformed (e,f) at room temperature, and (g,h) at 673 K, and a [$\bar{1}$26]-oriented sample deformed (i,j) at room temperature. The compression axis is in the vertical direction in the figure.

Figure 5(a) shows the typical dislocation structures observed in a [$\bar{1}$ 2 12]-oriented sample deformed at room temperature. Dislocations on the (111) slip plane tend to align along their screw orientation. Additionally, many cusps and prismatic loops are observed. These features represent that frequent cross-slip between {111} planes takes place. The dislocation structure observed for {111} slip in Pt-27 at.%Al is quite similar to that of bcc metals. It is considered that the core of dislocations on the {111} slip plane is non-planar, causing a high Peierls stress for their motion.

The dislocation dissociation scheme was confirmed by the weak-beam dark-field method. Two-fold dissociated dislocations are visible under the reflection vectors $g = \bar{2}02$ and $20\bar{2}$ as shown in figures 6(a) and (b), respectively. The dissociation width between the two superpartials is constant when inverting the g vectors. On the other hand, the two superpartials are completely out of contrast under $g = 020$ and $1\bar{1}1$ as shown in figures 6(c) and (d), respectively. From these observations, it turned out that dislocations with $b = [\bar{1}01]$ on (111) slip plane are dissociated into two collinear superpartials with $b = 1/2[\bar{1}01]$ separated by an APB, which is different from the dislocation dissociation scheme that has been believed to occur in Pt$_3$Al previously, that is to say, SISF-type dissociation [6,10-12].

Figure 5(b) show the typical dislocation structure observed in a $[\bar{2}34]$-oriented sample deformed at room temperature. Dislocations on the (001) slip plane are edge-oriented and fairly wavy on a local scale. Therefore, it is considered that the core of dislocations on the {001} slip plane is planar, causing a relatively low Peierls stress. Contrast analysis of weak-beam dark-field images indicates that the dislocations with $b = [\bar{1}10]$ on (001) plane are also dissociated into two collinear superpartials with $b = 1/2[\bar{1}10]$ separated by an APB. First-principles calculations indicated that the core of superpartial dislocations bounding an APB on the (001) plane is spatially extended onto the (111) and (11$\bar{1}$) planes, which have a common axis with the (001) slip plane, when the complex stacking fault (CSF) energy is too high [6]. However, our observations of the fairly wavy dislocations on the (001) plane suggest that the core of the superpartial dislocations might be spread on the (001) plane.

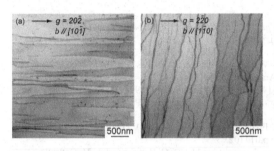

Figure 5. Dislocation structures observed in (a) a $[\bar{1}\,2\,12]$-oriented sample (BD: ~[111], $g = 20\bar{2}$), and (b) a $[\bar{2}34]$-oriented sample (BD: ~[001], $g = 2\bar{2}0$) deformed at room temperature. Thin foils were cut parallel to (111) and (001), respectively.

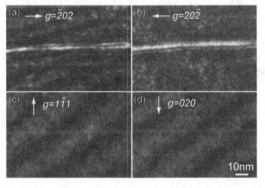

Figure 6. Weak-beam images of dislocations observed in a $[\bar{1}\,2\,12]$-oriented specimen deformed at room temperature. The reflection vectors used for imaging are indicated in each of the images.

CONCLUSIONS

Deformation behaviour of single crystals of Pt_3Al with the chemical composition of Pt-27 at.%Al was investigated. Thermal expansion measurement and TEM observation indicate that the $L1_2$ structure is no longer stable below around 220 K. Pt_3Al shows a negative (normal) temperature dependence of yield stress at low temperatures, which is quite different from other $L1_2$ intermetallics such as Ni_3Al. Dislocations on the $\{111\}$ slip plane tend to align along their screw orientation and are dissociated into two collinear superpartials with $b = 1/2[\bar{1}01]$ separated by an APB, which suggests the non-planar core of dislocations on $\{111\}$, causing a high Peierls stress for their motion. On the other hand, dislocations on the $\{001\}$ slip plane are fairy wavy on a local scale, inferring the planar core of dislocations on $\{001\}$.

ACKNOWLEDGMENTS

This work was supported by JSPS KAKENHI grant number 24246113 and the Elements Strategy Initiative for Structural Materials (ESISM) from the Ministry of Education, Culture, Sports, Science and Technology (MEXT) of Japan.

REFERENCES

1. J. H. Westbrook, in *Dislocations in Solids Vol. 10*, edited by F. R. N. Nabarro and M. S. Duesbery (Elsevier, Amsterdam, 1996), p. 1.
2. T. Takasugi, S. Watanabe, O. Izumi and N. K. Fat-Halla, Acta Metall. 37 (1989) p.3425.
3. T. Takasugi, S. Hirakawa, O. Izumi, S. Watanabe, Acta metall. 35(1987) p.2015.
4. D. M. Wee, D. P. Pope and V. Vitek, Acta Metall. 32 (1984) p.829.
5. F. E. Heredia, G. Tichy, D. P. Pope and V. Vitek, Acta metal. 37 (1989) p.2755.
6. V. Vitek and V. Paidar, in *Dislocations in Solids Vol. 14*, edited by J. P. Hirth (Elsevier. Amsterdam, 2008), p. 441.
7. Y. N. Gornostyrev, Y. Kontsevoi, A. J. Freeman, M. I. Katsnelson, A.V. Trefilov and A.I. Lichtensthtein, Phys. Rev. B 70 (2004) p.014102.
8. A. T. Paxton, in *Electron Theory in Alloy Design*, edited by D. G. Pettifor and A. H. Cottrell (Institute of Materials, London, 1992), p. 158.
9. H. R. Chauke, B. Minisini, R. Drautz, D. Nguyen-Manh, P.E. Ngoepe and D. G. Petiffor, Intermtallics 18 (2010) p.417.
10. V. Vitek, D. P. Pope and M.Yamaguchi, Scripta Metall. 15 (1981) p.1029.
11. M. Yamaguchi, V. Paidar, D. P. Pope and V. Vitek, Philos. Mag. A 45 (1982) p.867.
12. V. Paidar, M. Yamaguchi, D. P. Pope and V. Vitek, Philos. Mag. A 45 (1982) p.883.

Mater. Res. Soc. Symp. Proc. Vol. 1516 © 2013 Materials Research Society
DOI: 10.1557/opl.2013.188

Crystal Structure Evolution of La$_2$Ni$_7$ during Hydrogenation

Yuki Iwatake, Kyosuke Kishida and Haruyuki Inui
Department of Materials Science and Engineering, Kyoto University,
Sakyo-ku, Kyoto 606-8501, Japan

ABSTRACT

Atomic scale characterization of the La$_2$Ni$_7$ hydrides by high-angle annular dark-field scanning transmission electron microscopy (HAADF-STEM) revealed that not only the anisotropic expansion of the La$_2$Ni$_4$ unit layer previously reported but also the shearing on the basal plane of the La$_2$Ni$_4$ unit layers occur during one-cycle of hydrogen absorption/desorption process. Two different types of orthorhombic La$_2$Ni$_7$ hydrides with the same atomic arrangement of La and different atomic arrangement of Ni were observed depending on the maximum hydrogen concentration achieved during one hydrogen absorption/desorption cycle.

INTRODUCTION

(RE,Mg)$_2$Ni$_7$-based hydrogen-absorbing alloys (RE: Rare Earth) have attracted a great deal of interest as negative electrode materials in nickel-metal-hydride (Ni-MH) batteries with very low self-discharge property and high capacity. Such attractive properties are closely related to the crystal structure of (RE,Mg)$_2$Ni$_7$ phase and its stability upon hydrogen absorption/ desorption cycles. For further development of a new class of hydrogen-absorbing materials for negative electrodes of Ni-MH batteries, it is essential to understand the details of crystal structure evolution during hydrogen absorption/desorption cycles. Most of La-Ni binary intermetallic compounds such as LaNi$_3$, La$_2$Ni$_7$ and La$_5$Ni$_{19}$ possess common block-stacking structures, in which each block layer is composed of one La$_2$Ni$_4$ (Laves-type) unit layer and one or more LaNi$_5$ (Haucke-type) unit layers stacking along the c-axis (figure 1) [1]. La$_2$Ni$_7$ with the Ce$_2$Ni$_7$-type hexagonal structure (space group: $P6_3/mmc$, #194) is made with two block layers, each consisting of one La$_2$Ni$_4$-unit layer and two LaNi$_5$-unit layers [2]. It has long been believed that reversibility of the hydrogen absorption/desorption behavior of La$_2$Ni$_7$ is poor because of its amorphization upon hydrogenation, therefore structural variation during hydrogen absorption/desorption cycles has not been studied in detail for a long time [3]. Recently, good reversibility of hydrogen storage for La$_2$Ni$_7$ has been confirmed and crystal structural variation during hydrogenation has been studied by powder x-ray and neutron diffraction, respectively [4-6]. Common characteristics suggested by the previous studies is that hydrogen is preferentially absorbed in the La$_2$Ni$_4$ unit layer at low pressure and the hydride formed at this stage is stable after desorption. The crystal structure of the hydride has been proposed to be characterized by the preferential and anisotropic expansion of the La$_2$Ni$_4$ unit layers along the stacking direction [5,6]. However, the details of the crystal structure of the La$_2$Ni$_7$ hydride are still controversial. In this study, the structural change of La$_2$Ni$_7$ phase during hydrogenation was investigated by atomic-resolution high-angle annular dark-field scanning transmission electron microscopy (HAADF-STEM) and transmission electron microscopy (TEM).

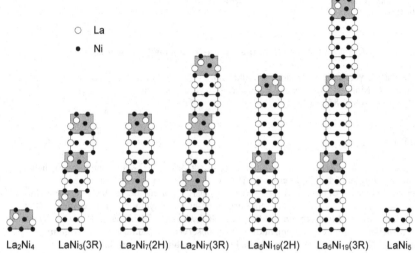

Figure 1. Block-stacking structure of some intermetallic compounds in the La-Ni binary system. Open and filled circles indicate La and Ni atoms, respectively.

EXPERIMENTAL PROCEDURES

Ingots of binary stoichiometric La_2Ni_7 were prepared by arc melting high purity lanthanum (99.9%) and nickel (99.99%). The as-prepared ingots were heat-treated at 930 °C for 168 h in vacuum and then furnace-cooled. Pressure-composition (P-C) isotherms were measured with a Sieverts apparatus at 30 °C after crushing these ingots into powder with a particle size of approximately 20-50 μm. Powder specimens were preheated in vacuum at 80 °C for 2 h in prior to the measurement of P-C isotherms. Two types of specimens absorbed hydrogen up to two different hydrogen concentrations of 0.5H/M and 1.2H/M and desorbed were prepared for TEM/STEM observations. Thin foils for TEM observations were prepared by a mechanical polishing powder specimens dispersed in epoxy resin and finished by Ar-ion milling. Microstructures of specimens were examined with a JEM-2000FX transmission electron microscope (TEM) and a JEM-2100F scanning transmission microscope (STEM) equipped with a field emission gun, both operated at 200kV.

RESULTS AND DISCUSSION

Figure 2 shows a typical P-C isotherm of one cycle of hydrogen absorption/desorption of La_2Ni_7 up to 4.5MPa as a function of the hydrogen-to-metal atomic ratio (H/M). The P-C isotherm exhibit at least three plateaus, among which those in a region from 0 to 0.5 H/M could not be detected since the pressure level is lower than the detection limit of our apparatus used in

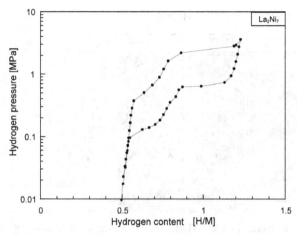

Figure 2. P-C isotherm of La$_2$Ni$_7$ for the 1cycle of hydrogen absorption-desorption at 30 °C.

this study. We hereafter assumed the plateau in this region as the first plateau and the hydride formed at this stage is assigned as a β-hydride. La$_2$Ni$_7$ absorbed hydrogen up to 1.2H/M in total and about 0.5H/M of hydrogen remained after desorption since the equilibrium pressure on the first plateau is too low. Considering the amount of remaining hydrogen, the β-hydride is likely to exist after 1cycle hydrogen absorption/ desorption up to 1.2H/M at 30 °C. However, detailed TEM analysis, which will be described later, revealed that the hydride formed at the first plateau and that remained after 1 cycle of hydrogen absorption/desorption up to 1.2H/M possess different crystal structures. Thus, the latter hydride remained after 1 cycle of hydrogen absorption/desorption up to 1.2H/M is assigned as a β'-hydride throughout this paper.

TEM and STEM observations were carried out for both the β and β'-hydrides in order to determine their crystal structures. Atomic-resolution high-angle annular dark-field (HAADF)-STEM images taken from the La$_2$Ni$_7$ phase and its β'-hydride are shown in figure 3. In the atomic resolution HAADF-STEM images, bright spots directly correspond to the position of the atomic columns with the strong Z-dependence of the contrast [7,8]. Since the atomic number of La ($Z = 57$) is much larger than that of Ni ($Z = 28$), only the positions of La atomic columns can be directly determined from the atomic resolution HAADF-STEM images obtained with the TEM used in this study. Careful inspection revealed that the β'-hydride possesses a unit cell with the c-axis about 20 %-longer than that of parental La$_2$Ni$_7$ phase, while the a-axis remains almost unchanged. In details, the c-axis of the La$_2$Ni$_4$ unit layer in the β'-hydride is confirmed to be about 57% larger than that of La$_2$Ni$_7$. Such characteristics of anisotropic expansion of the unit cell upon hydrogenation are well-consistent with those reported previously [5]. However, it should be noted that two different types of atomic arrangements were observed in the atomic resolution HAADF-STEM images projected along two directions both corresponding to <11$\bar{2}$0> of the parental hexagonal La$_2$Ni$_7$.This clearly indicates that the crystal system of the β'-hydride is not hexagonal but orthorhombic (or lower), which is in marked contrast with the previously proposed models based on powder X-ray and neutron diffraction experiments [5,9].

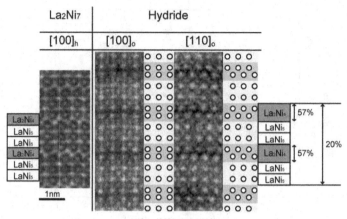

Figure 3. Atomic resolution HAADF-STEM images of the La$_2$Ni$_7$ phase and its β'-hydride. Open circles indicate the positions of La atomic columns.

A new crystal structure model was constructed based on the projected positions of La atomic columns observed in various HAADF-STEM images taken along some low-index orientations of the β'-hydride (figure 4). In the figure, the atomic arrangements of La atoms for two models, namely model 1 previously reported and model 2 determined in this study, are indicated for comparison. If only the atomic positions of La are taken into account, the previously reported structure (model 1) can be described as an alternate stacking of structural blocks in A and B positions as in the case of the parental La$_2$Ni$_7$ with a hexagonal symmetry. In contrast, the model 2 determined in this study possesses a continuous stacking of AA'-type, where the stacking positions A' and B are related with one of three crystallographically identical displacement vectors t_1, t_2 and t_3 indicated in the figure. The relative relations between models 1 and 2 indicate that the shearing of the La$_2$Ni$_4$ unit layers represented by one of the displacement vectors occurs in addition to the anisotropic expansion of the La$_2$Ni$_2$ unit layers during the hydrogenation process of the La$_2$Ni$_7$ phase. Preliminary study has confirmed that the β'-hydride is stable after several cycles of hydrogen absorption/desorption up to 1.2 H/M, which suggests that the high stability of the β'-hydride is likely to be a key factor for good reversibility in hydrogen absorption/desorption of La$_2$Ni$_7$.

The same atomic arrangement of La atoms was confirmed also for the β-hydride observed after 1 cycle of hydrogen absorption/desorption up to 0.5H/M by careful inspection of many atomic resolution HAADF-STEM images taken along various low- index orientations perpendicular to the c-axis. Although the both β- and β'-hydrides are found to possess the same arrangement of La atoms, apparent differences were seen in some selected area diffraction patterns. Figure 5 shows a typical example of the differences in the SAED patterns taken along direction corresponding to <11$\bar{2}$0> of the parental La$_2$Ni$_7$ phase. It is obvious that the diffraction spots of [0kl]* ($k = 2n + 1$, n: integers) are observed only in figure 5(b) taken from the β-hydride. This clearly indicates that the crystal structures of the β- and β'-hydrides are different from each

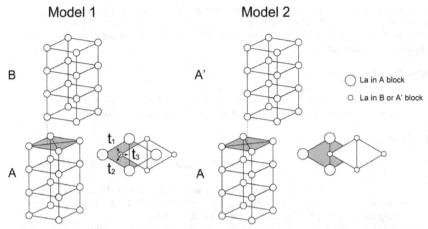

Figure 4. Block-stacking sequences for two model structures of the β- and β'-hydrides of La$_2$Ni$_7$. Only the atomic positions of La atoms are indicated.

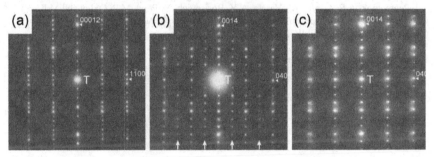

Figure 5. SAED patterns taken from (a) La$_2$Ni$_7$, (b) β-hydride and (c) β'-hydride.

other. Since only the atomic positions for La atoms were considered above, such differences observed in some SAED patterns should be related to the differences in atomic arrangement of Ni atoms for the β- and β'-hydrides. Detailed crystal structure analysis including the atomic positions of Ni atoms is currently on-going and will be published elsewhere.

CONCLUSIONS

Crystal structures of two types of hydrides of La$_2$Ni$_7$ phase after 1 cycle of hydrogen absorption/ desorption up to two different hydrogen concentrations has been investigated by atomic resolution HAADF-STEM and TEM. The preferential expansion about 57% along the *c*-axis as well as the shearing on the basal plane of the La$_2$Ni$_4$ unit layers was found to occur simultaneously upon hydrogenation, which results in the hexagonal to orthorhombic transformation. Two hydrides formed after 1 cycle of hydrogen absorption/desorption up to 0.5H/M and 1.2H/M possess the same atomic arrangement of La atoms, while those of Ni atoms are different from each other.

ACKNOWLEDGEMENTS

This work was supported by Grant-in-Aid for Scientific Research from the Ministry of Education, Culture, Sports, Science and Technology (MEXT), Japan (No. 24656409).

REFERENCE

1. T. Yamamoto, H. Inui, M. Yamaguchi, K. Sato, S. Fujitani, I. Yonezu, and K. Nishio, Acta Mater., **45**, 5213 (1997).
2. K.H.J. Buschow and A.S. Van Der Goot, J. Less-Comm. Met., **22**, 419 (1970).
3. H. Oesterreicher, J. Clintonn, and H. Bittner, Mater. Res. Bull., **11**, 1241 (1976).
4. U.I. Chung and J.Y. Lee, J. Non-Cryst. Solids, **110**, 203 (1989).
5. V.A. Yartys, A.B. Riabov, R.V. Denys, M. Sato, and R.G. Delaplane, J. Alloys Compd., **408-412**, 273 (2006).
6. K. Iwase, K. Sasaki, Y. Nakamura, and E. Akiba, Inorg. Chem., **49**, 8763 (2010).
7. S.J. Pennycook, A.R. Lupini, M. Varela, A.Y. Borisevich, Y. Peng, M.P. Oxley and M.F. Chisholm, in *Scanning Transmission Electron Microscopy for Nanostructure Characterization*, edited by W. Zhou and Z.L.Wang, (Springer, New York, 2006) p.152.
8. K. Kishida and N.D. Browning, Physica C **351**, 281 (2001).
9. K.H.J. Buschow, J. Magn. Magn. Mater., **40**, 224 (1983).

Mater. Res. Soc. Symp. Proc. Vol. 1516 © 2013 Materials Research Society
DOI: 10.1557/opl.2013.63

Effects of Ternary Additions on the Microstructure of Directionally-Solidified
MoSi$_2$/Mo$_5$Si$_3$ Eutectic Composites

Kosuke Fujiwara, Yuta Sasai, Kyosuke Kishida and Haruyuki Inui
Department of Materials Science and Engineering, Kyoto University,
Sakyo-ku, Kyoto 606-8501, Japan

ABSTRACT

Microstructural variations of directionally-solidified (DS) MoSi$_2$/Mo$_5$Si$_3$ eutectic
composites caused by ternary additions of V, Cr, Nb, Ta and W were investigated. Ternary
addition of relatively large amount of V resulted in the apparent modification of the morphology
as well as the macroscopic inclination angle of one type of the interfaces extending nearly along
the growth direction in the DS ingots. Ternary addition of Cr was found to change the
microstructures of the DS ingots from a cellular-type with coarse boundary regions to a plate-like
cellular-type with increasing the amount of the ternary addition. No significant change of script-
lamellar morphology was observed for the DS ingots containing Nb, Ta and W irrespective of
the amount of the ternary addition.

INTRODUCTION

The intermetallic compound MoSi$_2$ with the C11$_b$ structure is one of the promising
structural materials for ultra-high temperature applications due to its very high melting point
(2020 °C) and excellent oxidation resistance [1-5]. However, poor fracture toughness at room
temperature and insufficient high-temperature strength and creep strength have hindered the
wide-spread practical use of the material. Extensive studies have been carried out to overcome
such drawbacks by forming an in-situ composite with a secondary phase [4-9]. Among possible
candidates of the secondary phase, the MoSi$_2$/Mo$_5$Si$_3$ eutectic composites are quite attractive
since they possess high eutectic temperature (1900 °C for binary alloys) and also because fine
microstructures of a script-lamellar type have been reported to be formed simply by various
unidirectional solidification processes (see figure 1) [5]. Only limited studies on microstructures
and mechanical properties of binary MoSi$_2$/Mo$_5$Si$_3$ eutectic composites have been done so far
and revealed that the creep strengths of directionally-solidified (DS) MoSi$_2$/Mo$_5$Si$_3$ eutectic
composites are superior to those of MoSi$_2$ single crystals [8]. Very recently, we prepared various
DS ingots of binary eutectic composite with different thicknesses of both MoSi$_2$ and Mo$_5$Si$_3$
phases by controlling the growth rate during DS process and revealed that high-temperature
strength drastically increases with decreasing the thickness of MoSi$_2$ phase [10]. In addition, the
microstructure and mechanical properties of two-phase composites are generally influenced by
interface properties, such as lattice misfits and segregation behavior of additional ternary
elements. It is thus quite important to establish ways to optimize interface properties of the
MoSi$_2$/Mo$_5$Si$_3$ eutectic alloys. However, the effects of ternary additions on the microstructures
and mechanical properties have not been studied extensively yet. In the present study, we
investigate the effects of some ternary additions on the microstructure of MoSi$_2$/Mo$_5$Si$_3$ eutectic
composites especially focusing on the variation of lattice misfits by ternary additions.

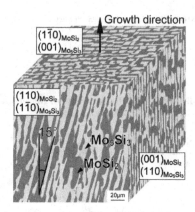

Figure 1. SEM backscattered electron micrograph of a directionally-solidified $MoSi_2/Mo_5Si_3$ eutectic composite grown at a rate of 10 mm/h.

EXPERIMENT

Rod ingots of binary alloys with a $MoSi_2/Mo_5Si_3$ eutectic composition of Mo-54 at.% Si and some ternary alloys with chemical compositions of Mo - 54 at.% Si - x at.% X (X = V, Cr, Nb, Ta and W, x = 1, 2, 5) were prepared by arc-melting. DS ingots were grown from the as-prepared rod ingots by an optical floating zone method at a growth rate of 10 mm/h. Microstructures of DS ingots of the eutectic composites were examined by scanning electron microscopy (SEM) with a JEOL JSM-7001FA equipped with a field emission gun and scanning transmission electron microscopy (STEM) with a JEOL JEM-2100F equipped with a field emission gun operated at 200 kV, respectively. Chemical compositions were analyzed by energy dispersive X-ray spectroscopy (EDS) in the SEM. Lattice parameters for both phases in each alloy were determined by X-ray diffraction (XRD) method in order to estimate the lattice misfits at various interfaces.

RESULTS AND DISCUSSION

Microstructures and lattice misfits

Typical script-lamellar structure of a DS ingot for a binary $MoSi_2/Mo_5Si_3$ eutectic composite grown at a rate of 10 mm/h is shown in figure 1. Detailed XRD and TEM analysis confirmed two component phases were grown so as to maintain the orientation relationships that $[1\bar{1}0]_{MoSi_2}//[001]_{Mo_5Si_3}$ and $(110)_{MoSi_2}//(1\bar{1}0)_{Mo_5Si_3}$ with a minor deviation about 2 to 3 ° in the script-lamellar structure, which are consistent with those previously reported [7]. Growth directions for $MoSi_2$ and Mo_5Si_3 phases in the eutectic composite were nearly parallel to

$[1\bar{1}0]_{MoSi_2}$ and $[001]_{Mo_5Si_3}$. Fine script-lamellar microstructure composed of a continuous $MoSi_2$ matrix and an interconnected network of Mo_5Si_3 was developed with a macroscopic inclination about 15° from the growth direction as clearly seen on the $(110)_{MoSi_2}//(1\bar{1}0)_{Mo_5Si_3}$ surface of the composite. The macroscopic inclination has been reported to be achieved microscopically by so-called ledge-terrace structure, whose terrace parts correspond to the $(001)_{MoSi_2}//(110)_{Mo_5Si_3}$ interfaces [7]. Thus, the lattice misfits for two types of major interfaces, which are of $(001)_{MoSi_2}//(110)_{Mo_5Si_3}$ and $(110)_{MoSi_2}//(1\bar{1}0)_{Mo_5Si_3}$, are considered in the present study. Two kinds of lattice misfits for each interface should be taking into account because of the orientation relationships and tetragonality of both component phases (figure 2). The $(001)_{MoSi_2}//(110)_{Mo_5Si_3}$ interface possesses lattice misfit A (between d_{110} of $MoSi_2$ and d_{330} of Mo_5Si_3) and B (between d_{110} of $MoSi_2$ and d_{002} of Mo_5Si_3), while the $(110)_{MoSi_2}//(1\bar{1}0)_{Mo_5Si_3}$ interface has misfit B and C (between d_{002} of $MoSi_2$ and d_{220} of Mo_5Si_3). The lattice misfits can be expressed by the following equation.

$$\delta = \frac{d_{hkl\ of\ MoSi_2} - d_{hkl\ of\ Mo_5Si_3}}{d_{hkl\ of\ MoSi_2}} \times 100 \tag{1}$$

The amount of lattice misfit A, B and C of binary eutectic composite were determined to be -0.36, -8.1 and 13 %, respectively.

Effects of ternary additions

For the five elements (V, Cr, Nb, Ta and W) investigated in this study, no significant change of script-lamellar morphology was observed for DS ingots containing Nb, Ta and W irrespective of the amount of the ternary addition as shown in figure 3(a). In the case of V-alloyed ingots, an apparent change of morphology for a certain type of interface was observed only for a relatively large amount of ternary addition of 5 at.% V. Curved interfaces were developed and consequently no apparent macroscopic inclination was seen on the

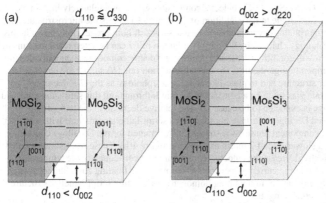

Figure 2. Schematic illustration of lattice misfits in two types of interfaces; (a) $(001)_{MoSi_2}$ // $(110)_{Mo_5Si_3}$ interface and (b) $(110)_{MoSi_2}$ // $(1\bar{1}0)_{Mo_5Si_3}$ interface.

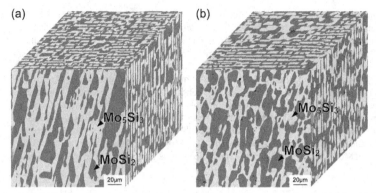

Figure 3. SEM backscattered electron micrographs showing the morphology of (a) 5 at.% Ta-alloyed and (b) 5 at.% V-alloyed DS ingots grown at a rate of 10 mm/h.

$(110)_{MoSi_2}//(1\bar{1}0)_{Mo_5Si_3}$ surface of the 5 at.% V alloyed DS ingot (figure 3(b)), which is in good contrast to the relatively flat interfaces inclined from the growth direction observed on $(110)_{MoSi_2}//(1\bar{1}0)_{Mo_5Si_3}$ surface for the binary and ternary DS ingots with Nb, Ta or W addition. Such variations of the interface morphology are considered to reflect the different variations of the lattice misfits as described later.

A ternary addition of Cr was found to change microstructures of the DS ingots significantly. For the DS ingots containing 1 or 2 at.% Cr, a coarse cellular structure with script-lamellar structure at the central part of each columnar cell and coarse and irregular shaped lamellar at the cell boundary regions was observed (figure 4(a)). Such a cellular structure is also observed in binary DS $MoSi_2/Mo_5Si_3$ eutectic composites grown at a relatively high rate of 100 mm/h. Further increase of the ternary addition of Cr (5 at.% Cr) alters the microstructure into plate-like cellular morphology with relatively coarse grains of both phases (not shown in this paper). Such morphology changes by the ternary addition of Cr can be interpreted in terms of the variations in the growth conditions. Microstructures of eutectic composites generally depend on the ratio of the temperature gradient (G) to the solidification rate (R) so that the lamellar structure changes to cellular structure and to plate-like cellular morphology as the G/R decreases [11]. Since the growth rate was set to be a constant to 10 mm/h throughout this study, the observed morphology change is expected to be caused by the relative decrease in the temperature gradient G for the Cr-alloyed DS ingots. In fact, homogeneous script-lamellar structure with a coarser distribution of the component phases was successfully obtained for the DS alloys containing 2 at.% Cr grown at a slower rate of 2 mm/h as shown in figure 4(b). These results indicate that the Cr addition drastically decrease the eutectic temperature and consequently the temperature gradient G is lowered in the current experimental setting of the floating zone furnace, which results in the relative difficulty in obtaining homogeneous and fine script-lamellar structure for the Cr-alloyed DS eutectic ingots.

Figure 5 shows variations of lattice misfits for V, Cr, Nb, Ta and W alloyed ingots estimated with the equation (1) based on lattice parameters determined by XRD method plotted

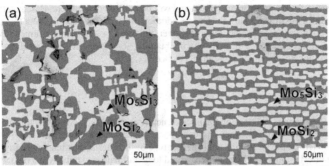

Figure 4. SEM backscattered electron micrographs of the transverse sections of the DS ingots with a nominal composition of Mo - 54 at.% Si - 2 at.% Cr alloy grown at rates of (a) 10 mm/h and (b) 2 mm/h.

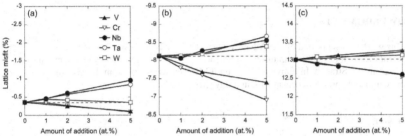

Figure 5. Variations of lattice misfit by V, Cr, Nb, Ta and W additions. (a) Misfit A (between d_{110} of $MoSi_2$ and d_{330} of Mo_5Si_3), (b) misfit B (between d_{110} of $MoSi_2$ and d_{002} of Mo_5Si_3) and (c) misfit C (between d_{002} of $MoSi_2$ and d_{220} of Mo_5Si_3).

as a function of the amount of ternary addition to the DS ingots. For the W-alloyed DS eutectic composites, for which ternary element of W are known to be partitioned into both phases, variations of lattice misfits are relatively small compared to those for the other ternary systems. For the remaining ternary systems containing V, Cr, Nb and Ta, SEM-EDS analysis confirmed that ternary elements are preferentially partitioned into Mo_5Si_3, however, the effects of the ternary elements on the lattice misfits are different from each other. The absolute values of the lattice misfits A and B for the Nb- and Ta-alloyed ingots increase with increasing the amount of the ternary addition, while that of the misfit C decreases. For the V- and Cr-alloyed ingots, the absolute values of the lattice misfits A and B decrease with the increase of the ternary addition, whereas that of the misfit C almost unchanged. Considering the fact that only the morphology of the $(001)_{MoSi_2}//(110)_{Mo_5Si_3}$ interfaces viewed on the $(110)_{MoSi_2}//(1\bar{1}0)_{Mo_5Si_3}$ surface is changed apparently by the 5 at.% addition of V, the variation of the misfit B, which corresponds to the difference between d_{110} of $MoSi_2$ and d_{002} of Mo_5Si_3, is likely to be a controlling factor for the interface morphology. Detailed influences of the reduction of misfit B on the interface microstructures are currently under investigation and will be published elsewhere. Based on the similarity in the lattice misfit variations, the ternary addition of Cr is expected to exhibit the similar influence on the interface morphology. However, the trend has not been confirmed yet for the Cr-alloyed ingots because of the difficulty in obtaining homogeneous script-lamellar structure for the DS ingots containing more than 5 at.% of Cr.

CONCLUSIONS

The effects of ternary additions on the microstructures of DS $MoSi_2/Mo_5Si_3$ eutectic composites and the lattice misfits between two component phases were investigated. The morphology and macroscopic inclination angle of one type of the interfaces extending nearly along the growth direction in the DS ingots were found to be influenced mainly by the reduction of the lattice misfit corresponding to the difference between the interplanar distance of (110) of $MoSi_2$ and that of (002) of Mo_5Si_3. No significant change of script-lamellar morphology was observed for the DS ingots containing ternary elements of Nb, Ta or W.

ACKNOWLEDGMENTS

This work was supported by the Advanced Low Carbon Technology Research and Development Program (ALCA) from the Japan Science and Technology Agency (JST) and in part by the Elements Strategy Initiative for Structural Materials (ESISM) from the Ministry of Education, Culture, Sports, Science and Technology (MEXT), Japan.

REFERENCES

1. K. Ito, H. Inui, Y. Shirai and M. Yamaguchi, *Philos. Mag. A*, **72**, 1075 (1995).
2. H. Inui, K. Ishikawa and M. Yamaguchi, *Intermetallics*, **8**, 1131 (2000).
3. H. Inui, K. Ishikawa and M. Yamaguchi, *Intermetallics*, **8**, 1159 (2000).
4. A. K. Vasudevan and J. J. Petrovic, *Mater. Sci. Eng.*, **A155**, 1 (1992).
5. K. Ito, T. Yano, T. Nakamoto, H. Inui and M. Yamaguchi, *Intermetallics*, **4**, S119 (1996)
6. R. Gibala, A. K. Ghosh, D. C. Van Aken, D. J. Srolovitz, A. Basu, H. Chang, D. P. Mason and W. Yang, *Mater. Sci. Eng.*, **A155**, 147 (1992).
7. D. P. Mason, D. C. Van Aken and J. F. Mansfield, *Acta metall. mater.*, **43**, 1189 (1995).
8. D. P. Mason and D. C. Van Aken, *Acta metal. mater.*, **43**, 1201 (1995).
9. S. Ueno, T. Fukui, R. Tanaka, S. Miura and Y. Mishima, Mater. Trans., **40**, 369 (1999).
10. Y. Sasai, A. Inoue, K. Fujiwara, K. Kishida and H. Inui, in *Intermetallics-Based Alloys – Science and Technology and Applications*, edited by I. Baker, M. Heilmaier, S. Kumar, and K. Yoshimi (Mater. Res. Soc. Symp. Proc., **1516**, Warrendale, PA, 2012), in this volume.
11. I. R. Hughes and H. Jones, *Journal of Materials science*, **11**, 1781 (1976)

Mater. Res. Soc. Symp. Proc. Vol. 1516 © 2013 Materials Research Society
DOI: 10.1557/opl.2013.128

Plastic Deformation of Directionally-Solidified MoSi$_2$/Mo$_5$Si$_3$ Eutectic Composites

Yuta Sasai, Atsushi Inoue, Kosuke Fujiwara, Kyosuke Kishida, and Haruyuki Inui
Department of Materials Science and Engineering, Kyoto University,
Sakyo-ku, Kyoto 606-8501, Japan

ABSTRACT

Deformation behavior of the directionally-solidified MoSi$_2$/Mo$_5$Si$_3$ eutectic composites has been investigated as a function of the average thickness of MoSi$_2$ phase over a temperature range from 900 to 1500°C. The average thickness of both MoSi$_2$ and Mo$_5$Si$_3$ phases in the directionally-solidified ingots with script-lamellar morphologies grown by optical floating zone method decreases with increasing the growth rate. Plastic deformation was observed above 1000°C for all the DS ingots grown at different growth rates when the loading axis is parallel to $[1\bar{1}0]_{MoSi_2}$ close to the growth direction. Yield stress decreases monotonically with increasing temperature. Yield stress at 1400°C increases drastically with decreasing the average thickness of MoSi$_2$ phase.

INTRODUCTION

In the last decades, there has been increasing demands for new structural materials which can be used in oxidizing environments at higher temperatures than the upper limit for Ni-based superalloys. MoSi$_2$ with the C11$_b$ structure has been recognized as one of the most attractive candidates because of its high melting point (2020°C), excellent oxidation resistance, and high thermal conductivity (figure 1a) [1,2]. For the practical application of this material, however, it is essential to improve its low-temperature fracture toughness and high-temperature strength. One possible way is to form an in-situ composite with a secondary phase. Mo$_5$Si$_3$ with the D8$_m$ structure (figure 1b) is one of the candidates because it has excellent creep strength at elevated temperatures and also because directionally-solidified MoSi$_2$/Mo$_5$Si$_3$ eutectic alloys possess very high eutectic temperature (1900°C), a fine script-lamellar microstructure composed of a continuous MoSi$_2$ matrix and an interconnected network of Mo$_5$Si$_3$, and better creep properties than the other MoSi$_2$-based composites [3,4]. The orientation relationship between MoSi$_2$ and Mo$_5$Si$_3$ phases in directionally-solidified eutectic compounds has been reported as $[1\bar{1}0]_{MoSi_2}$ and $[001]_{Mo_5Si_3}$ being approximately parallel with the growth direction, $(110)_{MoSi_2}$ // $(110)_{Mo_5Si_3}$ and $(001)_{MoSi_2}$ // $(1\bar{1}0)_{Mo_5Si_3}$ [3].

Mechanical properties of directionally-solidified (DS) eutectic composites are thought to be affected by the thickness of both component phases in the script-lamellar structures which are likely to be controlled by the growth rate in directional solidification. In the present study, we prepared various DS eutectic alloys grown at various growth rate and investigated the effects of the thickness of the script-lamellar structure on the high-temperature strength of the MoSi$_2$/Mo$_5$Si$_3$ DS eutectic alloys.

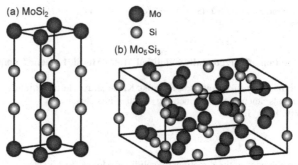

Figure 1. Crystal structures of (a) $MoSi_2$ and (b) Mo_5Si_3.

EXPERIMENT

Ingots of binary $MoSi_2/Mo_5Si_3$ eutectic alloys with a nominal composition of Mo -54 at.% Si were prepared by arc-melting high-purity Mo and Si. Directionally-solidified (DS) ingots were grown by the optical floating zone (FZ) method at various growth rates ranging from 5 to 200 mm/h under an Ar gas flow. Microstructures of these ingots were investigated by the scanning electron microscopy (SEM).

Specimens for compression tests with dimensions of $1.2 \times 1.2 \times 3 \ mm^3$ were sectioned from the as-grown DS ingots by electric discharge machining. Compression tests were carried out along $[1\bar{1}0]_{MoSi_2}$ close to the growth direction at a strain rate of $1 \times 10^{-4} \ s^{-1}$ in vacuum at temperatures ranging from 900 to 1500°C using an Instron-type testing machine. Deformation microstructures were examined by transmission electron microscopy (TEM).

RESULTS AND DISCUSSION

Microstructure

Figure 2 shows three-dimensional views of directionally-solidified binary eutectic alloys grown at two different growth rates of 10 and 100mm/h. At lower growth rates (5~ 50 mm/h), homogeneous script-lamellar structure was observed to be developed (figure 2a). In contrast, cellular microstructure with very fine script-lamellar structure at the central part of each columnar cell and coarse and irregular shaped lamellar at the cell boundary regions was evident in specimens grown at higher growth rates (100 ~ 200 mm/h) (figure 2b). As seen on the $(110)_{MoSi_2} \ // \ (1\bar{1}0)_{Mo_5Si_3}$ surfaces in figures 2a and 2b, Mo_5Si_3 rods with interconnecting branches elongate along directions inclined approximately ±15° from the growth direction in the script-lamellar structure, which is consistent with the previous report by Mason et al. [3]. Since backscattered Laue patterns revealed that the cellular structure also have the same orientation relationship with the homogeneous structure reported by Mason et al.[3], specimens with the cellular structure were also used for compression tests.

Quantitative characterization of the lamellar morphology was performed on the $(1\bar{1}0)_{MoSi_2} \ // \ (001)_{Mo_5Si_3}$ surface of each DS ingot. Volume fractions of Mo_5Si_3 phases were estimated by

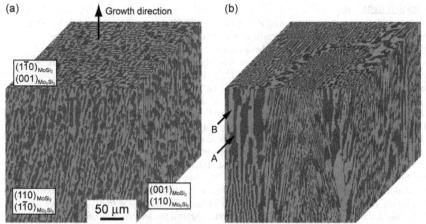

Figure 2. SEM backscattered electron images of directionally-solidified specimens grown at (a) 10 mm/h with a homogeneous script-lamellar microstructure and (b) 100 mm/h with a cellular structure. (Phase identification; A-MoSi$_2$, B-Mo$_5$Si$_3$)

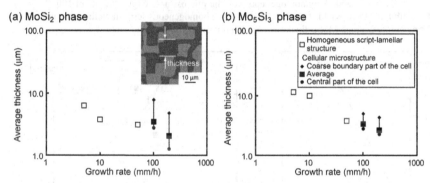

Figure 3. Average thicknesses of (a) MoSi$_2$ and (b) Mo$_5$Si$_3$ phases plotted as a function of the growth rate.

measuring the area fraction of brighter regions in SEM back scattered electron images taking account the script-lamellar morphology. Volume fractions of Mo$_5$Si$_3$ phase were about 47 % for all specimens prepared in this study. Figure 3 shows average thickness values of both the MoSi$_2$ and Mo$_5$Si$_3$ phases plotted as a function of growth rates. For specimens with the cellular structure, average thickness values of the central part of the cells, coarse boundary part and their average are indicated with circles, diamonds and squares, respectively in the figures. It is apparent from the figures that the average thickness values for both the phases decrease with increasing the growth rate. Thus it can be confirmed that finer microstructures can be obtained by increasing the growth rate through the unidirectional solidification process.

Deformation Behavior

Plastic deformation was observed above 1000 °C for all the specimens prepared from the various DS ingots grown at different growth rates. At 900 °C, the fracture of specimen was observed to occur in the elastic region without exhibiting any appreciable plastic flow. Typical stress - strain curves obtained for the specimens from the DS ingot grown at the growth rate of 5mm/h tested in compression at 1000 and 1400 °C up to about 2% plastic strain are shown in figure 4. The stress – strain curves for specimens deformed above 1000 °C generally exhibited yield drop behavior followed by a gradual decrease of flow stress and the amount of the yield drop decreased with increasing test temperature. Steady-state flow behavior from a relatively small amount of plastic strain was observed in all specimens compressed above 1300 °C. Such characteristics of the flow behavior are similar to those observed in many transition metal silicides [2,5-6].The yield stress exhibited strong temperature dependence, decreasing monotonously and drastically with increasing temperature. The yield stresses of the present $MoSi_2/Mo_5Si_3$ DS eutectic alloys are 3 ~ 8 times as high as those of binary $MoSi_2$ single crystals deformed along the corresponding loading axis orientation, which clearly indicates that $MoSi_2$ phase is highly strengthened by forming the fine script-lamellar structure with Mo_5Si_3 phase [2].

Figure 5 shows yield stresses at 1400 °C plotted against the average thickness of $MoSi_2$ phase in the $MoSi_2/Mo_5Si_3$ DS eutectic alloys. The yield stress values at 1400 °C are strongly dependent on the average thickness of $MoSi_2$ phase, exhibiting a steep increase with the decrease of the average thickness of $MoSi_2$ phase for the specimens with homogeneous script-lamellar structure. A specimen with the inhomogeneous cellular microstructure (growth rate of 100 mm/h) exhibited a lower yield stress than that with the homogeneous script-lamellar structure with comparable average $MoSi_2$ thickness. Such a relatively low value of yield stress is

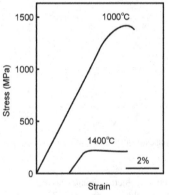

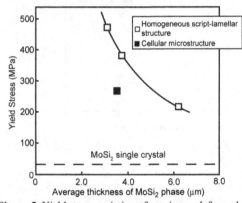

Figure 4. Stress – strain curves of the DS alloys grown at a growth rate of 5mm/h tested in compression at 1000 and 1400 °C.

Figure 5. Yield stress variation of specimens deformed at 1400 °C as a function of the average thickness of $MoSi_2$ phase in the $MoSi_2/Mo_5Si_3$ DS eutectic alloys.

considered to indicate the deformation behavior of coarse boundary regions plays a dominant role for the specimens with inhomogeneous cellular structures.

Figure 6 shows a bright-field TEM image of a coarse boundary region in a specimen grown at a rate of 100 mm/h and deformed at 1400 °C. Long and curved dislocations, which frequently form sub-boundaries (indicated by arrows in figure 6), were confirmed to exist only in $MoSi_2$ phase. This clearly indicates the preferential occurrence of plastic deformation in $MoSi_2$ phase and also the occurrence of dynamic recovery process accompanied by the climb motion of dislocations at this temperature. Detailed analyses of the deformation microstructures as a function of temperature as well as the average thickness of $MoSi_2$ phase are currently on going and will be published in elsewhere.

CONCLUSIONS

Deformation behavior of $MoSi_2/Mo_5Si_3$ directionally-solidified eutectic alloy at high temperatures has been investigated by compression tests along $[1\bar{1}0]_{MoSi_2}$ close to the growth direction. Yield stress values obtained at a temperature range from 1000 to 1500°C were several times as high as these of corresponding single crystals of $MoSi_2$ and decrease drastically with increasing temperature. The yield stress values at 1400 °C increased steeply with the increase of the growth rate of the $MoSi_2/Mo_5Si_3$ DS eutectic alloys.

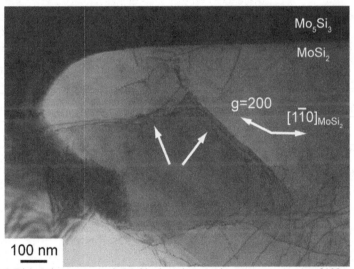

Figure 6. Dislocation structures in $MoSi_2$ phase in a specimen grown at a rate of 100 mm/h and deformed at 1400°C. The thin foil was cut parallel to $(001)_{MoSi_2}$, which is parallel to the compression axis. Some sub-boundaries are indicated by arrows.

ACKNOWLEDGMENTS

This work was supported by the Advanced Low Carbon Technology Research and Development Program (ALCA) from the Japan Science and Technology Agency (JST) and in part by the Elements Strategy Initiative for Structural Materials (ESISM) from the Ministry of Education, Culture, Sports, Science and Technology (MEXT), Japan.

REFERENCES

1. A. K. Vasudevan and J. J. Petrovic, Mater. Sci. Eng. A, **155**, 1 (1992).
2. K. Ito, H. Inui, Y. Shirai, and M.Yamaguchi, Philos. Mag. A, **72**, 1075 (1995).
3. D. P. Mason, D. C. Van Aken, and J. F. Mansfield, Acta metal. mater., **43**, 1189 (1995).
4. D. P. Mason and D. C. Van Aken, Acta metal. mater., **43**, 1201 (1995).
5. K. Kishida, M. Fujiwawra, H. Adachi, K. Tanaka, and H. Inui, Acta Mater., **58**, 846 (2010).
6. Y. Ochiai, K. Kishida, K. Tanaka, and H. Inui, in *Intermetallic-Based Alloys for Structural and Functional Applications*, edited by B. Bewlay, M. Palm, S. Kumar, and K. Yoshimi (Mater. Res. Soc. Symp. Proc., **1295**, Warrendale, PA, 2011), pp. 213-218.

Mater. Res. Soc. Symp. Proc. Vol. 1516 © 2012 Materials Research Society
DOI: 10.1557/opl.2012.1669

Structural and magnetic characterization of spark plasma sintered
Fe-50Co alloys

Mahesh Kumar Mani[1*], Giuseppe Viola[2,3], Mike J Reece[2,3], Jeremy P Hall[1], Sam L Evans[4]

1-Wolfson Centre for Magnetics, Cardiff School of Engineering, Cardiff University, UK
2- School of Engineering and Materials Science, Queen Mary University of London, London, UK
3-Nanoforce Technology Ltd., London, U.K
4-Institute of Medical Engineering and Physics, Cardiff University, UK

ABSTRACT

Fe-50 wt% Co alloy powders with average particle size of 10 μm were compacted by spark plasma sintering (SPS) at 700, 800, 900 and 950°C by applying 40, 80, 100 MPa uniaxial pressures for 2, 5, 10 minutes. The densities of the samples were found to increase with temperature from 700 to 900°C for constant sintering pressure and time and to decrease for the material sintered at 950°C. The effects of sintering time on density were more significant in samples sintered at 700°C and 800°C than those densified at 900°C. The consequences of small increases in mechanical pressure during sintering on density values were significant for samples sintered at 700°C. The coercivity (H_c) of the compacts decreased significantly with increasing sintering temperature, and with increasing dwell time at sintering temperatures lower than 700°C. The sample sintered at 950°C, which contains the largest grains among the prepared samples and porous microstructure, exhibited the minimum coercivity. Unlike H_c, the remanence (B_r) and saturation induction (B_{sat}) values were more strongly affected by the specimen density than by grain size. B_r and B_{sat} values were found to vary linearly with sintering temperature and pressure owing to increasing density. An increase in soaking time at 800 and 900 °C, although enabling higher density, exhibited contradicting effects on B_{sat} values. The SPS parameters to obtain maximum density and optimum magnetic properties for Fe-50% Co alloy were found to be 900°C, 80 MPa and 2-5 minutes.

INTRODUCTION

There is continuing interest to replace hydraulic and pneumatic systems in aircraft with reliable, easy to maintain and efficient electrical subsystems [1,2]. The soft magnetic material selected for such electric subsystems should retain good magnetic characteristics at high operating temperatures (500-700°C) and should exhibit high saturation induction (> 2 T) to maximise power density in weight sensitive applications [3]. Fe-(30-50) Co alloys, exhibiting high saturation magnetisation (2.3–2.45T) and Curie temperature (920–985°C) are suitable commercial soft magnetic alloys for such applications. In addition, Fe-Co alloy with near-equiatomic composition displays the highest permeability and zero magnetocrystalline anisotropy in the FeCo system. However, the ordered intermetallic Fe-50% Co alloys are brittle and are difficult to deform without any ternary addition [4-6]. Powder metallurgy (PM) is the only available alternate fabrication route to shape the brittle alloy into useful parts. PM technologies like metal injection moulding (MIM), cold compaction followed by sintering, hot isostatic pressing (HIP) have been employed to consolidate brittle FeCo and ductile FeCo-V magnetic powders [7-9]. The saturation induction of magnetic materials processed by PM route from pre-alloyed powder depends on the density of the compacts [10].

In the last two decades, spark plasma sintering (SPS) has been widely used to densify a wide variety of materials because of its unique characteristics such as low sintering temperatures, short holding time and cleaner grain boundaries. It has also been reported that SPS helped to attain density values close to theoretical density without any binder or prior cold compaction [11,12]. SPS was employed to consolidate nanocrystalline FeCo-based powders and it was reported that the electric field/pulsed current applied during sintering favours strong interparticle bonding and rapid densification of the nanocrystalline powders, without coarsening of the microstructure below 900°C [13]. Kim et al attained 95% of theoretical density and magnetisation of 230 emu/g in nanocrystalline Fe-30% Co alloy prepared by SPS at 900°C for 5 min under a pressure of 60 MPa [14]. A systematic study to clarify the effects of SPS parameters such as temperature, pressure and time on the densification of Fe-50%Co based alloy and their magnetic response is still missing. In the current study, the magnetic behaviour of the Fe-50%Co compacts prepared under different SPS conditions was investigated and correlated with their microstructure.

EXPERIMENTAL DETAILS

FeCo alloy powder with average particle size of 9.5- 11 μm was obtained from Sandvik Osprey Powder Group. The morphology, particle size distribution and composition of the procured powders were studied by scanning electron microscopy with EDX attachment (Nano Technology Systems, Carl Zeiss). The magnetic powders were compacted in a graphite die using a spark plasma sintering furnace (HPD 25/1 FCT, Germany). A schematic drawing depicting the main parts of the sintering system is shown in Fig.1., and the experimental details are described in greater detail by Milsom et al [15]. All the samples were heated to the sintering temperature at a constant rate of 100 °C/min. Sintering was conducted for 2, 5, 10 minutes at four different temperatures between 700 and 950°C in a vacuum of 10^{-2} torr under 40, 80 and 100 MPa pressure. Hereafter, the compacts prepared will be referred with the sintering parameters in the order: temperature-pressure-time. The samples were cooled in the furnace and removed from the die using a manual hydraulic press.

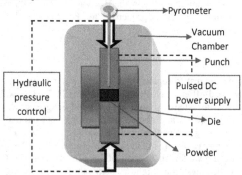

Fig.1 Schematic representation of SPS system

The density of the compact was measured using the Archimedes' immersion method in water. The cross section of the samples was ground and polished with abrasive discs with roughness values of 15, 9, 3 and 1μm and then etched with Nital for 30 sec. The microstructural features revealed after etching were examined using SEM and optical microscopy. Samples with rectangular cross-section (24 mm x 18 mm) were cut from 30 mm diameter sintered discs and their quasi DC magnetic response was measured using a fully automatic universal measurement system developed in the laboratory by varying the magnetic field up to 20 kA/m [16].

RESULTS AND DISCUSSION

Characterisation of FeCo powders

The powder prepared by the gas atomisation process exhibits spherical morphology and a wide size range from less than 1 μm to 25 μm as shown in Fig 2(a). EDX analysis confirmed the presence of Fe and Co (Fig. 2(b)). The average chemical composition of the powder particles was Fe= 51.14±0.10 wt% and Co=48.86±0.10 wt% (Fe-52.48±0.10 and Co 47.55±0.10 by atomic percentage).

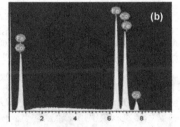

Fig.2 (a) SEM image of and (b) EDX spectrum of as-received Fe-Co powders

The effects of sintering parameters such as temperature, pressure and time on microstructure and magnetic properties will be discussed in the following three sections.

Effects of sintering temperature

As shown in the bar chart (Fig. 3(a)), an increase in sintering temperature up to 900°C improved the density of the Fe-Co alloy consolidated by SPS. Further increase in temperature promoted grain growth instead of densification as illustrated schematically in Fig. 3 (b) and a fall in the density value was observed. Fine particles were packed more effectively in the gaps of large particles under the influence of current and mechanical pressure at 900°C than at 700 and 800°C. Therefore, a significant reduction in the porosity content was realised when the sintering temperature was increased up to 900°C (Fig 4 (a) & (b)). However, at temperature higher than 900 °C, grain coarsening and an increased porosity was produced as shown in the micrographs, Fig. 4(c) & (d).

The density and grain size are the two main variables that can influence the magnetic performance of pre-alloyed powders sintered by SPS. An increase in the compact density with temperature up to 900°C determined an increasing saturation magnetisation and reduced coercivity as shown in Fig. 5 (a). The effects of pores and grain size on magnetic properties can be explained by comparing the hysteresis response of 950-40-5 and 900-40-5 (Fig 5(b)). The 950-40-5 sample with a porous microstructure and coarse grains exhibited lower coercivity and saturation magnetisation than the 900-40-5 sample with a dense microstructure and small grains. This clearly implies that grain size exerts a strong influence on coercivity while the density of the compact predominantly controls saturation magnetisation as observed in other PM processed magnetic materials [10]. The most favourable sintering temperature for SPS processing was found to be 900°C, which is less than the sintering temperature employed in MIM (1330°C) [7] and in sintering followed by cold compaction (1000-1400°C) [8] to process iron-cobalt powders.

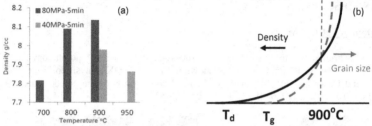

Fig.3 (a) Effect of sintering temperature on density of compacts at two different pressure levels (theoretical density of the alloy is 8.18 g/cc) (b) Schematic drawing demonstrating the T dependence of the density and grain size. T_d and T_g are the onset temperature of densification and grain growth.

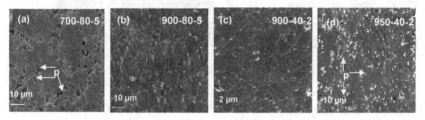

Fig.4. SEM micrographs of material sintered at different temperature-pressure-time combinations. Pores are marked "p" in the images.

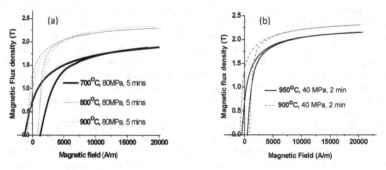

Fig.5 Upper half of the magnetic hysteresis curve of samples sintered at different temperatures

204

Effects of sintering pressure

The employment of graphite as the material for punches and dies limits the mechanical pressure to less than 100 MPa. At 700 and 800°C, the magnetic powders were compacted under 80 and 100 MPa. At 900 °C, the maximum pressure was limited to 80 MPa to avoid damage to the die. Table 1 summarises the effects of mechanical pressure on density and magnetic characteristics at different temperatures. An increase in the pressure values, irrespective of the incremental amount, always improved the density values and therefore the saturation induction of the sintered material. Additional pressure, with temperature and dwell time kept constant, promoted remanence reduction due to improvements in compact density. Other factors which may contribute for this reduction include long range order parameter, internal stress and grain size and their effects need to be studied in detail. The effect of variation of pressure addition on coercivity was not significant.

Table I Effect of sintering pressure on density and magnetic properties

Sample identity	Density	H_c (A/m)	B_r (T)	B_{sat} (T)
700-**80**-10	7.86	1130	1.19	2.14
700-**100**-10	8.01	1185	1.00	2.18
800-**80**-10	8.12	868	1.15	2.25
800-**100**-10	8.12	854	0.88	2.34
900-**40**-2	7.98	809	1.38	2.30
900-**80**-2	8.10	797	1.24	2.34
900-**40**-5	8.10	777	1.32	2.30
900-**80**-5	8.14	786	0.98	2.33

Effects of sintering time

The increase in density values with sintering time was more significant in samples sintered at 700 and 800°C than at 900°C as depicted in Fig. 6 (a). There was no improvement in density on increasing the soaking time beyond 5 min at 900°C. The effects of holding time increments on saturation magnetisation and coercivity are shown in Fig 7 (a) and (b).

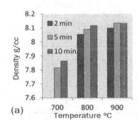
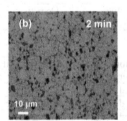
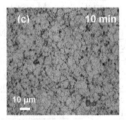

Fig. 6 (a) Effect of sintering times on density at different temperatures, optical micrographs of material sintered at 800 °C under 80 MPa pressure for (b) 2 min and for (c) 10 min

It has been observed that increasing the holding time, at temperatures as low as 700 °C, improves saturation magnetisation by enhancing the compact density. However, sintering the samples for longer duration at temperatures between 800-900°C has contrary effects on saturation magnetisation. The saturation magnetisation, which depends on density of the compact, showed a negative trend with holding times at 800°C and 900°C, even though there were improvements in density values with holding time.

The coercivity values, which strongly affect the hysteresis loss component of the compact, were reduced by increasing the holding time. The drop in the H_c values with soaking time was more substantial at 700°C than that at 800 and 900°C.

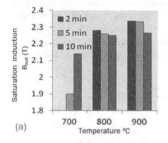 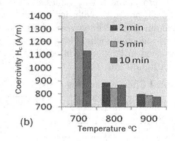

Fig.7 Effect of sintering times on (a) saturation induction (B_{sat}) and (b) coercivity (H_c) at different temperatures

FeCo based alloys processed by PM route

Table II summarises the attempts made to process FeCo based materials by powder metallurgy route and the saturation magnetisation and density values reported. It is evident that SPS helped to improve saturation induction values of the compacts formed from pre-alloyed powders significantly by improving the sintering density. FeCo materials processed under optimum SPS conditions, exhibit density values close to that of theoretical values. As a result, their magnetic induction is close to that of wrought alloys of similar composition, which is 2.4 T [3].

Table II Saturation induction and density of FeCo based alloy processed by PM route

Material	PM processing route	Density	Saturation induction (T)	Ref.
Fe-50 Co	MIM	8.03 g/cc	1.69	[7]
Fe-49Co-2V	MIM	7.82 g/cc	1.82	[7]
Fe-50 wt% Co	Cold compaction followed by sintering	95% relative density	2.15	[8]
Fe-Co alloy	Not mentioned	8.0 g/cc	2	[17]
Fe-49Co-2V	Not mentioned	> 7.4 g/cc	2.0	[18]
Fe-50 Co	Not mentioned	> 7.4 g/cc	1.7	[18]
Fe- 49 wt %Co	SPS	8.14 g/cc (99.5% relative density)	2.34	This work

CONCLUSIONS

1. Increasing the SPS sintering temperature up to 900 °C, improved the density of compacts and hence the magnetic properties. At temperatures as high as 950°C, grain coarsening is thermodynamically more favourable than densification.
2. Increasing the sintering pressure values had a positive effect on density, saturation magnetisation and remanence but mixed effects on coercivity.
3. Sintering Fe-Co alloy for longer soaking times yielded more favourable results at lower temperatures than at temperatures higher than 800°C.
4. The SPS parameters to obtain maximum density and optimum magnetic properties for Fe-50% Co alloy were found to be 900°C, 80 MPa and 2-5 minutes.
5. The saturation induction of Fe-Co alloy processed under optimum SPS conditions was determined to be at least 10% higher than that prepared by other PM processing routes.

REFERENCES

1. R.E Quigley, Proc. of IEEE Applied Power Electronics Conf. 'More electric aircraft', 906-911 (1993).
2. R.I Jones, Proc. of the Institution of Mech. Engineers, Part G: J of Aerospace Engineering **216**, 259-269 (2002).
3. R. S. Sundar and S. C. Deevi, *International Mater. Reviews* **50**, 157-192 (2005).
4. L. Zhao and I. Baker, *Acta Metall. Mater.* **42**, 1953-58 (1994).
5. T. Sourmail, *Progress in Mat. Sci.* **50**, 816–880 (2005).
6. K Kawahara, *J Mater. Sci.* **18**, 1709-18 (1983).
7. A. Silva, P. Wendhausen, R. Machado and W. Ristow Jr, *Mater. Sci. Forum* **534-536** 1353-56 (2007).
8. W. Yamagishi, K. Hashimoto, T. Sato, S. Ogawa and Z. Henmi, *IEEE Trans. on Magnetics* **Mag-22**, 641-643 (1986).
9. Z Turgut, M Huang, J.C. Horwath, and R.T. Fingers, *J. App. Physics* **103**, 07E7241-3 (2008).
10. F. Hanejko, H. Rutz and C. Oliver, *Advances in Powder Metallurgy & Particulate Materials* **6**, 375-404 (1992), Metal Powder Industries Federation, Princeton, NJ.
11. V.Mamedov, *Powder Metallurgy* **45**, 322-328 (2002).
12. M. Nygren and Z. Shen, *Key Engineering Materials* **264-268,** 719-724 (2004).
13. R. Nicula, V.D. Cojocaru, M. Stir, J. Hennicke and E. Burkel, *J. of Alloys & Compounds* **434–435**, 362–366 (2007).
14. Y.D. Kim, J.Y. Chung, J. Kim and H. Jeon, *Mater. Sci. & Engg.* **A291**, 17-21 (2000).
15. B Milsom, G Viola, Z Gao, F Inam, T Peijs and M.J Reece, *Journal of the European Ceramic Society* **32**, 4149-4156 (2012).
16. P Anderson, *J Mag. & Mag. Mater.* **320**, e589-e593 (2008).
17. J.A. Bas, J.A. Calero and M.J. Dougan, *Journal of Magnetism and Magnetic Materials* **254–255**, 391-398 (2003).
18. P.W. Lee, *ASM Handbook - Powder Metal Technologies and Applications* 7, ed. (American Society for Metals, 1998) p 2523.

Mater. Res. Soc. Symp. Proc. Vol. 1516 © 2013 Materials Research Society
DOI: 10.1557/opl.2013.64

Micro-pillar Compression of Ni-base Superalloy Single Crystals

Kabir Arora, Kyosuke Kishida, and Haruyuki Inui,
Department of Materials Science and Engineering, Kyoto University,
Sakyo-ku, Kyoto, 606-8501 JAPAN

ABSTRACT

Micro-compression behavior of single crystalline Ni-base superalloy CMSX-4 has been studied especially focusing on a specimen size range comparable to the size of Ni_3Al (γ') precipitates of about 500 nm using square cross-section micro-pillar specimens. The variation in stress levels exhibited a considerable increase for smaller micro-pillar specimens with edge lengths below 2 μm. Shearing of both the Ni (γ) channels and the γ' precipitates was observed indicating that stress levels were significant enough to cause the dislocations to cut into the γ' precipitates at room temperatures.

INTRODUCTION

Micro-compression testing as a technique to investigate the deformation behavior of materials is still in its nascent stages. The point to point variations in mechanical properties and their combined effect on overall properties with decreasing sample size can be investigated using micro-compression testing. A number of studies have been conducted on the size-scale effects ('smaller-is-stronger' phenomenon) and the mechanical response of micro-crystals of pure metals[1], intermetallic phases [2] and some two-phase alloys, under the application of uniaxial compressive stress with a few of them focusing particularly on Ni-base superalloys[3-4]. The microstructure of these alloys consists of ordered Ni_3Al (γ') precipitates with the $L1_2$ structure, coherently embedded in an FCC nickel (γ) matrix. The unique γ /γ' microstructure of Ni-base superalloys is the major reason for them being the materials of choice for high temperature applications. With the varying lattice parameter of both γ and γ', the internal stress conditions may also vary due to the change in the magnitude and sense of the lattice mismatch δ [5-6].

Figure 1 shows an SEM micrograph of a single crystalline superalloy CMSX-4, which is widely used as a blade material in industrial gas turbine applications. At size-scales comparable to the size of γ' precipitates of the order of 1μm and smaller, the effects due to the magnitude and sense of internal stresses may become rather significant. The previous micro-pillar compression studies for two-phase superalloys investigated the behavior above 2 μm only and have observed a size-scale dependence similar to those for single-phase materials, but no results have been reported so far for smaller size-scales comparable to the size of the γ' precipitates. At such small sizes, the exhaustion hardening and source truncation mechanism have been considered as possible causes for the size dependence of the flow stress for single phase materials [7], however, almost nothing has been revealed for two-phase materials. In this paper, we have studied micro-pillar compression behavior using square cross-section micro-pillar specimens with edge lengths below 2 μm to evaluate the effect of the presence of a γ' precipitate with dimensions comparable to the sample and to elucidate the controlling mechanisms of the size-scale dependence of the flow stress for two-phase superalloy CMSX-4.

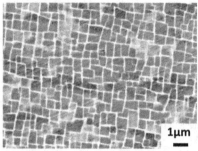

Figure 1. An SEM image of an as-received ingot of the CMSX-4.

EXPERIMENTAL DETAILS

Rectangular parallelepiped specimens with one pair of surfaces perpendicular to [001] were cut from the as-received ingot of the single crystalline CMSX-4 by electric discharge machining. The obtained surface was mechanically polished and then electropolished to remove the excessive damage. Electropolishing was carried out using a mixture of perchloric acid, methanol and butanol mixed in the ratio 1:12:6 by volume. Micro-pillars for compression tests were made on the polished surface using focused ion beam (FIB) milling system. Orientations of side surfaces of the micro-pillars was chosen to be perpendicular to [100] and [010] so as to align all edges of the micro-pillars to the cuboidal precipitates. Micro-pillars with various edge lengths ranging from 0.6 µm to 8 µm were prepared for the study. The average size of the γ' particles was confirmed to be between 400 – 800 nm as shown in figure 1. Micro-pillar compression tests were carried out using a micro hardness testing machine equipped with a flat diamond tip with a diameter of 20 µm. A scanning electron microscope equipped with a field emission gun was used to examine the micro-pillars before and after mechanical testing.

RESULTS & DISCUSSION

The stress-strain curves for micro-pillar specimens with edge lengths of 4.0 µm and 1.5 µm are shown in Figures 2a and 2b, respectively. The stress-strain curves for the 4.0 µm-edge specimens can be characterized by three parts, namely (i) an elastic region, followed by (ii) a transition to stable flow and a continually decreasing work hardening rate after the onset of plastic flow, and finally (iii) a large strain-burst after about 4 % plastic strain. The observed characteristics of the flow behavior are well consistent with those reported previously [1-2].

Figures 3a and 3b show SEM images of the micro-pillar specimens with an edge length of 4 µm deformed to two different plastic strain levels. The specimen with a lower amount of plastic strain of about 2.5 %, which corresponds to the work hardening stage, displayed two sets of slip traces on (11$\bar{1}$) (dominant) and (111) (minor), both of which extend continuously from one edge to another. This indicates that not only the γ channels but also γ' particles were shear already at this stage. The slip traces are evident only in the top half of the sample with the lower half being relatively free of any slip traces at this relatively early stage of plastic deformation. As the amount of total strain was increased, sheared area were confirmed to be developed by thickening of slip bands to fill the entire gauge length (figure 3b), indicating that the large strain-burst observed in the stress-strain curves for the 7.5 % and 14.0 % deformed specimens was mainly caused by rapid multiplication of dislocations with changing slip planes. Such a thickening behavior of slip bands accompanied by a large strain-burst has been observed in the compression of cylindrical micro-pillars of Ni₃Al with relatively large sizes above 5 μm in diameter oriented for single-slip activation.

In the stress-strain curves for the 1.5 μm-specimens, no apparent large strain-burst was observed at least up to about 10 % plastic strain, which is in marked contrast with the case of larger micro-pillar specimens shown in figure 2a. Size–scale effects become more significant with the reduction in size, with the flow stress at 0.2 % plastic strains being between 1000-1100 MPa for 1.5 μm samples.

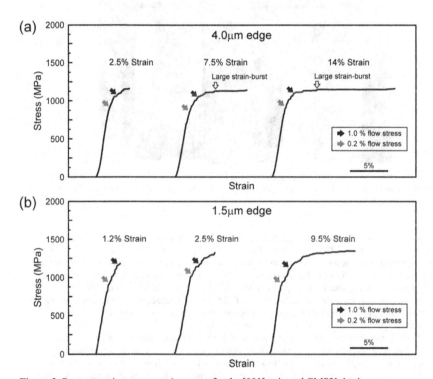

Figure 2. Representative stress-strain curves for the [001]-oriented CMSX-4 micro compression experiments for specimens with edge lengths of (a) 4.0μm and (b) 1.5μm.

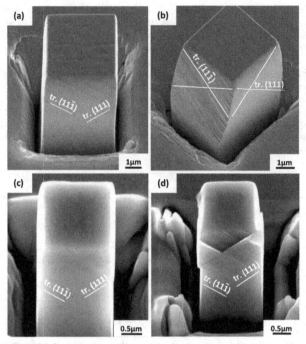

Figure 3. Images of deformed micro-pillars of the [001]-oriented CMSX-4 with edge lengths of (a,b) 4.0 μm and (c,d) 1.5 μm. Total plastic strains achieved are (a)2.5 %, (b)14.0 % , (c) 2.5 % and (d) 9.5 %.

Figures 3c and 3d show the 1.5 μm-edge specimens after deformation to total plastic strains of about 2.5 and 9.5 %, respectively. Fine slip traces are seen for the specimen at 2.5 % plastic deformation. Presence of intense shear along multiple slip planes is clearly evident for the specimen deformed to 9.5 % plastic strain. Such deformation behavior is also different from that in the 4.0 μm specimen. The 1.5 μm specimen showed intense shear at lesser total plastic strains, while the 4.0 μm specimen showed a thickening of slip lines at a higher strain. The absence of a large strain-burst in the stress-strain curve for this specimen may be explained on the basis of the simultaneous occurrence of the intensive shear on two slip planes

For sub-micro-sized specimens, relatively wide scatter of the stress-strain curves were observed. Figure 4a depicts the stress-strain behavior of specimens with and edge length of 0.7 μm deformed to 20 % plastic strain. As marked by gray and black arrows, values of 0.2 % and 1.0 % flow stresses differ markedly from specimen to specimen, while flow stress values become nearly identical with increasing the amount of plastic strain. Another difference observed in the stress strain curves is that intermittent occurrence of small strain-bursts and work hardening

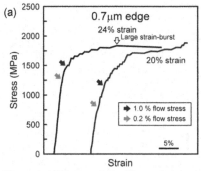 

Figure 4. (a) Representative Stress-strain curves and (b) an SEM image for the [001]-oriented CMSX-4 micro-pillars with an edge length of 0.7 µm deformed to about 20% plastic strain.

regions over much wider range of strain up to at least about 15 % plastic strain than in the larger specimens. These characteristics are becoming more significant as the size of the specimens approaches the sub-micron scales. The wider scatter observed in the stress-strain curves may be related to the fact that the deformation behavior approaches to that of single phase with decreasing the specimen size so that only one γ' precipitate exist in a cross section of a micro-pillar. Figure 4b shows an SEM image of a 0.7 µm-edge specimen after deformation exhibiting several shear steps on the surface including an intense shear step, which is similar to that observed in compression of single phase γ' micro-pillars. Thus, lower values of flow stresses at 0.2% and 1.0% plastic strain for the specimen deformed to 20 % strain is inferred to reflect the deformation behavior of single phase γ', which is caused by the so-called exhaustion hardening accompanied by micro-strain bursts in the elastic region according to the previously reported deformation behavior of single phase γ' [2]. Detailed analysis of deformation microstructures by transmission electron microscopy is currently in progress in order to clarify the actual cause of the observed variation in the deformation behavior.

The size-scale dependence of the resolved shear stress at 1.0% flow stress is shown in figure 5. The values of the resolved shear stress show a size-scale dependence with the power law exponent, $n = 0.087$ which is in correspondence, even at sub-micro scales to the results previously reported [4].

CONCLUSIONS

This study is an attempt towards understanding the behavior of nickel-base superalloys at scales, where the specimen edge dimensions become comparable to the size of the γ' precipitates. We have tried to formulate the flow behavior as a size-scale phenomenon, suggesting that the regions of alternating strain-bursts and work hardening are a product of the entrapment of dislocations in a finite volume. Thus, as the sizes of the sample for the micro-compression experiments are taken to the sub-micro levels (0.6-0.7 µm), the range of the unique alternating behavior increase, and the mode of deformation of the micro-pillars shifts to an intensive shear

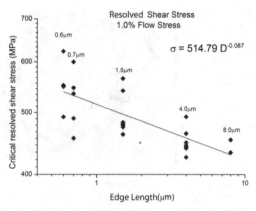

Figure 5. Log-log plot showing the relationship between the resolved shear stress and sample size at 1.0% yield stress

ACKNOWLEDGMENTS

This work was supported in part by the Elements Strategy Initiative for Structural Materials (ESISM) from the Ministry of Education, Culture, Sports, Science and Technology (MEXT), Japan.

REFERENCES

1. D.M. Dimiduk, M.D. Uchic and T.A. Parthasarathy, Acta Mater., **53**, 4065 (2005).
2. D.M. Dimiduk, M.D. Uchic, S.I. Rao, P.A. Shade, C. Woddward, G.B. Viswanathan, E.M. Nadgorny, S.Polasik, D.M. Norfleet and M.J. Mills, Philos. Mag. (2012) in press.
3. M.D. Uchic and D.M. Dimiduk, Mater. Sci. Eng. A, **400-401**, 268 (2005).
4. P.A. Shade, M.D. Uchic, D.M. Dimiduk, G.B. Viswanathan, R. Wheeler and H.L. Fraser, Mater. Sci. Eng. A, **535**, 53 (2012).
5. T.M. Pollock and A.S. Argon, Acta Metall. Mater., **42**, 1859 (1994).
6. F.R.N. Nabarro, Metall. Mater. Trans. A, **27A**, 513 (1996).
7. S. I. Rao, D.M. Dimiduk, T.A. Parthasarathy, M.D. Uchic, M. Tang and C. Woodward, Acta Mater., **56**, 3245 (2008).

Mater. Res. Soc. Symp. Proc. Vol. 1516 © 2013 Materials Research Society
DOI: 10.1557/opl.2012.1750

Diffusion brazing of γ-TiAl-alloys: Investigations of the joint by electron microscopy and high-energy X-ray diffraction

Katja Hauschildt, Andreas Stark, Uwe Lorenz, Norbert Schell, Torben Fischer, Malte Blankenburg, Martin Mueller, Florian Pyczak
Institute of Materials Research, Helmholtz-Zentrum Geesthacht, Centre for Materials and Coastal Research, Max-Planck-Str. 1, D-21502 Geesthacht, Germany

ABSTRACT

Diffusion brazing is a potential method to repair parts made from TiAl-alloys. Two different brazing materials with varying contents of titanium, iron and nickel were investigated. The phases present in the brazed zone were identified by high energy X-ray diffraction (HEXRD) at the material science beamline HEMS at the PETRA III synchrotron facility at DESY in Hamburg, Germany, and the microstructure was characterised by scanning electron microscopy (SEM). The braze zone itself is composed of one to two transitional layers from the substrate material to the middle of the joint. Near the substrate material the phase constitution reassembles a TiAl-alloy while the middle of the joint is similar to α/β-titanium alloys. Besides phases commonly encountered in TiAl-alloys such as γ, α₂ and β, additional phases, which are related to the presence of nickel or iron as melting point depressing elements are present. The microstructure of the brazed zone changes significantly during a subsequent heat treatment.

INTRODUCTION

Repairing parts made of TiAl-alloys is an interesting approach due to the increasing use of these alloys in aero engine and automotive applications. The closure of cracks (in noncritical or not highly loaded areas) in aero engine vanes and blades is an example for the potential use of such repair methods. To have mechanical properties on a similar level as the substrate material it is desirable to get a microstructure in the repaired region which does not differ strongly from the base material [1]. Here diffusion brazing could be an attractive method for TiAl. While diffusion bonding provides even better microstructural homogeneity with the substrate, this method is not suited for most repair problems as for example crack closure.

The idea of diffusion brazing is to use a solder with a lower melting point than the substrate material [2]. This is achieved by adding a melting point depressing element to the solder. At temperatures below the melting point of the substrate material the brazing solder melts. During a holding time at brazing temperature the melting point depressing element diffuses into the substrate material and the melting point of the material in the brazing zone increases continuously until it is fully solidified.

In this work two different brazing materials, one with nickel and one with iron, are investigated. Both are studied in as-brazed state and after a heat treatment at 1000 °C. The phases existing in the joint were identified by space-resolved high energy X-ray diffraction (HEXRD) and the microstructure in the brazing zone was characterized by scanning electron microscopy.

EXPERIMENT

As substrate material the γ-TiAl alloy TNB-V5 (Ti-45Al-5Nb-0.2B-0.2C, all atomic percent) is used. Two brazing solders of the compositions Ti-29Fe [3] and Ti-24Ni [4] are applied. The constituents of the brazing solders are chosen due to the strong effect of both elements Fe and Ni on the melting point of titanium and the composition is near the eutectic point for both binary systems. The theoretical melting point of Ti-24Ni lies at 942 °C and the one of Ti-29Fe at 1078 °C.

For brazing the γ-TiAl substrate was cut into blocks of approximately 10x10x5 mm³ and a foil of brazing solder with a thickness of 300 to 350 μm was placed between two blocks. The material of the brazing solder was produced in an arc furnace. Furthermore, these stacks of substrate and solder foil were brazed in a vacuum furnace for 24 hours at a temperature of 1110 °C. The heating rate was 21.7 K/min. Afterwards the temperature was hold for 24 hours followed by furnace cooling. In addition to the as-brazed states also specimens were produced which experienced a subsequent heat treatment for 168 hours at 1000 °C and then furnace cooling.

The specimens were cut, ground and electrolytically or vibration polished for scanning electron microscopy (SEM) investigations with a Leo Gemini 1530 with field emission gun using backscattered electron (BSE) contrast.

For the high energy X-ray diffraction (HEXRD) a slice of circa 1 mm was cut from each specimen state. The HEXRD investigations were done at the HZG run high energy material science beamline HEMS at PETRA III at DESY in Hamburg, Germany. While a thicker slice (e.g. 4 mm) would have contained a higher number of grains and provided better grain statistics, it was decided to use the rather thin slice to avoid uncertainties about the position of the diffracting volume. This effect would have been more severe for a thicker specimen. The measurements were taken with an energy of 87.1 keV and corresponding wavelength of 0.014235 nm. The diffraction patterns were recorded with a mar345 image plate detector for the as-brazed and a PerkinElmer flatpanel detector for the heat treated samples. A beam of 100 μm width and 50 μm height was scanned in 25 μm steps across the joint. The joint was positioned horizontally in this setup. During the measurement of one pattern, the sample was turned by 60° around the vertical axis to get diffraction information from a higher number of grains. The detected Debye-Scherrer rings were integrated over 360° and displayed as a function of 2Θ.

RESULTS AND DISCUSSION

Microstructure

Looking at the microstructure of the different samples, it is noticeable, that there are great differences between both brazing solders and between the as-brazed and after heat treated states. In figure 1 a BSE picture of each sample is shown. Figures 1a and 1b are the as-brazed states and 1c and 1d display the heat treated states. In general it is recognizable, that there are always one or two transition zones between the substrate material and the middle of the joint. The better homogeneity of the heat treated samples is also obvious. Additionally the substrate material is fine grained with grain sizes of 5 to 10 μm and shows two phases with a lighter and a darker gray.

The as-brazed state of the Ti-29Fe solder (fig. 1a) shows needles of dark gray embedded in a bright phase in the middle of the joint. In addition grains with light gray colour of a size of 20 μm are found at the transition of the joint to the substrate material. These grew probably vertical from the substrate material into the joint. It is evident, that the grains of the substrate material are bigger at the junction. When comparing with figure 1c, Ti-29Fe specimen after heat treatment at 1000 °C, it is recognizable that the width of the joint zone nearly doubled. Furthermore the needles of the dark gray phase in the joint grew shorter and thicker and some globular grains of this phase are now found, which coexist with a second phase appearing brighter in the SEM micrograph.

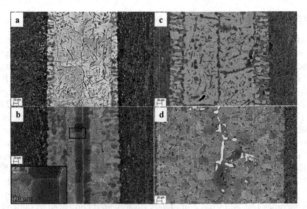

Figure 1. BSE pictures of the different samples: a) Ti-29Fe as-brazed state, b) Ti-24Ni as-brazed state, c) Ti-29Fe after heat treatment and d) Ti-24Ni after heat treatment.

In figure 1b, showing the as-brazed state of Ti-24Ni, one can also notice a brighter phase while the middle of the joint is composed of darker grains with a size of 30 to 50 µm. Furthermore, at the transition between substrate material and joint zone grains in a range of 15 to 20 µm are found, which appear light gray in the SEM micrographs. In addition in the joint one can distinguish small, bright particles. For better visibility a respective area is shown in an inset with higher magnification in figure 1b. Due to the bright appearance in BSE contrast this phase probably contains a higher amount of heavy elements like nickel. In the very centre of the joint zone a small layer consisting of a white and a lighter gray phase is found. When comparing with the heat treated state (figure 1d), it is obvious, that the joint got almost two times wider. Also coarse white particles are present in the middle of the joint. It is interesting to note, that the layer of grains, which appeared darker in the middle of the joint in the as-brazed state, has dissolved. In addition the grains in the whole brazing zone have coarsened. This significant alteration during the heat treatment indicates that the microstructure of the as-brazed state is far from the thermodynamic equilibrium at 1000 °C.

Comparing both as-brazed states it is remarkable, that in the middle of the Ti-24Ni specimen two distinct layers with bigger grains are found, which is not seen in the Ti-29Fe specimen. Here the structure is formed by small brighter, spicular grains. A rather unexpected finding is the similar width of the joints, which is remarkable, since the melting point of Ti-24Ni is with 942 °C by more than 100 °C lower than for Ti-29Fe with 1078 °C. Nevertheless, the most striking difference is the appearance of small, bright precipitates in the Ti-24Ni specimen. It will be shown in the next section that these particles belong to a new phase forming in the joint zone.

Furthermore, the two heat treated samples show, that the specimen with Fe-solder does not change as pronounced during heat treatment compared with the as-brazed state than the one with the Ni-solder. Obviously both joints widened during the heat treatment but the Ti-24Ni specimen much stronger. This is probably due to the higher diffusion coefficient of Ni in TiAl at 1000 °C compared to Fe ($D_{Fe} \sim 1.29 \times 10^{-16}\, m^2 s^{-1}$ and $D_{Ni} \sim 2.02 \times 10^{-16}\, m^2 s^{-1}$, calc. after [5]).

Phase analysis with HEXRD

The phases present in the material can be identified with HEXRD. Due to the space-resolved measurement we additionally get information of their distribution within the joint. Figure 2 shows four representative diffractograms of the as-brazed states (their positions are marked in figure 3). In the substrate material TNB-V5 (fig. 2a), two phases, the α_2-Ti_3Al-phase (hexagonal $D0_{19}$ structure, $P6_3/mmc$) and the γ-TiAl-phase (tetragonal $L1_0$ structure, P4/mmm) can be clearly distinguished [6]. In figure 2b a XRD pattern of the transition zone of the Ti-24Ni specimen is displayed. Therein a third phase can be observed, the ω_0-Ti_4Al_3Nb-phase (hexagonal $B8_2$ structure, $P6_3/mmc$), which transforms during cooling from the ordered β_0-TiAl-phase (cubic B2 structure, Pm-3m) [7]. Additional peaks can be noticed at a 2Θ-angle of 3.8° to 3.9° which can not be related to the three above mentioned phases. They finally could be identified as the τ_3-TiNiAl-phase (hexagonal C14 structure, $P6_3/mmc$) with the lattice parameters a = 5.0 Å and c = 8.1 Å [8,9]. Besides, the volume fraction of the α_2-phase increased compared to the substrate material and there is almost no γ-phase present. In the middle of the Ti-24Ni joint (fig. 2c), ω_0 is almost lacking, τ_3 becomes less and γ has completely disappeared compared to figure 2b. In figure 2d, the middle of the Ti-29Fe joint, it is obvious to see that at the position of the ω_0 double-peak (2$\Theta \sim 3.6°$) one broad peak exists. This can be attributed to the presence of the phases ω_0 and β_0 together. Thus, one can assume that iron retards the transformation from ordered β_0 to ω_0 during cooling. Compared to the Ti-24Ni joint, a little volume fraction of γ-phase exists, but no α_2-phase and no τ_3-phase can be detected, which is in accordance with ternary phase diagrams [8,10].

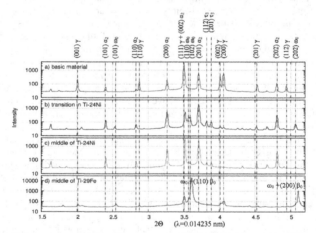

Figure 2. X-ray diffractograms of a) the substrate material TNB-V5, b) the transition from substrate material to the middle of the joint of Ti-24Ni, c) the middle of the joint from Ti-24Ni and d) the middle of the joint from Ti-29Fe. Each position is marked in fig 3.

Comparison of space-resolved synchrotron measurements and SEM pictures

Figure 3 shows the BSE pictures of the four samples combined with the step-resolved synchrotron XRD data. The diffraction patterns of the 21 steps are aligned side by side as diffraction angle vs. distance diagrams with the peak intensity coded in grayscale. The first row shows both Ti-29Fe specimens (fig. 3a,c) and the second both Ti-24Ni specimens (fig. 3b,d).

The Ti-29Fe specimen in the as-brazed state (fig. 3a) shows a bright phase in the joint, which is indentified as ω_o- and β_o-phase together. The α_2-phase almost lacks and the small volume fraction of γ can be attributed to the needles. At the transition from joint to substrate material a small amount of bright phase is visible infiltrating the substrate. In the heat treated state of the Ti-29Fe specimen (fig. 3c) it is noticeable, that the α_2-phase now is also identified in the middle of the joint, which can be explained by a homogenization of the chemical composition. In addition it should be noticed that the peak positions of the γ double-peak show a small 2Θ shift over the joint. This can be attributed to small changes of the lattice parameters due to variations in the chemical composition of γ.

The Ti-24Ni specimen in the as-brazed state is shown in figure 3b. Here one can identify the ω_o- and τ_3-phase as bright phases in the joint. In the middle of the joint no γ-phase is visible. The big, dark grains are probably α_2-phase. In the heat treated state (fig. 3d) it is evident, that the

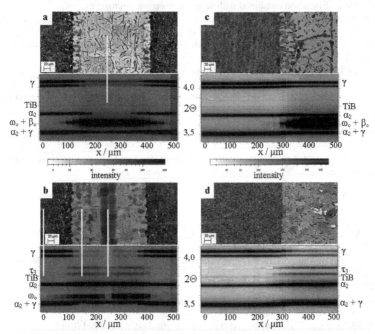

Figure 3. Measurements with synchrotron in comparison to the BSE pictures of a) Ti-29Fe in as-brazed state, b) Ti-24Ni in as-brazed state, c) Ti-29Fe after heat treatment and d) Ti-24Ni after heat treatment. The lines mark the position of the diffractograms in fig. 2.

ω_o-phase completely dissolved. In the middle of the joint only α_2, γ and τ_3 are observable. Except for the presence of τ_3 this is near to the constitution of the substrate. Further studies are planed to check the influence of the τ_3-phase on the strength.

It is notable that in all states the TiB-peak extends in the joint. This means, that either solid borides are flushed into the molten region during the beginning of the brazing process or boron is diffused into the region of the joint.

Comparing both as-brazed states it is remarkable, that the Ti-29Fe-joint is almost exclusively composed of ω_o- and β_o-phase and small amounts of γ. In contrast, the Ti-24Ni specimen shows the three phases α_2, ω_o and τ_3. This can be explained by the fact that in Ti-based alloys Fe is a stronger β-stabilizer than Ni, since in the binary Ni-Ti phase diagram β exists down to a temperature of 765 °C [4] whereas in the Fe-Ti phase diagram it is stable down to 568°C [3].

The most remarkable point is that by an additional heat treatment the ω_o- or β_o-phase can be fully removed in Ti-24Ni but not in Ti-29Fe. This means the chemical homogenization of the joint and the substrate is much faster in the Ti-24Ni than in the Ti-29Fe specimen.

CONCLUSIONS

We studied the joining of TiAl parts by diffusion brazing with a Fe and a Ni solder using high energy X-ray diffraction combined with scanning electron microscopy.

In the brazing joint one to two transition states occur with different phase constitutions. The Fe-joint consists almost exclusively of ω_o and β_o with a small amount of γ whereas the Ni-joint consists of α_2, ω_o and τ_3. Remarkable in the Ni-joint is the existence of a new phase in comparison to γ-TiAl, the τ_3-phase.

After an additional heat treatment the ω_o-phase is completely dissolved in the Ni-joint, whereas in the Fe-joint it still exists. Du to the higher diffusion coefficient of Ni in TiAl the chemical homogenization between the joint region and the substrate is better in the Ni-joint than in the Fe-joint.

ACKNOWLEDGMENTS

Experimental support is gratefully acknowledged to Stefan Eggert, Dirk Matthiessen and Bernhard Eltzschig.

REFERENCES

1. S. Neumeier, M. Dinkel, F. Pyczak, M. Göken, *Mater. Sc. Eng.* **A528**, 815 (2011)
2. H. Duan, M. Koçak, K.-H. Bohm, V. Ventzke, *Sci. Technol. Weld. Joi.* **9**, 513 (2004)
3. H. Okamoto, *J. Phase Equilib.* **17**, 369 (1996)
4. *Phase Diagrams of Binary Nickel Alloys*, edited by P. Nash (ASM International, Materials Park, OH, USA, 1991) p. 342
5. C. Herzig, T. Przeorski, M. Friesel, F. Hisker, S. Divinski, *Intermetallics* **9**, 461 (2001)
6. F. Appel, J. D. H. Paul, M. Oehring, in Gamma Titanium Aluminide Alloys (Wiley-VCH, Weinheim, Germany, 2011) chapter 1-2
7. A. Stark, M. Oehring, F. Pyczak, A. Schreyer, *Adv. Eng. Mater.* **13**, 700 (2011)
8. J.C. Schuster, Z. Pan, S. Liu, F. Weitzer, Y. Du, *Intermetallics* **15**, 1257 (2007)
9. P. Nash and W. W. Liang, *Metall. Trans. A* **16A**, 319 (1985)
10. M. Palm and J. Lacaze, *Intermetallics* **14**, 1291 (2006)

Mater. Res. Soc. Symp. Proc. Vol. 1516 © 2012 Materials Research Society
DOI: 10.1557/opl.2012.1683

Microstructural evolution of monocrystalline Co–Al–W-based superalloys by high-temperature creep deformation

Takahiro Sumitani[1], Katsushi Tanaka[1] and Haruyuki Inui[2]

[1] Department of Mechanics, Kobe University, Rokkodai-cho, Nada-ku, Kobe 657-8501, Japan
[2] Department of Materials Science and Engineering, Kyoto University, Sakyo-ku, Kyoto 606-8501, Japan

ABSTRACT

Creep tests of monocrystalline Co–Al–W-based alloys with a tensile stress of 137 MPa at 1000 °C were carried out. The microstructures of the crept specimens were investigated by scanning electron microscope (SEM) and transmission electron microscope (TEM). The γ' phase in the specimens was not only elongated along the stress direction as expected by the sign of the lattice misfit but also elongated in one of the <100> directions perpendicular to the stress direction. As a result, the shape of the γ' phase is not a rod but a plate. In the TEM images, it was observed that many SISFs are induced in the γ' phase by creep. A similar microstructure is also observed in Ni-based superalloys, but the microstructure was formed under relatively lower temperatures and higher applied stresses. The observation of numerous stacking faults in the γ' phase is a clear indication that the γ' phase precipitated in the present alloy is weaker than that in many modern Ni-based superalloys.

INTRODUCTION

Because the efficiency of turbine engines for aerospace propulsion and land-based power generation increases generally with the increase in the combustion gas temperature, there have been ever-increasing demands for structural materials that can withstand severe oxidizing environments and high operating temperatures. Currently, Ni-based superalloys containing an $L1_2$-ordered intermetallic phase are the most widely used high-temperature structural materials in aircraft engines and power generation systems. Although the temperature capability of Ni-based superalloys has increased gradually year by year through modification of their alloy chemistry, this capability has reached almost the melting temperatures of these alloys, and thus the development of new high-temperature alloys is eagerly awaited.

The recent discovery of stable $L1_2$-ordered intermetallic compounds, $Co_3(Al, W)$, coexisting in a solid solution based on fcc-Co [1] has opened up a pathway to the development of high-temperature structural materials based on cobalt. The compounds exhibit the formation of γ+γ' two-phase microstructures that resemble those in Ni-based superalloys. The discovery has stimulated extensive research [2-17] on the development of a new class of high-temperature structural materials based on cobalt, which are named "Co-based superalloys." Because the melting temperature of Co is higher than that of Ni by 40 K, it may be possible for Co-based superalloys to achieve a temperature capability higher than that of Ni-based superalloys. Studies on phase equilibria of the ternary and higher-order systems indicate that the γ' solvus temperature in the ternary system is quite low (<1000 °C) when compared to the Ni–Al binary system [1, 2, 9]. The addition of transition metals such as Ta, Ti, and Nb is beneficial for increasing the γ' solvus temperature of Co–Al–W-based alloys [14], possibly leading to a better high-temperature strength. In contrast to Ni-based superalloys, the measured values of lattice misfit between the two phases in Co–Al–W-based alloys are all positive (i.e., the lattice constant of the γ' phase is larger than that of the γ phase). According to the mechanism of rafting of the γ' phase operating in creep, a different type of raft structure is expected to form. The results of measurements of single-crystal elastic constants of the γ' phase in the ternary system suggest the ductile nature of the L1₂ compound. The magnitude of the anisotropic factor and elastic stiffness constants also suggests the formation of γ+γ' two-phase microstructures with cuboidal γ' precipitates well aligned parallel to <100> and well faceted parallel to {100} [4]. The plastic deformation behavior has also been investigated in both γ+γ' two-phase [2, 3, 6] and γ' single-phase [2, 15] systems in compression. Single crystals of Co–Al–W-based alloys with γ+γ' two-phase microstructures exhibit a promising high-temperature strength in compression, especially when Ta is added [1, 3, 6]. However, polycrystals of the ternary L1₂ compound exhibit a modest positive dependence of yield stress on temperature only at high temperatures above 700 °C [15], in contrast to L1₂ compounds based on Ni₃Al [18-20]. The creep properties of Co–Al–W-based alloys were first reported by Bauer et al. [13] in polycrystalline specimens alloyed with boron. They also reported both the creep properties of polycrystalline specimens under different temperatures (850 and 950 °C) and the microstructural changes, in which the alloys tend to form a rod-like parallel raft structure [16]. Similar microstructural changes during creep at 900 °C have been reported in the case of single crystals [17].

In the present study, we investigate the creep deformation behavior of single crystals of Co–Al–W-based alloys with γ+γ' two-phase microstructures. Because we pay special attention to the positive values of lattice misfit between the γ and γ' phases in analyzing the creep properties and resultant microstructures, the temperature for the creep tests was chosen as 1000 °C in order to clearly observe the changes in microstructure during creep.

EXPERIMENTAL PROCEDURE

Alloys with four different nominal compositions, listed in table I, were prepared by Ar arc-melting and appropriate mixtures of pure elements under an argon atmosphere. These alloys are designated as alloy-1 to alloy-4 throughout this paper. Alloy-1 is a ternary alloy that serves as a base for the Co-based superalloys. In order to increase the γ' solvus temperature, tantalum was added to alloy-2 and alloy-3, in which different amounts of tungsten are alloyed in order to control the volume fraction of the γ' phase. Both tantalum and nickel were alloyed to alloy-4 to further increase the γ' solvus temperature and to improve the phase stability.

Monocrystal bars with diameters of 15 mm and lengths of approximately 150 mm were grown by a modified Bridgman technique. These single crystals were subjected to solution heat-treatment at 1150 °C for 15 ks, followed by aging treatment at 1000 °C for three days in order to obtain homogeneously distributed cuboidal γ' precipitates in the γ matrix. Specimens for tensile creep tests were machined by grinding a part of the single crystals. The tensile axis orientations for all four specimens were within 15° from the [001] direction. Creep tests were performed in tension under a constant stress of 137 MPa in air at 1000 °C. The microstructures of the specimens were examined before and after creep testing by scanning electron microscopy with a back-scattered electron detector (SEM-BSE) and transmission electron microscopy (TEM).

Table I. Nominal compositions (at%) of alloys investigated, volume fraction of the γ′ precipitates at 1000 °C, and γ′ solvus temperature.

Alloy	Co	Al	W	Ta	Ni	V_f (%)	T_s/°C
1	82	9	9	-	-	0	977
2	79	9	10	2	-	85	1096
3	81	9	8	2	-	50	1087
4	62	9	7	2	20	43	1098

RESULTS AND DISCUSSION

Microstructures after heat treatment

The SEM-BSE microstructures of single crystals of the alloy samples after solution and aging heat treatments for precipitation of the γ′ phase are shown in figures 1(a)–(d). The observations were made on the (001) plane, almost perpendicular to the growth direction. Because the γ′ phase contains heavier elements than the γ phase, γ′ precipitates and the γ matrix are observed as bright and dark contrasts, respectively. Alloy-2 to alloy-4 contain cuboidal γ′ precipitates with dimensions of approximately 400 nm in edge length arranged along the <100> directions in the γ matrix. Because the γ′ solvus temperature of alloy-1 is lower than the heat treatment temperature, only small (secondary) γ′ precipitates that formed during cooling after heat treatment are observed for the ternary alloy-1. Small (secondary) γ′ precipitates formed in the channels of the γ matrix are also observed in alloy-2 to alloy-4. The volume fraction of the γ′ phase at the heat-treatment temperature of 1000 °C was evaluated for each alloy from an area fraction of the γ′ precipitates, as tabulated in table I. The γ′ solvus temperature is determined by measurements of thermal expansions of the alloys. The results are also tabulated in table I.

Creep properties

The creep strains of single crystals of alloy-1 to alloy-4 obtained at 1000 °C under a constant load of 137 MPa in tension are plotted in figure 2 as a function of time. Rupture occurs

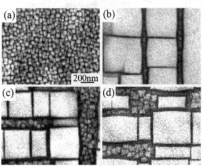

Figure 1. SEM back-scattered electron images of the alloys investigated. (a) alloy-1, (b) alloy-2, (c) alloy-3, and (d) alloy-4.

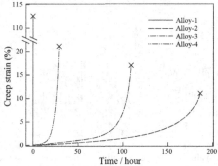

Figure 2. Creep curves for alloy-1 to alloy-4 under a constant tensile stress of 137 MPa at 1000 °C.

immediately after loading for alloy-1. The other alloys were able to survive for some time before rupture occurred. This significant difference is considered to result from the fact that alloy-1 exhibits a microstructure of a γ single phase at the test temperature. This is because the γ′ solvus temperature of this alloy is lower than the test temperature (table I), in contrast to the γ+γ′ two-phase microstructures that are retained in the other alloy samples. The rupture time for alloy-4 is less than 30 h, while it is almost 190 h for alloy-2.

In modern Ni-based superalloys in which the *n* (normal)-type raft structure is formed, the increase in the creep rate in the accelerating region is usually rather high. The increase in the creep rate in the accelerating region tends to be accelerated when the alloy forms the raft structure with higher regularity and interface flatness, leading to instantaneous rupture. However, the increase in the creep rate in the accelerating region is generally very slow for the present Co–Al–W-based alloys. This property is beneficial to the practical use of Co–Al–W-based alloys, since rupture occurs very slowly. This may be related to how a raft structure is formed (if formed) and with what regularly it is formed.

Deformation microstructures

Figure 3 shows SEM-BSE images of the microstructures of alloy-2 to alloy-4 after creep testing at 1000 °C and 137 MPa. Observations were made in areas at a distance from the rupture surface and on both (001) and (100) planes, which are almost perpendicular to and parallel to the tensile axis orientation, respectively, for each alloy. As is the case with figure 1, the γ and γ′ phases are imaged as dark and bright contrasts, respectively. For all these alloys, the γ′ phase is directionally coarsened along the [100] and [010] directions when imaged on (001), whereas it is

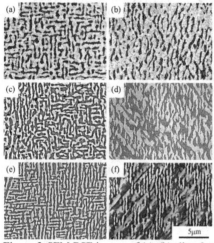 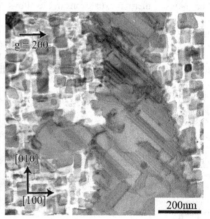

Figure 3. SEM-BSE images of (a)–(b) alloy-2, (c)–(d) alloy-3, and (e)–(f) alloy-4 after creep testing. Note that (a), (c), and (e) show the (001) plane, and (b), (d), and (f) show the {100} plane.

Figure 4. TEM bright-field image showing stacking faults in the rafted γ′ precipitates.

directionally coarsened along the [001] direction when imaged on (100). The latter is a clear indication of the formation of a p (parallel)-type raft structure [5] in which the γ' phase is directionally coarsened along the tensile axis direction, which is consistent with what is expected from the positive values of the lattice misfit between the γ and γ' phases. In these images, it should be noted that the γ' phase in this case seems to have plate shapes or combination of plates lying on (100) and (010) rather than just rod shapes.

Figure 4 shows a typical TEM bright-field image of alloy-3 after creep testing at 1000 °C and 137 MPa. The thin foil for the TEM observation was cut parallel to the (001) planes, which is almost perpendicular of the tensile stress orientation. Large γ' precipitates coarsened along the [100] and [010] directions are observed, which is consistent with the SEM-BSE images shown in figure 3. Small (secondary) γ' precipitates distributed in the γ channels are believed to re-precipitate during cooling after creep testing. Therefore, it is practically impossible to observe dislocations in the γ channels and γ/γ' interfaces that might carry strains during creep. Instead, numerous stacking faults are observed to lie on the {111} planes in large γ' precipitates. A similar microstructure is also observed in the close vicinity of the rupture surface of Ni-based superalloys subjected to creep under higher applied stresses of approximately 350 MPa [21]. This fact implies that an applied stress of 137 MPa for Co–Al–W-based alloys corresponds to higher applied stresses for Ni-based superalloys resulting from that the high-temperature strength of the γ' phase [Co$_3$(Al, W)] in the present Co–Al–W-based alloys is much weaker than that of the γ' phase (Ni$_3$Al) in Ni-based superalloys.

CONCLUSIONS

The creep deformation behavior of and microstructural changes in single crystals of Co–Al–W-based alloys with $\gamma+\gamma'$ two-phase microstructures were investigated under a constant load of 137 MPa in tension at 1000 °C. The results obtained are summarized as follows:
(1) The creep rupture life is significantly enhanced in the presence of cuboidal γ' precipitates. When the testing temperature is higher than the γ' solvus temperature, as in the case of ternary alloys, the specimen ruptures immediately upon loading, whereas other alloys with γ' precipitates can survive for a certain amount of time before rupture.
(2) The so-called p (parallel)-type raft structure, in which the γ' phase is elongated along the tensile axis direction, is formed during creep in Co–Al–W-based alloys, unlike in Ni-based superalloys. This microstructural change is consistent with what is expected from the positive values of lattice misfit between the γ and γ' phases.

ACKNOWLEDGMENTS

This work was supported by Grant-in-Aid for Scientific Research (A) from the Ministry of Education, Culture, Sports, Science and Technology (MEXT), Japan, and in part by the Global COE (Center of Excellence) Program of International Center for Integrated Research and Advanced Education in Materials Science from the MEXT, Japan.

REFERENCES

1. J. Sato, T. Omori, K. Oikawa, I. Ohnuma, R. Kainuma and K. Ishida, *Science* **312**, 90 (2006).

2. S. Miura, K. Ohkubo and T. Mohri, *Mater. Trans.* **48**, 2403 (2007).
3. A. Suzuki, G. C. DeNolf and T. M. Pollock, *Scripta Mater.* **56**, 385 (2007).
4. K. Tanaka, T. Ohashi, K. Kishida and H. Inui, *Appl. Phys. Lett.* **91**, 181907 (2007).
5. K. Shinagawa, T. Omori, J. Sato, K. Oikawa, I. Ohnuma, R. Kainuma and K. Ishida, *Mater. Trans.* **49**, 1474 (2008).
6. A. Suzuki and T. M. Pollock, *Acta Mater.* **56**, 1288 (2008).
7. M. Osaki, S. Ueta, T. Shimizu, T. Omori and K. Ishida, *Denkiseikou* **79**, 197 (2008).
8. H. Chinen, T. Omori, K. Oikawa, I. Ohnuma, R. Kainuma and K. Ishida, *J. Phase Equi. Diffusion* **30**, 587 (2009).
9. S. Kobayashi, Y. Tsukamoto, T. Takasugi, H, Chinen, T. Omori, K. Ishida and S. Zaefferer, *Intermetallics* **17**, 1085 (2009).
10. K. Shinagawa, T. Omori, K. Oikawa, R. Kainuma and K. Ishida, *Scripta Mater.* **61**, 612 (2009).
11. T. M. Pollock, J. Dibbern, M. Tsunekane, J. Zhu and A. Suzuki, *JOM* **62**, 58 (2010).
12. M. Chen, C. Y. Wang, *Physics Letters* **A374**, 3238 (2010).
13. A. Bauer, S. Neumeier, F. Pyczak and M. Göken, *Scripta Mater.* **63**, 1197 (2010).
14. M. Ooshima, K. Tanaka, N. L. Okamoto, K. Kishida and H. Inui, *J. Alloys Comp.* **508**, 71 (2010).
15. H. Inui, T. Oohashi, N. L. Okamoto, K. Kishida and K. Tanaka, *Mater. Sci. Forum* **638-642**, 1342 (2010).
16. S. M. Copley and B. H. Kear, *Trans. Metall. Soc. AIME* **239**, 977 (1967).
17. A. Bauer, S. Neumeier, F. Pyczak, R. F. Singer and M. Göken, *Mater. Sci. Eng.* **A550**, 333 (2012).
18. M. S. Titus, A. Suzuki and T. M. Pollock, *Scripta Mater.* **66**, 574 (2012).
19. P. H. Thornton, R. G. Davis and T. L. Johnston, *Metall. Trans.* **1**, 207 (1970).
20. V. Paidar, D. P. Pope, V. Vitek, *Acta Metall.* **32**, 435 (1984).
21. D. M. Knowles and Q. Z. Chen, *Mater. Sci. Eng.* **A356**, 352 (2003).

Mater. Res. Soc. Symp. Proc. Vol. 1516 © 2013 Materials Research Society
DOI: 10.1557/opl.2013.392

Effect of Ball Milling on Magnetic Properties of Nb Substituted $R_2Fe_{16}Nb_1$ (R: Gd and Er) Alloys

B. K. Rai and S. R. Mishra
Department of Physics, The University of Memphis, Memphis, TN 38152.

S. Khanra and K. Ghosh
Department of Physics, Astronomy, and Materials Science, The Missouri State University, Springfield, MO 65897

Abstract
In this work, we report the effect of high energy ball milling (HEBM) on Nb doped $R_2Fe_{16}Nb_1$ (R= Gd, Er) compounds. The focus of the work is to bring enhancement in magnetic properties of R_2Fe_{17} (2:17) compounds with the ball milling. Specifically, we find that the ball milling increases saturation magnetization, coercivity, and Curie temperature. The increase in the magnetization and Curie temperature upon ball milling is related to the lattice expansion and microstrains while the increase in coercivity is related to the grain refinement.

Introduction
Iron-rich rare-earth intermetallics (R_2Fe_{17}) and their compounds have drawn considerable attention because of their potential use as hard magnetic materials. Various interesting changes in these compounds have been observed upon the insertion of interstitial atoms or substitutional atoms. These changes include modification of the lattice parameter, magnetization, Curie temperature (T_c), and crystal field experience by rare-earth atoms. Although interstitial compounds are promising candidates for permanent magnets, they are thermodynamically unstable. Alternately, replacement of Fe by non-magnetic atoms such as Al, Ga, or Si [1,2,3] or refractory metals such as Mo and Nb [4,5] is preferred which produce dramatic changes in the magnetic properties of the material. The substitution of non-magnetic elements enhances T_c by changing the unit cell volume which enhances the Fe-Fe exchange interaction [6]. Furthermore, substitution of non-magnetic dopants in R_2Fe_{17} also leads to spin reorientation transition [7,8] suggesting the role of non-magnetic atoms in modifying magnetocrystalline anisotropy. However, the substitution of non-magnetic atoms in R_2Fe_{17} drastically reduces the saturation magnetization due to the magnetic dilution. The substitution of Nb in these compounds has proven to reduce free iron. The presence of excess iron is often detrimental to the coercivity [9].

In view of these facts, the present work is focused on increasing T_c of R_2Fe_{17} intermetallics with a low level of non-magnetic substitution for Fe and thus expecting a minimum deterioration in the saturation magnetization. We use HEBM to mill Nb-doped $Gd_2Fe_{16}Nb_1$ and $Er_2Fe_{16}Nb_1$; crystallographically different compounds with Th_2Zn_{17} (*R-3m*) and Th_2Ni_{17} (*P63/mmc*) structure respectively. The magnetic properties of a nanostructure depend on the intrinsic material properties, as well as on grain size, grain size distribution, the magnetic character of the interphase (the region between grains), and the intergrain magnetic coupling. The potential of the HEBM method in material processing has been extensively explored for amorphous alloys [10], grain refinement [11], and nanocomposite mixing [12]. The HEBM can be used to produce a variety of effects in intermetallic alloys due to the complex dependence of the nanostructure on milling intensity, temperature, and other factors [13].

Experiment

$Gd_2Fe_{16}Nb_1$ and $Er_2Fe_{16}Nb_1$ samples were prepared by arc melting pure elements (99.99%) in high purity argon atmosphere. The ingots were remelted at least four times to ensure their homogeneity. Approximately 1 gram of sample was then subjected to HEBM (Across International, High energy vibratory ball mill) in the presence of ethanol for different time viz. 0, 30, 60, and 90 minutes. Samples were sealed in 80 ml stainless steel jar with stainless steel balls. Balls to sample mass ratio was 10:1. Samples were milled at 1200 rpm oscillating frequency. The phase and structure of the samples were determined using x-ray powder diffraction (XRD) (Bruker D8 Advance) with Cu Kα radiation. The magnetic properties of the samples were investigated using vibrating sample magnetometer (VSM) (Dexing Magnet, China) at 80 and 300 K. The Curie temperature, T_c, was determined using the VSM in a field of 40 kA/m between 300-600K.

Results and Discussions

Figures 1(a) and (b) show the XRD patterns of the $R_2Fe_{16}Nb$ alloy with a function of ball milling (BM) time. The starting material has Th_2Zni_{17}-type structure (R= Gd) and Th_2Ni_{17}-type structure (R= Er) with a small amount of secondary $NbFe_2$ and Fe phases respectively. With the increase in the BM time, the peak intensity of R_2Fe_{17} (2:17) phase and $NbFe_2$ phase diminishes whereas the peak intensity of α-Fe increases in $Gd_2Fe_{16}Nb_1$. This implies insertion of additional Nb in $Gd_2Fe_{16}Nb_{1+δ}$. The diminishing peak intensity of the 2:17 phase in both 2:17 alloys indicates gradual transformation of the micro-crystalline phase into an amorphous phase upon ball milling [14,15]. The observed XRD line broadening with the milling time could be due to the crystalline size reduction and microcrystalline stresses arising from defects [16]. Due to peak broadening α-Fe peaks at higher angle are not visible.

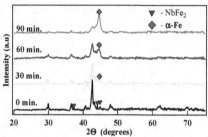

Figure 1(a): XRD patterns of $Gd_2Fe_{16}Nb_1$ as a function of ball milling time. Unindexed peaks correspond to 2:17 intermetallic phase.

Figure 1(b): XRD patterns of $Er_2Fe_{16}Nb_1$ as a function of ball milling time. Unindexed peaks correspond to 2:17 intermetallic phase.

Table I list the crystal lattice parameters viz. a and c, as a function of x, Nb. The lattice parameters are calculated using following equation:

$$d_{(hkl)} = \left(\frac{4}{3} \frac{h^2 + hk + k^2}{a^2} + \frac{l^2}{c^2} \right)^{-\frac{1}{2}}$$

........eq.(1)

where, d(hkl) is the inter-planar spacing and (hkl) are the Miller indices for Bragg peaks at $2\theta=29.87°$ (112) and $36.57°$ (300). The unit-cell volume is calculated using the formula: $V = a^2c \sin 60°$. The presence of secondary phases (Fe and $NbFe_2$) in starting powders brings different trend in the lattice parameters which can be seen in the Fig 2(a) and (b). In the $Gd_2Fe_{16}Nb_1$ sample, a linear expansion of unit-cell volume was observed as a function of BM time. This observation is consistent with the diminishing peak intensity of the $NbFe_2$ phase with increase in BM time. As per the nature of ball milling, it introduces defects and induces microstructural refinement in the alloys. In our understanding, the presence of defects and refinement enhances the diffusivity of Nb (out of $NbFe_2$ phase) into 2:17 phase. This diffusion causes the linear increase in the unit cell volume. This interpretation is further corroborated by observed increase in the α-Fe peak intensity due to dissociation of α-Fe from $NbFe_2$ phase upon ball milling. However in $Er_2Fe_{16}Nb_1$ samples, unit cell volume expansion was insignificant. Thus observed differences in crystalline parameters of $Gd_2Fe_{16}Nb_1$ and $Er_2Fe_{16}Nb_1$ might be attributed to the presence of secondary $NbFe_2$ phase in the $Gd_2Fe_{16}Nb_1$ starting powder. These secondary phases were also noticed in Mössbauer spectra (results not shown).

Table I: Lattice parameters and unit cell volume of $R_2Fe_{16}Nb_1$ compounds.

BM Time (min)	$Gd_2Fe_{16}Nb_1$			$Er_2Fe_{16}Nb_1$		
	a, (Å)	c, (Å)	V, (Å³)	a, (Å)	c, (Å)	V, (Å³)
0	8.4613	8.3718	519.06	8.4422	8.3914	517.91
30	8.4698	8.3807	520.04	8.4392	8.3652	515.94
60	8.4672	8.3993	521.49	8.4384	8.3736	516.36
90	8.4661	8.4158	522.37	8.4429	8.3775	517.15

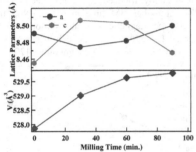 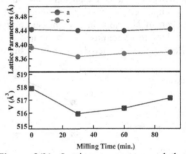

Figure 2(a): Lattice parameters and the unit cell volume of $Gd_2Fe_{16}Nb_1$ as a function of ball milling time.

Figure 2(b): Lattice parameters and the unit cell volume of $Er_2Fe_{16}Nb_1$ as a function of ball milling time.

The magnetic properties of $Gd_2Fe_{16}Nb_1$ and $Er_2Fe_{16}Nb_1$ at 80 and 300 K are listed in the Table II. Figure 3(a) and (b) show the saturation magnetization (M_s) and the coercivity (H_c) of the $Gd_2Fe_{16}Nb_1$ and $Er_2Fe_{16}Nb_1$ samples as a function of BM time. An increase in the H_c and M_s is evident in $R_2Fe_{16}Nb_1$ alloy upon ball milling. The increase in M_s is partly due to lattice expansion and from α-Fe. The trend of lattice parameters expansion and increase in the M_s in $Gd_2Fe_{16}Nb_1$ samples are in agreement with the Zhou et. al [17] whereas the increment in magnetization of $Er_2Fe_{16}Nb_1$ samples might be coming from the addition of α-Fe in the compound. An overall improvement in coercivity is observed for samples ball milled for 30 min. We attribute the increase in coercivity to the milling-induced defects and strain, which are metastable enough to be removed by room temperature annealing [18]. However, the coercivity of ball milled samples decreases for long milling times. This is due to (i) amorphization of 2:17 phase, (ii) agglomeration of the nanoparticles after long hours of ball milling, and (iii) increment of soft α-Fe content [19,15].

Table II: Magnetic Properties of $R_2Fe_{16}Nb_1$ compounds.

BM time (min)	Gd$_2$Fe$_{16}$Nb$_1$					Er$_2$Fe$_{16}$Nb$_1$				
	80 K		300 K			80 K		300 K	
	Tc(K)	Ms (emu/g)	Hc (Oe)	Ms (emu/g)	Hc (Oe)	Tc(K)	Ms (emu/g)	Hc (Oe)	Ms (emu/g)	Hc (Oe)
0	510	88	200	62	170	410	53	126	50	44
30	520	90	228	66	180	485	72	194	66	120
60	530	116	195	81	168	482	105	178	67	112
90	550	145	174	81	166	480	82	157	80	129

The Curie temperature of milled alloys is listed in the Table II. An increase in T_c is evident upon ball milling $R_2Fe_{16}Nb_1$ alloys. The maximum in T_c was observed at ~ 485 K for 30 min BM $Er_2Fe_{16}Nb_1$ and at ~ 550 K for 90 min BM $Gd_2Fe_{16}Nb_1$. The maximum in T_c for $Gd_2Fe_{16}Nb_1$ is ~ 70 K higher than that of Gd_2Fe_{17} alloy [20]. Generally, the T_c in rare-earth transition-metal intermetallic compounds is governed by three kinds of exchange interactions, namely, the 3d–3d exchange interaction between the magnetic moments of the Fe sublattice, the 4f–4f exchange interaction between the magnetic moments within the R sublattice, and the intersublattice the 3d–4f exchange interaction [21]. Again, introduction of Nb (from NbFe$_2$) in the unit cell causes dumbbell Fe-Fe distance to expand which improves the strength of exchange interaction which eventually brings in improvement in the T_c. On the other hand, the T_c increase of $Er_2Fe_{16}Nb_1$ could be related to induced defects as lattice expansion is not observed for this alloy. Induced defect have been known to bring improvement in T_c of materials [15,22,23]. At higher BM time in $Gd_2Fe_{17}Nb_1$, the introduction of Nb atom in 2:17 phase compensate with the magnetic moment dilution in 2:17 phase. It does not cause much variation in T_c. The observed T_c ~ 550 K in $Gd_2Fe_{16}Nb_1$ is superior to any series (R_2Fe_{17} intermetallic) of rare-earth transition metal-rich intermetallic compounds [20].

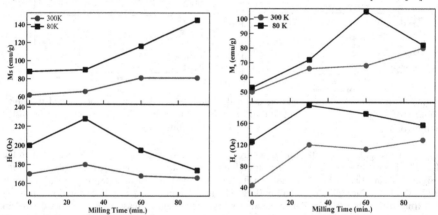

Figure 3 (a): Coercivity and saturation magnetization of Gd$_2$Fe$_{16}$Nb$_1$ alloy.

Figure 3(b): Coercivity and saturation magnetization of Er$_2$Fe$_{16}$Nb$_1$ alloy.

Conclusion

The effects of HEBM on structural and magnetic properties of arc melted $R_2Fe_{16}Nb_1$ (R = Er and Gd) compounds were assessed. A substantial improvement in Tc was observed for both compounds. The insertion of Nb in $Gd_2Fe_{16}Nb_1$ causes an increase in unit cell volume which enhances the Fe-Fe exchange interactions. In $Er_2Fe_{16}Nb_1$ lattice defects could be responsible for T_c enhancement. It is to be noted that the Tc enhancement in these compounds is obtained at a significantly lower level of Nb doping as compared to Al, Ga or Si doped R_2Fe_{17} compounds [1,2,3,24]. The overall improvement in Ms, Hc, and Tc by simple ball milling process is of a significant advantage in terms of energy density to cost ratio of the permanent magnet.

Acknowledgement
This research was supported in part by the U.S. National Science Foundation through Grant No. 1029780 (NSF-CMMI) and 0965801 (NSF-EAGER).

References

1. F. Weitmr, K. Iliehl, and P. Rogl, J. Appl. Phys. **65**, 4963 (1989).
2. M. Venkatesan, K. V. S. Rama Rao, and U. V. Varadraju, Physica B **291**, 159 (2000).
3. E. E. Alp, A. M. Umarji, S. K. Malik, G. K. Shenoy et.al, J. Magn. Magn. Mater. **68**, 305 (1987).
4. M. Daniil, Z. M. Chen, G. C. Hadjipanayis, A. Moukarika, V. Papaefthymiou, J. Magn. Magn. Mater. **234**, 375 (2001).
5. J. A. Chelvane, M. Palit, S. Pandian, M. M. Raja, and V. Chandrasekaran, Hyperfine Interact **187**, 87 (2008).
6. R. Xua, L. Zhen, D. Yang, J. Wu, X. Wang, Q. Wang, C. Chen, and L. Dai, Mater. Lett. **57**, 146(2002).
7. K. Ohno, T. Urakabe, T. Takei, T. Saito, K. Shinagawa, and T. Tsushima, J. Magn. Magn. Mater. **147**, 279 (1995).
8. Y. D. Zhang, W. A. Hines, J. I. Budnick, D. P. Yang, B. G. Shen, and Z. H. Cheng, IEEE Trans. MAGN. **37**, 2546 (2001).
9. S. A. Sinan, D. S. Edgley, and I. R. Harris, J. Alloys. Comp. **226**, 170 (1995).
10. C. C. Koch, O. B. Cavin, C. G. Mc Kamey, and J. O. Scarbrough, Appl. Phys. Lett., **43** (1983) 1017.
11. H. J. Fecht, NanoStructure Materials, **1**, 125(1992).
12. G. B. Schaffer, Scripta Metall., **27**, 1(1992).
13. C.C. Koch, Scripta Met. **34**, 21 (1996).
14. X. C. Kou, R. Grössinger, T. H. Jacobs, and K. H. J. Buschow, Physica B, **168**, 181(1991).
15. C. Suryanarayan, Progress in Materials Science 46 (2001) 1.
16. Y. H. Wang, Basics of X-ray diffraction (Atomic Energy publishing House, Beijing, 1993).
17. G. F. Zhou and H. Bakker, Scripta Mater. **34**, 29(1996).
18. G. Herzer, IEEE Trans. MAGN. **26**, 1397 (1990).
19. X. Rui , J. E. Shield, Z. Sun, L. Yue, Y. Xu, D. J. Sellmyer, Z. Liu, and D.J. Miller, J. Magn. Magn. Mater. **305**, 76 (2006).
20. K. J. Strnat, Handbook on Ferromagnetic Materials, vol 4, eds. E. P. Wohlfarth and K. H. J. Buschow (North-Holland, Amsterdam, 1988) chap. 2.
21. K. H. J. Buschow, Rep. Prog. Phys. **40**, 1179 (1977).
22. I. Nehdi, L. Bessais, C. Djega-Mariadassou, M. Abdellaoui, H. Zarrouk, J. Alloys and Comp. **351**, 24 (2003).
23. A. Iwase, Y. Hamatani, Y. Mukumoto, N. Ishikawa, Y. Chimi, T. Kambara, C. Muller, R. Neumann, and F. Ono, Nuclear Instr. Meth. Phys. Res. B **209**, 323 (2003).
24. R.van Mens, J. Magn. Mag. Mater. **61**, 24(1986).

Shape Memory Alloys

Mater. Res. Soc. Symp. Proc. Vol. 1516 © 2012 Materials Research Society
DOI: 10.1557/opl.2012.1728

Internal Friction and Dynamic Modulus in High Temperature Ru-Nb Shape Memory Intermetallics

Laura Dirand[1], Maria L. Nó[1], Karine Chastaing[2], Anne Denquin[2], Jose San Juan[3]

[1] Dpt. Física Aplicada II, Facultad de Ciencia y Tecnología, Universidad del País Vasco, Apdo. 644, 48080 Bilbao, Spain.
[2] ONERA (DMMP), 29 avenue de la Division Leclerc, F-92322 Châtillon Cedex, France.
[3] Dpt. Física Materia Condensada, Facultad de Ciencia y Tecnología, Universidad del País Vasco, Apdo. 644, 48080 Bilbao, Spain.

ABSTRACT

Nowadays, aeronautic and aerospace are the more demanding sectors for shape memory alloys (SMA) after the bio-medical one. In particular the interest has been recently focused on very high temperature SMA, which would be able of working as sensors and actuators in the hot areas of the engines and exaust devices.
In the present work we undertook a study of the Ru-Nb SMA Intermetallics, which undergo two succesive martensitic transformations around 1050 K and 1180 K respectively, depending on composition. This study has been focused on measurements of internal friction spectra and dynamic modulus variation up to 1700 K, which have been carried out in a sub-resonant torsion mechanical spectrometer.
The internal friction and dynamic modulus have been studied as a function of the heating-cooling rate and the frequency in order to compare experimental behaviour with theoretical models for martensitic transformations. In addition to the internal friction peaks linked to both martensitic transformations we have also observed a complex relaxation process around 950 K, which seems to be linked to the interaction of the martensite interfaces with structural defects. An analysis and discusion of the potential microscopic mechanisms are also presented.

INTRODUCTION

Shape memory alloys (SMA) are functional intermetallics that can perform both sensing and actuating functions thanks to a reversible first order phase transition characterized mainly by an atomic shearing of the lattice planes from the high temperature and high symmetry phase, called Austenite, to the low temperature and lower symmetry phase called Martensite. In SMA the transition between both phases can be induced thermally or by the application of a stress promoting crystallographic shearing during the transformation, and consequently the shape memory and superelastic properties are controlled by the behavior of this thermo-elastic martensitic transformation (MT). See the textbooks [1, 2] for a general overview. The most developed and commercially distributed SMA is binary Ti-Ni, which unfortunately has his transformation temperatures limited to below 100°C. However there are many potential applications requiring high temperature SMA (HT-SMA) and several families of materials have been developed with transformation temperatures between 100°C and 600°C, as recently reviewed by Ma et al. [3]. Nevertheless there are very few potential candidates to fulfill the requirements to be used above 600°C, in hot areas of aero-engines and aerospace devices, which constitutes a scientific and technological challenge. Indeed, at present only the Ti-Pt(Ir) [4-7],

Ru-Nb [8-13] and Ru-Ta [8, 12-16] systems seems to be able of operating in the temperature range from 600°C to 1200°C, and can be considered as ultra-high temperature SMA (UHT-SMA). The present work is focused on Ru-50Nb (at%) with the highest transformation temperatures in this system and exhibiting two successive martensitic transformations from a high temperature β austenite, cubic B2 ordered (Pm$\bar{3}$m), to a tetragonal β' martensite (P4/mmm) and then from β' to another monoclinic β'' martensite (P2/m) at lower temperature. The lattice cells are shown in Table 1, which includes the lattice parameters found in the literature [9, 17-19].

Phase	Phase Structure	Space Group	Lattice Parameter (Å)
Cubic β austenite (high temperature)		Pm$\bar{3}$m	a = 3.185 [17] a = 3.176 [18]
Tetragonal β' martensite (intermediate temperature)		P4/mmm	a = 3.11 [17] a = 3.106 [18] c = 3.332 [17] c = 3.307 [18]
Monoclinic β'' martensite (low temperature)		P2/m	a = 8.06 [9] a = 8.10 [19] b = 4.28 [9] b = 4.23 [19] c = 5.48 [9] c = 5.22 [19] β(°) = 97 [9,19]

Table 1. Crystalline structures of different phases in Ru-50Nb (at%) intermetallics.

These martensites correspond to the equilibrium phases of the Ru-Nb phase diagram [8, 11] in contrast with the martensites from most SMA; which are meta-stable phases appearing in non-equilibrium conditions from austenite. This fact together with the low strain recovery associated to these martensites, particularly to β'' [12], offer some doubts about the thermo-elastic behavior of such martensites. Then, the aim of the present work is to investigate the internal friction and dynamic modulus behavior of Ru-50Nb (at%) as a function of temperature, in order to elucidate whether the β–β' and β'–β'' martensitic transformations exhibit the expected behavior for a thermo-elastic martensitic transformations [20, 21].

MATERIALS AND TECHNIQUES

Ru-Nb alloys were produced from commercially pure elements by arc-melting under an argon atmosphere and re-melted several times to ensure homogeneity [19]. Then the material was heat-treated in vacuum at 1673 K for 5.7×10^5 s to obtain a homogeneous microstructure. The ingot was electro-spark machined and final samples were mechanically polished allowing the observation of a grain size of about 500 μm. The microstructure of the martensite was characterized at room temperature by scanning electron microscopy in back-scattered mode at 10 kV in a JEOL 7000F. The microstructure of the monoclinic β'' martensite, constituted mainly by shelf-accommodated groups of variants, is shown in Figure 1. To study the martensitic transformation as a function of temperature internal friction (IF) and dynamic modulus (DM) were measured in a high temperature mechanical spectrometer based on a forced inverted torsion pendulum working between 300 K and 1800 K, at different frequencies from 10^{-3} Hz to 10 Hz [22]. Measurements were performed in high vacuum, $\approx 10^{-5}$ mbar, to avoid oxidation of the sample.

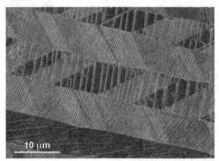

Figure 1. Scanning electron micrograph of monoclinic martensite in Ru-50Nb shape memory alloy.

RESULTS AND DISCUSSION

Internal friction and dynamic modulus curves during cooling and heating between 600 K and 1330 K were measured at 0.1 Hz with a heating/cooling rate of 1 K/min and an oscillating amplitude (ε_0) of $\varepsilon_0 = 10^{-5}$. The results are shown in Figure 2 where sharp peaks are observed at 1158 K and 1027 K during cooling and at 1171 K and 1046 K during heating, corresponding to the forward (TP$_F$) and reverse (TP$_R$) martensitic transformations from β–β' and β'–β'' respectively. Taking as reference the peak temperature for the β–β' MT, our results agree quite well with those from the literature measured by dilatometry and differential thermal analysis (DTA): 1159 K [11, 19] and 1158 K [8] for TP$_F$ as well as 1172 K [11, 19] and 1168 K [8] for TP$_R$.

One interesting point that can be noticed in Fig. 2 is the clear softening of the dynamic modulus just before and during the martensitic transformation, which is recovered after the transformation. The drop of the dynamic modulus between 1300 K and 1180 K is clear evidence of a pre-martensitic softening of the elastic constants in β phase, which has been discussed in previous work [23].

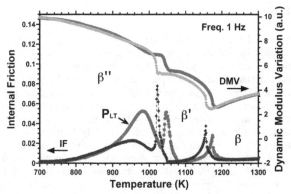

Figure 2. Internal friction spectrum and dynamic modulus variation curve (at 1 Hz) during heating (red and magenta dots) and cooling (blue and cyan diamonds). The sharp peaks of both transformation β–β' and β'–β'', as well as the low temperature peak P_{LT}, are plotted.

It should be also noted that the softening of the dynamic modulus is observed in Fig. 2 not only for the β–β' transformation but also for the β'–β'' transformation, which gives a clear indication of the thermo-elastic nature of these martensitic transformations. In addition a broad peak P_{LT} is observed in Fig. 2 at lower temperature than the β'–β'' transformation peak. This peak is rather unexpected because such an anomaly has been never reported by dilatometry [8] or by DTA [11, 19]. Consequently we may disregard the possibility of attributing it to a martensitic transformation and it may be associated to some relaxation process linked to defects or interfaces mobility.

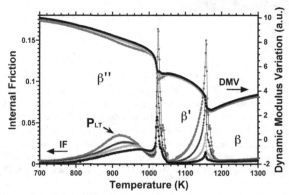

Figure 3. Internal friction and dynamic modulus measured during cooling, for different frequencies from 0.1 Hz (red dots) to 0.3 (blue diamonds), 1 Hz (cyan triangles) and 3 Hz (black dots), with the same heating rate of 1 K/min.

To elucidate this matter, IF and DM have been measured at different frequencies between 3 Hz and 0.1 Hz and in Figure 3 the corresponding cooling runs are plotted. In this series of spectra we may also observe that the strength of both transformation peaks β–β' and β'–β'' exhibit a dependence on $1/\omega$ according the dependence on $\dot{N}/\omega$ predicted by the original model of Gremaud et al. [24]. This behavior is characteristic of thermo-elastic martensitic transformations, being verified on many SMA, and this fact support the idea of the thermo-elastic nature of both martensitic transformations β–β' and β'–β'', as indicated by the softening of the dynamic modulus.

In what concerns the low temperature peak P_{LT}, spectra of Fig. 3 show a shift of the temperature of the maximum with frequency as expected from a relaxation process. Although the analysis of this peak is out of the scope of the present paper and will be presented elsewhere [25], we may advance the idea that the apparent activation enthalpy measured from this temperatures shift is about 7 eV, i.e. too high to be attributed to a relaxation process with some specific physical meaning. In fact P_{LT} is not a real relaxation peak because it disappears at constant temperature exhibiting a strong dependence on time. Indeed the internal friction and dynamic modulus are plotted in Figure 4 at 1 Hz during heating up to 962 K (close to the maximum of P_{LT} peak) and then they are measured as a function of time at constant temperature. The strong decrease of the IF can be interpreted as a decrease of the number of the relaxing elements given place to IF, and the evolution on time of the DM provides evidence of a simultaneous hardening as a function of time.

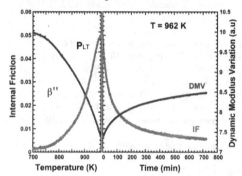

Figure 4. Internal friction (red dot) and dynamic modulus evolution (black cross) during heating (1 K/min) from 600 K up to 962 K (temperature of the maximum of the P_{LT} peak), and then holding isothermally 12 h at this temperature. The frequency is 1 Hz.

Taking into account that the strength of the peak is completely regenerated when heating just above A_f of the β'–β'' martensitic transformations [25], a relationship between the peak P_{LT} and this transformation seems to be clear. In light of the present results a mechanism of pinning of the β'' martensite interfaces by point defects mobiles in the temperature range from 800 K to 1000 K is proposed. At this point it is difficult to determine the nature of such point defects, but our samples as well as other Ru-50Nb alloys from the literature [8, 11] are expected to contain some hundreds ppm of substitutional impurities and about 200 to 700 ppm of oxygen, which could constitute the mobile defects able to migrate towards the interfaces of the martensite. The above interpretation is consistent with the observed dependence on time and the disappearance of β'' interfaces during the reverse transformation β''–β', which will re-dissolve in solid solution the point defects pinned at the interfaces, regenerating the P_{LT} peak.

CONCLUSIONS

Internal friction and dynamic modulus have been measured in Ru-50Nb (at%) intermetallics. We may conclude that both martensitic transformations β–β' and β'–β'' in Ru-50Nb SMA exhibit a sharp internal friction peak and a dynamic softening evidencing a clear thermoelastic behavior. The β'–β'' transformation is influenced by a mechanism of interfaces pinning by point defects.

ACKNOWLEDGMENTS

This work has been supported by the Spanish Ministry of Science and Innovation, MICINN, project MAT2009-12492 and the CONSOLIDER-INGENIO 2010 CSD2009-00013, by the Consolidated Research Group IT-10-310 from the Education Department and by the project ETORTEK ACTIMAT-10 from the Industry Department of the Basque Government.

REFERENCES

1. *Shape Memory Materials*, 1999, edited by K. Otsuka and C.M. Wayman (Cambridge University Press, Cambridge, 1998).
2. *Shape Memory and Superelastic Alloys*, 2011, edited by K. Yamauchi, I. Ohkata, K. Tsuchiya and S. Miyazaki (Woodhead Publishing, Cambridge, 2011).
3. J. Ma, I. Karaman and R.D. Noebe, Int. Mat. Rev. **55**, 257 (2010).
4. T. Biggs, M.B. Cortie, M.J. Witcomb and L.A. Cornish, Metall. Mater. Trans. **A 32**,1881 (2001).
5. Y. Yamabe-Mitarai, T. Hara and H. Hosoda, Mat. Sci. For. **426**, 2267 (2003).
6. Y. Yamabe-Mitarai, T. Hara, S. Miura and H. Hosoda, Mater. Trans. **47**, 650 (2006).
7. Y. Yamabe-Mitarai, T. Hara, S. Miura and H. Hosoda, Intermetallics **18**, 2275 (2010).
8. R.W. Fonda, H.N. Jones and R.A. Vandermeer, Scripta Mater. **39**, 1031 (1998).
9. R.W. Fonda and H.N. Jones, Mat. Sci. Eng. **A 273-275**, 275 (1999).
10. X. Gao, W. Cai, Y.F. Zheng and L.C. Zhao, Mat. Sci. Eng. **A 438-440**, 862 (2006).
11. K. Chastaing, A. Denquin, R. Portier and P. Vermaut, Mat. Sci. Eng. **A 481**, 702 (2008).
12. A. Manzoni, K. Chastaing, A. Denquin, P. Vermaut and R. Portier, ESOMAT-2009, 05021 (2009).
13. A. Manzoni, K. Chastaing, A. Denquin, P. Vermaut, J. Van Humbeeck and R. Portier, Scripta Mater **64**, 1071 (2011).
14. R.W. Fonda and R.A. Vandermeer, Phil. Mag. **A 76**, 119 (1997).
15. Z. He, J. Zhon and Y. Furuya, Mat. Sci. Eng. **A 348**, 36 (2003).
16. Z. He, F. Wang and J. Zhon, J. Mater. Sci. Technol. **22**, 634 (2006).
17. B.H. Chen and H.F. Franzen, J. Less Common Metals **253**, 13 (1989).
18. S.M. Shapiro, G. Xu, G. Gu, J. Gardner and R.W. Fonda, Phys. Rev. **B 73**, 214114 (2006).
19. K. Chastaing, PhD Theses, Université Pierre et Marie Curie, Paris (2007).
20. R.B. Perez-Saez, V. Recarte, M.L. Nó and J. San Juan, Phys. Rev. **B 57**, 5684 (1998).
21. J. San Juan and R.B. Perez-Saez, Mat. Sci. For. **366-368**, 416 (2001).
22. P. Simas, J. San Juan, R. Schaller and M.L. Nó, Key Eng. Mater. **423**, 89 (2010).
23. L. Dirand, M.L. Nó, K. Chastaing, A. Denquin and J. San Juan, Appl. Phys. Lett. **101**, 161909 (2012).
24. G. Gremaud, J.E. Bidaux and W. Benoit, Helv. Phys. Acta **60**, 947 (1987).
25. L. Dirand, M.L. Nó, K. Chastaing, A. Denquin and J. San Juan, to be published.

L1$_2$, B2 and DO$_3$ Compounds

Mater. Res. Soc. Symp. Proc. Vol. 1516 © 2013 Materials Research Society
DOI: 10.1557/opl.2013.65

Effect of Interstitial Carbon Atoms on Phase Stability and Mechanical Properties of E2₁ (L1₂) Ni₃AlC₁₋ₓ Single Crystals

Yoshisato Kimura, Masato Kawakita, Hiroyasu Yuyama and Yaw-Wang Chai

Tokyo Institute of Technology, Interdisciplinary Graduate School of Science and Engineering, Department of Materials Science and Engineering, 4259-J3-19 Nagatsuta, Midori-ku, Yokohama 226-8502, Japan.

ABSTRACT

Single crystals of E2₁ (L1₂) Ni₃AlC₁₋ₓ were prepared by the unidirectional solidification using the optical floating zone melting method to determine their mechanical properties. Particularly the effects of interstitial carbon atoms on mechanical properties were evaluated by compression tests at room temperature. Operative slip system of E2₁ Ni₃AlC is {111}<011> type which is the same as that of L1₂ Ni₃Al. Strength of Ni₃AlC single crystals increases with carbon concentration due to the solid solution effect, though the stress relief of yielding behavior is enhanced at the intermediate carbon content at around 3at%. A large gap appears in the carbon concentration dependence of critical resolved shear stress (as well as yield stress) at almost the same carbon content. This discontinuity in strengthening is attributed to the interaction between multiple solute carbon atoms and mobile dislocations.

INTRODUCTION

The E2₁ type intermetallic compounds M_3AlC_{1-x}, where M is Co, Fe, Ti and so forth, have been investigated with expectation that they can be promising strengthener for a new class of heat resistant alloys [1-14]. It is because the E2₁ type ordered crystal structure quite resembles to that of the L1₂ type Ni₃Al which is well-known strengthener of Ni-base superalloys. Phase stability of E2₁ type compounds has been investigated for a long time in their history. However their mechanical properties had not been examined until Hosoda et al. reported the mechanical properties of Co_3AlC_{1-x} based multi-phase alloys [2], except for the cases in which E2₁ type compounds were used as strengthening precipitates. The present authors, Kimura et al., fabricated E2₁ Co_3AlC_{1-x} single crystals and reported their mechanical properties [7]. Not only as the strengthener but also as the matrix, E2₁ type compounds M_3AlC_{1-x}, can be promising candidate materials for high temperature structural application.

Two types of ordered structures, E2₁ and L1₂, can be differentiated only by an interstitial carbon atom occupying at the cell center of the E2₁ type as shown in Fig. 1. In the chemical formula of E2₁ type M_3AlC_{1-x}, x stands for the deficiency of carbon concentration as a vacancy at the C-site of E2₁. We suggest that the E2₁ type intermetallic compounds can be classified into two groups. One is the ternary compounds which are interstitially stabilized by a ternary element carbon atom occupying at its cell center. Typical examples of this group are Co_3AlC_{1-x} and Fe_3AlC_{1-x}. Note that neither L1₂ Co₃Al nor L1₂ Fe₃Al exists at least as a stable phase in the binary system. The other is the binary L1₂-based compounds such as Ni_3AlC_{1-x}. To minimize the elastic energy, a carbon atom is allowed to occupy at the octahedral interstice only at the cell center which is surrounded by six Ni atoms octahedron so that the cell is kept as cubic. Hence, a value of x is very large in this case, in other words, the solubility of carbon is restricted small.

Thus, subscript of $_{1-x}$ is omitted hereafter and solute carbon concentration in Ni_3AlC_{1-x} single crystals is denoted by at% C (see table 1).

Objective of the present work is to understand the effect of interstitial carbon atoms on mechanical properties of $E2_1$ Ni_3AlC single crystal from the viewpoint of the resemblance and the difference between $L1_2$ and $E2_1$ type ordered crystal structures. Our research group has been carrying out a series of systematic works aiming to establish the heat resistant alloy design based on $E2_1$ type intermetallic compounds.

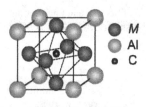

Figure 1. Unit cell of the $E2_1$ type ordered crystal structure M_3AlC_{1-x}.

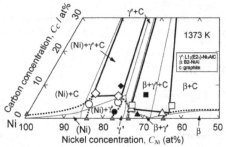

Figure 2. The isotherm of the Ni-Al-C ternary system at 1373 K [12].

EXPERIMENT

Several Ni_3AlC alloys were firstly prepared by induction heat melting and arc-melting under an argon gas atmosphere using high purity elements. Subsequently, single crystals of Ni_3AlC alloys were fabricated by unidirectional solidification using optical floating zone (OFZ) melting method. In order to prevent oxidation of alloys, flowing argon gas atmosphere was applied under a slightly positive pressure. The solidification rate was controlled in a range from 2.0 to 5.0 mm/h. To suppress the compositional super cooling at the liquid-solid interface, the melt was stirred by rotating feed and seed rods in the opposite direction at the rate of 45 rpm. Chemical compositions of single crystals were measured by electron probe microanalysis (EPMA). Carbon concentration was measured using LDE2 with the accelerating voltage of 15kV.

Compression test was conducted using a universal mechanical testing machine, Autograph 10TE, to measure the compressive mechanical properties of Ni_3AlC single crystals in the temperature range from 300 K to 1273 K. Two different loading axes were selected; one is near [-123] orientation and the other is near [001] orientation. Crystallographic orientations of single crystals were measured by X-ray analysis using back Laue. Pillar shaped compression test specimens were cut into the dimension of about 3 x 3 x 6 mm^3 or 2 x 2 x 4 mm^3 depending on the size allowance of single crystals. The initial strain rate was at around 1.6 x 10^{-4} s^{-1}. Transmission electron microscopy (TEM) was also conducted on a few compression test specimens after the test to observe deformed microstructure and to evaluate burgers vector of mobile dislocations.

RESULTS AND DISCUSSION

Preparation of Ni₃AlC single crystals and phase stability

Chemical compositions of Ni₃AlC single crystals measured using EPMA are summarized in Table 1 together with the compositional ratio of Ni to Al (Ni:Al). We intended to gradually vary the carbon concentration in a range from 1 at.% up to about 6 at.% with a step of 1 at.%, while trying to fix the compositional ratio of Ni to Al as the stoichiometric composition of 3 to 1. However the lowest carbon content was about 2.1 at.% C. The difficulty of controlling chemical composition is according to the crystal growth based on the traveling solvent effect due to the zone melting method. Chemical composition of Ni₃AlC phase is given as the equilibrium composition due to the tie-line between Ni₃AlC phase and the liquid phase at the holding temperature during the unidirectional solidification. It is not possible to control the chemical composition of Ni₃AlC phase by modifying the nominal composition of the mother alloy.

The isotherm of the Ni-Al-C ternary phase diagram [15], which was partially reconfirmed at 1373 K in this work, is shown in Fig. 2. Solubility of carbon in E2₁ Ni₃AlC phase is about 8 at% on the Ni-rich side and about 4 at% on the Al-rich side, respectively. Since excess Ni atoms substitute for the Al-site, the total number of octahedral interstitial sites which are surrounded by six Ni atoms increases without raising elastic energy. Ni₃AlC single crystals with carbon concentration higher than 4.6 at% have slightly Ni-rich compositions (the ratio of Ni:Al is 3.1:1). Solubility limit of carbon in E2₁ Ni₃AlC was evaluated by Hosoda et al. based on theoretical calculation [4], and it was explained that the phase stability of E2₁ M₃AlC₁₋ₓ should be attributed to the attractive covalent bonding between 3d transition metals M and interstitial carbon.

Table 1 Chemical alloy composition of Ni₃AlC₁₋ₓ by EPMA

Alloys	Compressive axis	Chemical composition (at.%)			
		Ni	Al	C	Ni : Al
2.1C[-123]	[-123]	73.4	24.5	2.1	3.0 : 1
3.2C[-123]		72.0	24.8	3.2	2.9 : 1
4.1C[-123]		71.2	24.7	4.1	2.9 : 1
4.6C[-123]		71.9	23.5	4.6	3.1 : 1
2.5C[001]	[001]	73.0	24.5	2.5	3.0 : 1
3.6C[001]		72.1	24.3	3.6	3.0 : 1
4.6C[001]		71.9	23.5	4.6	3.1 : 1
5.2C[001]		71.7	23.1	5.2	3.1 : 1

Compressive Mechanical Properties of Ni₃AlC single crystals

To evaluate the effect of interstitial carbon atoms on the mechanical properties of Ni₃AlC single crystal with various carbon concentrations, compression tests were conducted mainly at room temperature. Strain rate dip tests, ten times faster and one tenth lower than the initial rate, were also carried out during the test. Typical stress-strain curves are shown in Fig. 3 for two orientations of compressive loading axis; (a) near [-123] and (b) near [001]. Schmid factors S_o, S_c and S_r, and their ratios N and R are shown together in each figure. Note that both octahedral slip and cube slip can be operative in [-123] orientation, while only octahedral slip can be

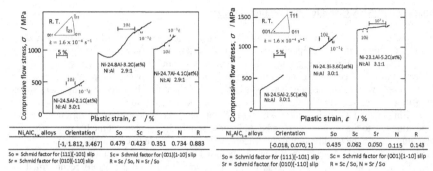

Ni$_4$AlC$_{1-x}$ alloys	Orientation	So	Sc	Sr	N	R
	[-1, 1.812, 3.467]	0.479	0.423	0.351	0.734	0.883

So = Schmid factor for (111)[-101] slip   Sc= Schmid factor for (001)[1-10] slip
Sr = Schmid factor for (010)[-110] slip   R = Sc / So, N = Sr / So

Ni$_3$AlC$_{1-x}$ alloys	Orientation	So	Sc	Sr	N	R
	[-0.018, 0.070, 1]	0.435	0.062	0.050	0.115	0.143

So = Schmid factor for (111)[-101] slip   Sc= Schmid factor for (001)[1-10] slip
Sr = Schmid factor for (010)[-110] slip   R = Sc / So, N = Sr / So

Figure 3. Stress-strain curves of compression tests conducted for E2$_1$ Ni$_3$AlC single crystals at room temperature; (a) [-123] orientation and (b) [001] orientation.

operative in [001] orientation. Compression tests were terminated at about 10 to 20 % plastic strain and all the specimens show excellent plastic deformability. Yield stress increases remarkably with carbon concentrations in both orientations, which indicates the solid solution effect due to interstitial carbon atoms. Yielding is clearly observed in all the specimens. It is interesting that yielding behavior becomes remarkably large, just like the Johnston-Gilman type yielding, particularly at intermediate carbon concentrations of 3.2C for [-123] and 3.6C for [001]. Changes in compressive flow stress as responses to the strain rate dip tests are about 4 ~ 5 MPa in low carbon specimens and about 9 ~ 16 MPa in high carbon counterparts. We interpret the rather small stress change to be governed by the density of mobile dislocations, and a large stress change is explained to be controlled by the mobility of dislocations. It indicates that the controlling factor of plastic deformation in E2$_1$ Ni$_3$AlC changes from the density of dislocations to the mobility of dislocations as the carbon concentration increases from the binary L1$_2$ Ni$_3$Al.

Carbon concentration dependence of critical resolved shear stress (CRSS) on (111)[-101] is represented in Fig. 4 for all the specimens of [-123] and [001] orientations. The result of the stoichiometric L1$_2$ Ni$_3$Al single crystal, reported by Demura et al., is represented for a comparison [16]. They explained that the orientation dependence of CRSS is very small in the

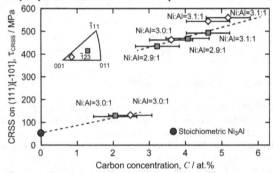

Figure 4. Carbon concentration dependence of critical resolved shear stress, τ_{CRSS}, on (111)[-101] measured for E2$_1$ Ni$_3$AlC single crystals at room temperature.

stoichiometric L1$_2$ Ni$_3$Al. We confirmed that the operative slip system of Ni$_3$AlC single crystal at room temperature is {111}<011> type which is the same as that of L1$_2$ Ni$_3$Al. Burgers vectors of mobile dislocations were analyzed using the visible-invisible criterion by TEM observation, and the possible slip plane was determined by the slip line trace analysis performed on two sides of compression test specimen. It is not possible to compare the difference of CRSS between two loading orientations exactly at the same carbon content here. Nevertheless, it can be said that the Schmidt law seems to be satisfied judging from Fig. 4. However, values of CRSS of [-123] seem to be slightly smaller than those of [001] in high carbon specimens. It may be attributed to that the cube slip somehow assists the major octahedral slip which would be effectively prohibited by interstitial carbon atoms. The solid solution hardening effect of off-stoichiometry should also be taken into account since the compositional ratio of Ni:Al is not exactly 3:1 (though close to 3:1).

It is very interesting that there appears a quite large gap in values of CRSS at around 3 at% C in carbon concentration dependence. In other words, CRSS increases drastically from about 100 MPa at low carbon contents around 2.5 at% jump up to about 400 MPa at intermediate carbon contents around 3.2 at%. The difference of resolved shear stress is about 300 MPa. Activation volume v^* was estimated by calculation based on the result of strain rate dip tests using the equation, $v^* = kT\{\ln(\gamma_2/\gamma_1)/(\tau_2-\tau_1)\}$, where k is Boltzmann constant, T is temperature, τ and γ are shear stress and shear strain, respectively. Carbon concentration dependence of the activation volume is shown in Fig. 5. Corresponding to a huge gap in CRSS, the activation volume drastically decreases as carbon concentration increases at about 3 at%. In the range of carbon concentration higher than 3 at%, activation volume tends to decrease gradually with increasing carbon concentration. A carbon atom induces the elastic stress filed around itself in the lattice. In the range of low carbon concentration, a mobile dislocation is affected by the elastic stress field of a single carbon atom. On the other hand, in the range of high carbon concentration, the elastic stress field becomes overlapped with each other. Thus, a mobile dislocation is affected by the overlapped elastic stress fields induced by multiple carbon atoms. A huge gap appearing in CRSS should be attributed to the change in the interaction between solute carbon atoms and mobile dislocations. The elastically affected region induced by solute carbon atoms should change from single atom contribution to multiple atoms overlapping at about 3 at% C. Specimens with [-123] orientation seem to have smaller activation volume than those with [001] orientation in high carbon concentration. It may be indicated that the cube slip

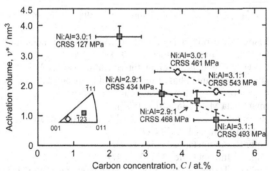

Figure 5. Carbon concentration dependence of Activation volume, v^*, calculated for E2$_1$ Ni$_3$AlC single crystals using the results of strain rate dip tests.

is somehow related to the primary octahedral slip, or the effect of solid solution hardening due to the off-stoichiometry regarding the Ni to Al ratio is involved.

CONCLUSIONS

Aiming to establish the basis of heat resistant alloy design based on $E2_1$ type intermetallic compound Ni_3AlC, the effect of interstitial carbon atoms on phase stability and mechanical properties have been investigated using single crystals fabricated in the present work. The following concluding remarks can be drawn in the present work.

1. Single crystal of $E2_1$ Ni_3AlC can be grown by the unidirectional solidification using the optical floating zone melting method.
2. Relatively large stress drop is observed in yielding behavior in the Ni_3AlC single crystal only with intermediate carbon content around 3 at% for both cases of loading axes [-123] and [001].
3. CRSS (yield stress) of $E2_1$ Ni_3AlC single crystal increases with carbon concentration due to solid solution effect of interstitial carbon, and there appears a huge gap in the carbon concentration dependence of CRSS at about 3 at% C.
4. Activation volume drastically decreases at about 3 at% C corresponding to a huge gap in CRSS. It should be explained by the change in the interaction between mobile dislocations and solute carbon atoms, i.e., the affected region induced by solute carbon atoms changes from single atom contribution to multiple atoms overlapping at about 3 at% C.

ACKNOWLEDGMENTS

The present work was partially supported by Grants-in-Aid for Scientific Research No. 23656453 JSPS, Japan.

REFERENCES

1. L. J. Huetter and H. H. Stadelmaier, Acta Metall. **6**, 367 (1958).
2. H. Hosoda, M. Takahashi, T. Suzuki and Y. Mishima, (Mater. Res. Soc. Symp. Proc. **288**, Pittsburgh, PA, 1993), pp. 793-798.
3. H. Hosoda, K. Suzuki and S. Hanada, (Mater. Res. Soc. Symp. Proc. **522**, Pittsburgh, PA, 2011), pp. KK8.31.1-6.
4. H. Hosoda and T. Inamura, (Mater. Res. Soc. Symp. Proc. **1295**, Pittsburgh, PA, 2011), pp. 77-82.
5. Y. Kimura, M. Takahashi, H. Hosoda, S. Miura and Y. Mishima, Intermetallics **8**, 749 (2000).
6. Y. Kimura, C.T. Liu and Y. Mishima, Intermetallics **9**, 1069 (2001).
7. Y. Kimura, K. Sakai, F. G. Wei and Y. Mishima, Intermetallics **14**, 1262 (2006).
8. Y. Kimura, K. Sakai and Y. Mishima, Phase Equilibria and Diffusion **27**, 14 (2006).
9. H. Ishii, K. Ohkubo, S. Miura and T. Mohri, Materials Transactions **44**, 1679 (2003).
10. L. Pang and K. S. Kumar, Acta Mater. **46**, 4017 (1998).
11. K. Ishida, H. Ohtani, N. Satoh, R. Kainuma and T, Nishizawa, ISIJ Int. **30**, 680 (1990).
12. M. Palm and G. Inden, Intermetallics **3**, 443 (1995).
13. W. Sanders and G. Sauthoff, Intermetallics **5**, 361 (1997).
14. W. Sanders and G. Sauthoff, Intermetallics **5**, 377 (1997).
15. G. Petzow and G. Effenberg, Ternary Alloys **3**, VHC, Weinheim, pp. 519-524 (1990).
16. M. Demura, Doctor Thesis, Univ. of Tokyo, (2003).

Mater. Res. Soc. Symp. Proc. Vol. 1516 © 2013 Materials Research Society
DOI: 10.1557/opl.2012.1751

The Mechanical Properties of Near-equiatomic B2/f.c.c. FeNiMnAl Alloys

Xiaolan Wu[1], Ian Baker[1], Hong Wu[2] and Paul R. Munroe[3]
[1]Thayer School of Engineering, Dartmouth College, Hanover NH 03755, U.S.A
[2]State Key Laboratory of Powder Metallurgy, Central South University, Changsha 410083, P.R. China
[3]Electron Microscope Unit, University of New South Wales, Sydney NSW 2052, Australia

ABSTRACT

Two types of as-cast microstructures have been observed in a series of near-equiatomic FeNiMnAl alloys: 1) an ultrafine microstructure in $Fe_{30}Ni_{20}Mn_{20}Al_{30}$ [1] and $Fe_{25}Ni_{25}Mn_{20}Al_{30}$, which consists of (Fe, Mn)-rich B2-ordered (ordered b.c.c.) and (Ni, Al)-rich $L2_1$-ordered (Heusler) phases aligned along <100>; and 2) a fine two-phase microstructure in $Fe_{30}Ni_{20}Mn_{30}Al_{20}$ and $Fe_{25}Ni_{25}Mn_{30}Al_{20}$, which consists of alternating (Fe, Mn)-rich f.c.c. and (Ni, Al)-rich B2-ordered platelets with an orientation relationship close to f.c.c (002) // B2 (002); f.c.c. [011] // B2 [001] [2]. The phases in $Fe_{25}Ni_{25}Mn_{20}Al_{30}$ coarsened upon annealing with no significant change in the chemical partitioning. The hardness behavior was studied as a function of the annealing time at 823 K. An $L2_1$-to-B2 transition, which occurred at 573-623K, was observed using in-situ heating in a TEM. After annealing at 973 K for 100 h, needle-shaped clusters of (Fe, Mn)-rich precipitates were observed along the grain boundaries and in the matrix. The temperature dependence of the yield strength of as-cast $Fe_{25}Ni_{25}Mn_{20}Al_{30}$ was also studied.

INTRODUCTION

Several two-phase alloys have been recently discovered in the FeNiMnAl system [1-3]. Because of their combination of attractive mechanical properties, especially high room temperature strength, inexpensive elemental components, and potentially high corrosion resistance due to the aluminum content, these FeNiMnAl alloys have been of great interest.

The primary phases observed in the as-cast alloys exhibit b.c.c., B2, $L2_1$ or f.c.c structures. The microstructure of two alloys with a higher Mn concentration and a lower Al concentration, i.e. $Fe_{30}Ni_{20}Mn_{30}Al_{20}$ and $Fe_{25}Ni_{25}Mn_{30}Al_{20}$, consists of alternating (Fe, Mn)-rich f.c.c and (Ni, Al)-rich B2 platelets with an orientation relationship close to f.c.c (002) // B2 (002); f.c.c. [011] // B2 [001] [2]. On the other hand, alloys with a lower Mn concentration and a higher Al concentration, such as $Fe_{30}Ni_{20}Mn_{30}Al_{30}$ show a very fine $B2/L2_1$ two-phase microstructure with interphase interfaces aligned along <100>.

The microstructure and mechanical properties of $Fe_{30}Ni_{20}Mn_{20}Al_{30}$ have been studied in some detail [1]. An $L2_1$-to-B2 transition occurred at 750 ± 25 K, resulting in a microstructure consisting of two B2 phases [1]. The hardness of as-cast $Fe_{30}Ni_{20}Mn_{20}Al_{30}$ was 514 ± 7 VPN [1]. Apart from a ~6% increase, the hardness showed no significant change upon subsequent annealing up to 72 h, although the microstructure coarsened considerably [1]. The temperature dependence of the compressive strength of as-cast $Fe_{30}Ni_{20}Mn_{20}Al_{30}$ showed three distinct regions: 1) fracture by transgranular cleavage before yielding below 623 K, the brittle-to-ductile transition temperature (BDTT); 2) a rapid decline in strength from ~1500 MPa to ~ 250 MPa between the BDTT and 873 K; and 3) a slower decrease to ~150 MPa above 873 K [1].

Since $Fe_{30}Ni_{20}Mn_{20}Al_{30}$ is quite brittle at the low temperatures, which limits its potential usage, another $B2/L2_1$ two-phase alloy, $Fe_{25}Ni_{25}Mn_{20}Al_{30}$, was studied because of its coarser phase size and higher hardnss in the as-cast state. This paper describes the microstructure and mechanical properties of $Fe_{25}Ni_{25}Mn_{20}Al_{30}$, and compares the behavior with other alloys examined in the FeNiMnAl system, such as $Fe_{30}Ni_{20}Mn_{20}Al_{30}$, $Fe_{30}Ni_{20}Mn_{30}Al_{20}$ and $Fe_{25}Ni_{25}Mn_{30}Al_{20}$.

EXPERIMENT

Buttons (~5 cm in diameter, ~50 g in mass) of nominal composition of $Fe_{25}Ni_{25}Mn_{20}Al_{30}$ were arc-melted in a water-chilled copper crucible under argon from pure high-purity elemental pieces. In order to produce a homogeneous alloy, the buttons were flipped and re-melted twice after the initial melting.

Microstructural analysis was performed using an FEI Tecnai F20 field emission gun (FEG) transmission electron microscope (TEM). Thin foils for the TEM were prepared by electropolishing in 20% nitric acid, 10% butoxyethanol and 70% methanol using a Struers Tenupol 5 at a voltage of ~10 V with a current of ~40 mA at ~260 K.

The microstructure of $Fe_{25}Ni_{25}Mn_{20}Al_{30}$ after a 100 h anneal at 973 K was characterized using an FEI XL-30 FEG scanning electron microscope (SEM) equipped with an electron backscattered electron (BSE) detector. Cross-sectional TEM specimens, used to examine the chemical composition, were prepared using a FEI Nova 200 Nanolab FIB. The specimens were examined using a Philips CM200 TEM operating at 200 kV. Elemental X-ray maps were collected in scanning transmission electron microscope (STEM) mode using energy dispersive X-ray spectroscopy (EDS).

The Vickers hardness of the $Fe_{25}Ni_{25}Mn_{20}Al_{30}$ after different heat treatments was determined with a Leitz MINIload tester at a load of 200 g. Samples were mounted in phenolic resin and polished to a mirror finish using 0.3 μm alumina powder before testing. The data reported are averages taken from 10 measurements, and the errors reported are one standard deviation.

Compression tests were performed on as-cast $Fe_{25}Ni_{25}Mn_{20}Al_{30}$ at temperatures up to 873 K at a strain rate of 5×10^{-4} s^{-1} using a hydraulic MTS with Inconel 718 platens. The specimens (~3×3×8 mm) were cut from the ingots using an abrasive saw followed by a grinding with 600 grit silica paper.

RESULTS AND DISCUSSION

Figure 1 shows bright field TEM images of the four two-phase FeNiMnAl alloys noted above in the as-cast state. Further TEM analysis shows that the microstructure of $Fe_{30}Ni_{20}Mn_{20}Al_{30}$ [1] and $Fe_{25}Ni_{25}Mn_{20}Al_{30}$ consists of (Fe, Mn)-rich B2 and (Ni, Al)-rich $L2_1$ phases aligned along <100>; and the microstructure of $Fe_{30}Ni_{20}Mn_{30}Al_{20}$ and $Fe_{25}Ni_{25}Mn_{30}Al_{20}$ consists of alternating (Fe, Mn)-rich f.c.c and (Ni, Al)-rich B2 platelets with an orientation relationship close to f.c.c (002) // B2 (002); f.c.c. [011] // B2 [001] [2]. The phase size for as-cast $Fe_{25}Ni_{25}Mn_{20}Al_{30}$ was ~10 nm, which is coarser than that of $Fe_{30}Ni_{20}Mn_{20}Al_{30}$ (~5 nm) [1] and increased to ~25 nm after a 240 h anneal at 823 K. In-situ TEM heating of $Fe_{25}Ni_{25}Mn_{20}Al_{30}$ up to 1173 K showed that an $L2_1$ to B2 transition started around 573 K and was complete by 623 K. A similar transition was observed in $Fe_{30}Ni_{20}Mn_{20}Al_{30}$, but at 673 K - 773 K [1].

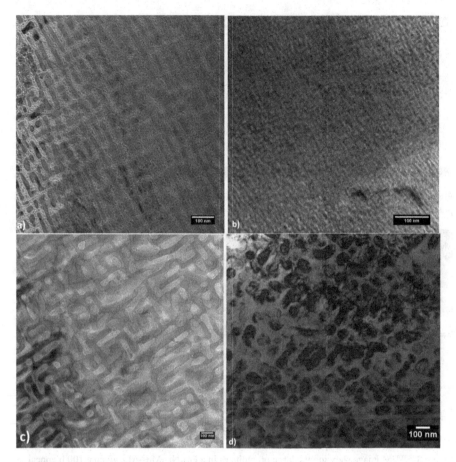

Figure 1. Bright field TEM images of as-cast (a) $Fe_{25}Ni_{25}Mn_{20}Al_{30}$, (b) $Fe_{30}Ni_{20}Mn_{20}Al_{30}$, (c) $Fe_{30}Ni_{20}Mn_{30}Al_{20}$, and (d) $Fe_{25}Ni_{25}Mn_{30}Al_{20}$.

After annealing at 973 K for 100 h, needle-shaped precipitates were observed both along the grain boundaries and in the matrix of $Fe_{25}Ni_{25}Mn_{20}Al_{30}$, see Figure 2. Figure 3 is a STEM image of the specimen produced by FIB lift-out from an area containing a high density of precipitates together with the corresponding X-ray elemental maps. The chemical composition of the precipitates is $Fe_{53}Ni_2Mn_{36}Al_9$, while the matrix composition is $Fe_{30}Ni_{26}Mn_{18}Al_{26}$. Similar precipitates were observed in other FeNiMnAl alloys [4, 5].

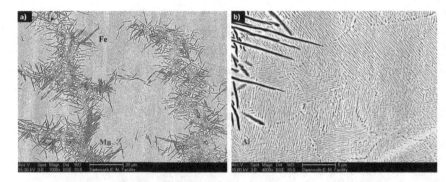

Figure 2. BSE images of $Fe_{25}Ni_{25}Mn_{20}Al_{30}$ after a 100 h anneal at 973 K: (a) showing large precipitates, and (b) higher magnification image showing the coarsened two-phase matrix.

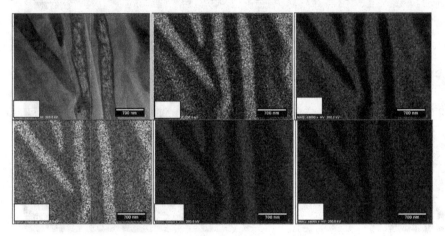

Figure 3. STEM image showing the large precipitates in a $Fe_{25}Ni_{25}Mn_{20}Al_{30}$ given a 100 h anneal at 973 K and corresponding X-ray maps using Fe, Ni, Mn and Al peaks.

The hardness behavior of $Fe_{25}Ni_{25}Mn_{20}Al_{30}$ as a function of annealing time at 823 K is shown in Figure 4 along with data for $Fe_{30}Ni_{20}Mn_{20}Al_{30}$ [1], $Fe_{30}Ni_{20}Mn_{30}Al_{20}$ and $Fe_{25}Ni_{25}Mn_{30}Al_{20}$ [2] for comparison. $Fe_{25}Ni_{25}Mn_{20}Al_{30}$ displayed a hardness of 524 ± 11 VPN in the as-cast state, and showed no significant change upon annealing, similar to the hardness behavior of $Fe_{30}Ni_{20}Mn_{20}Al_{30}$. Since the phase coarsened with no significant change in chemical partitioning upon annealing, the strength is independent of phase size, but correlated with the chemical amplitude, which could be explained using the models developed by Dahlgren [6] and Kato et al. [7]. Due to the presence of the f.c.c. phases, the as-cast hardness of two B2/f.c.c.

alloys, i.e. $Fe_{30}Ni_{20}Mn_{30}Al_{20}$ and $Fe_{25}Ni_{25}Mn_{30}Al_{20}$, is about ~100 VPN lower than that of two B2/L2$_1$ alloys, i.e. $Fe_{25}Ni_{25}Mn_{20}Al_{30}$ and $Fe_{30}Ni_{20}Mn_{20}Al_{30}$.

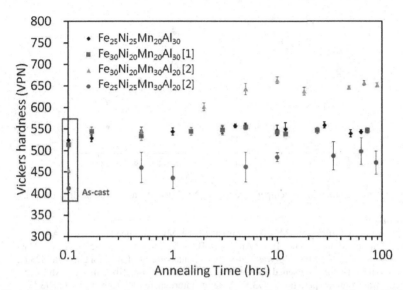

Figure 4: Room temperature hardness as a function of annealing time (log. scale) at 823 K for $Fe_{25}Ni_{25}Mn_{20}Al_{30}$, $Fe_{30}Ni_{20}Mn_{20}Al_{30}$ [1], $Fe_{30}Ni_{20}Mn_{30}Al_{20}$ and $Fe_{25}Ni_{25}Mn_{30}Al_{20}$ [2].

The compressive fracture strength of $Fe_{25}Ni_{25}Mn_{20}Al_{30}$ at room temperature was ~1700 MPa, which is ~21% higher than that of $Fe_{30}Ni_{20}Mn_{20}Al_{30}$ (~1400 MPa) [1]. Similar to $Fe_{30}Ni_{20}Mn_{20}Al_{30}$, the fracture strength of $Fe_{25}Ni_{25}Mn_{20}Al_{30}$ was independent of annealing time and all the specimens showed brittle fracture in compression at room temperature.

The temperature dependence of the compressive strength of as-cast $Fe_{25}Ni_{25}Mn_{20}Al_{30}$ is shown in Figure 5. Data for as-cast $Fe_{30}Ni_{20}Mn_{20}Al_{30}$ are plotted for comparison [1]. $Fe_{25}Ni_{25}Mn_{20}Al_{30}$ showed yielding before fracture when the temperature was higher than 373 K. The yield stress decreased from ~1700 MPa at 473 K to ~1600 MPa at 573 K, followed by a sharp drop to ~500 MPa at 873 K. Interestingly, $Fe_{25}Ni_{25}Mn_{20}Al_{30}$ showed a lower brittle-to-ductile transition temperature than $Fe_{30}Ni_{20}Mn_{20}Al_{30}$. Both alloys experience a dramatic decrease in the yield strength with the increase of temperature after the brittle-to-ductile transition temperature. This behavior at intermediate temperatures ($\sim 0.4T_m < T < \sim 0.6T_m$) is broadly similar to those observed in many B2 compounds, like FeAl [8] and NiAl [9], and is due to the onset of diffusive processes. The possible reason for lower BDTT for $Fe_{25}Ni_{25}Mn_{20}Al_{30}$ could be the coarser microstructure in $Fe_{25}Ni_{25}Mn_{20}Al_{30}$ (~10 nm in phase size) compared with that in $Fe_{30}Ni_{20}Mn_{20}Al_{30}$ (~5 nm in phase size) [1].

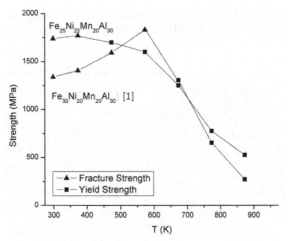

Figure 5. Strength as a function of temperature for $Fe_{25}Ni_{25}Mn_{20}Al_{30}$ and $Fe_{30}Ni_{20}Mn_{20}Al_{30}$ [1].

CONCLUSIONS

The microstructure of $Fe_{25}Ni_{25}Mn_{20}Al_{30}$ consists of (Fe, Mn)-enriched B2 and (Ni, Al)-enriched $L2_1$ phases aligned along <100>. An $L2_1$-to-B2 transition occurred at 573-623 K. The hardness of as-cast $Fe_{25}Ni_{25}Mn_{20}Al_{30}$ was 524 VPN and independent of annealing time at 823 K, although the microstructure coarsened upon annealing. The as-cast $Fe_{25}Ni_{25}Mn_{20}Al_{30}$ exhibited a brittle ductile transition temperature at 473 K. As temperature increased from BDTT to 873 K, the yield strength of as-cast $Fe_{25}Ni_{25}Mn_{20}Al_{30}$ decreased from ~1700 MPa to ~500 MPa.

ACKNOWLEDGMENTS

This research was supported by U.S. Department of Energy (DOE), Office of Basic Energy Science Award #DE-FG02-10ER46392. The views and conclusions contained herein are those of the authors and should not be interpreted as necessarily representing official policies, either expressed or implied of the DOE or the U.S. Government.

REFERENCES

1. X. Wu, I. Baker, H. Wu, M.K. Miller and P.R. Munroe, submitted to J. Mat. Sci.
2. I. Baker, H. Wu, X. Wu, M.K. Miller and P.R. Munroe, Mat. Char. 62, 952 (2011).
3. M.W. Wittmann, I. Baker, J.A. Hanna and P.R. Munroe PR, MRS Proc 842, S5.17, 35 (2005).
4. X. Wu, I. Baker, M.K. Miller, K.L. More, H. Bei and H. Wu, Intermet. 32, 1 (2013).
5. J.A. Loudis and I. Baker, Philos. Mag. 87, 5639 (2007).
6. S.D. Dahlgren, Met. Mater. Trans. A8, 347 (1977).
7. M. Kato, T. Mori and L.H. Schwartz, Acta Metall. 28, 285 (1980).
8. I. Baker, H. Xiao, O. Klein, C. Nelson and J. D. Whittenberger, Acta Metall. 43, 1723 (1995).
9. I. Baker, Mater. Sci. Eng. A 192, 1 (1995).

Mater. Res. Soc. Symp. Proc. Vol. 1516 © 2012 Materials Research Society
DOI: 10.1557/opl.2012.1564

Mechanical properties of NiAl-Mo composites produced by specially controlled directional solidification

G. Zhang[1], L. Hu[1*], W. Hu[1], G. Gottstein[1], S. Bogner[2], A. Bührig-Polaczek[2]

[1]Institute for Physical Metallurgy and Metal Physics, RWTH Aachen University, Kopernikusstraße 14, D-52074 Aachen, Germany
[2]Foundry Institute, RWTH Aachen University, Intzestraße 5, 52056 Aachen, Germany
*Correspondent author. Tel.: +49 (0) 241 80 20 25 7; fax: +49 (0) 241 80 22 30 1.
E-mail address: hulei@imm.rwth-aachen.de

ABSTRACT

Mo fiber reinforced NiAl in-situ composites with a nominal composition Ni-43.8Al-9.5Mo (at.%) were produced by specially controlled directional solidification (DS) using a laboratory-scale Bridgman furnace equipped with a liquid metal cooling (LMC) device. In these composites, single crystalline Mo fibers were precipitated out through eutectic reaction and aligned parallel to the growth direction of the ingot. Mechanical properties, i.e. the creep resistance at high temperatures (HT, between 900 °C and 1200 °C) and the fracture toughness at room temperature (RT) of in-situ NiAl-Mo composites, were characterized by tensile creep (along the growth direction) and flexure (four-point bending, vertical to the growth direction) tests, respectively. In the current study, a steady creep rate of $10^{-6}s^{-1}$ at 1100 °C under an initial applied tensile stress of 150MPa was measured. The flexure tests sustained a fracture toughness of 14.5 MPa·m$^{1/2}$ at room temperature. Compared to binary NiAl and other NiAl alloys, these properties showed a remarkably improvement in creep resistance at HT and fracture toughness at RT that makes this composite a potential candidate material for structural application at the temperatures above 1000 °C. The mechanisms responsible for the improvement of the mechanical properties in NiAl-Mo in-situ composites were discussed based on the investigation results.

INTRODUCTION

In-situ NiAl composites reinforced by refractory metallic phase like Mo through eutectic reaction has been considered to be an effective way to improve both the fracture toughness at RT and the creep resistance at HT of binary NiAl. Some substantial progresses have been achieved since the early 1990s. Joslin *et al.* [1] measured creep strength of 80 MPa at 1300K and a strain rate of $10^{-6}s^{-1}$ on directionally solidified (DS) NiAl-9Mo, compared to the creep strength of 20 MPa at 1300K and $10^{-6}s^{-1}$ on [001] NiAl. Misra and Heredia reported a RT fracture toughness of around 15 MPam$^{1/2}$ on DS NiAl-9Mo [2, 3], compared to the RT fracture toughness of [001] NiAl of about 8 MPam$^{1/2}$. However, due to the low temperature gradient (3~5 K/mm) of the Bridgeman furnace they used for directional solidification of NiAl-9Mo, the microstructure of as-produced materials was mainly comprised of cellular structures with irregularly aligned Mo fibers. In 2005, Bei *et al.* has successfully produced a unidirectional Mo fiber reinforced NiAl composite with regular alignment of fibers using an optical floating zone furnace, in which an ultimate tensile strength of about 260MPa at 1000 °C was measured, indicating the mechanical properties of *in-situ* NiAl composites like NiAl-9Mo can be remarkably improved by optimizing the alignment of reinforcing fibers with growth direction [4]. Present research work addressed the fracture toughness at RT and the tensile creep behavior at HT of as-produced composites with well aligned Mo fibers which were parallel to the growth direction.

EXPERIMENTAL METHODS

The Ni-44.5Al-9.2Mo (in at.%) buttons were produced by inductive melting furnace with high purity Ni, Al, and Mo (the purity >99.99%). For simplification, in what follows this composition will be designated NiAl-9Mo. Thereafter, the NiAl-9Mo buttons were loaded in an alumina crucible and directionally solidified in a Bridgman furnace with a specially controlled liquid metal cooling under argon protective atmosphere. The growth rate of the ingot varied from 80 mm/h to 20 mm/h. The local temperature gradient produced by metal cooling was around 8.9°C/mm. The NiAl-9Mo rods after DS were 8 mm in diameter and 40 mm in length. The samples for microstructural observation were prepared by mechanical polishing with diamond paste (particle size 1 μm) and then observed along longitudinal and transversal direction of the ingot by SEM (LEO1530).

For four-point bending tests, a U-shaped pre-notch was firstly machined by electronic discharge machining (EDM) on the specimens. Then a sharp V-shaped notch (Single Edge V-Shaped Notch Beam, SEVNB) was prepared by polishing the bottom of the U-shaped pre-notch with a razor blade using diamond paste (1μm).The geometry of the specimens and the spacing of the outer and inner span are shown in Figure 1a. The depth and bottom radius of the V-shaped notch after polishing were about 600 μm and 5-10 μm, respectively (Figure 1b). Bending tests were performed at room temperature in a servo-hydraulic materials testing machine (Schenk Hydropuls PSB250) using a head cross displacement rate of 5×10^{-4} mm/s, which was equivalent to an initial strain rate of 1.2×10^{-5} s^{-1}. Mode I fracture toughness was calculated according to the K calibration for pure bending [5].

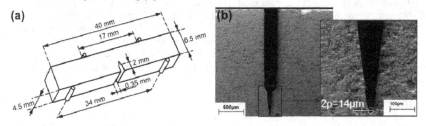

Figure 1. SEVNB NiAl-9Mo bending specimen showing, a) sample geometry with a U-shaped pre-notch and b) sharpened V-shaped notch with a root radius of approximately 7μm.

The creep behavior of DS NiAl-9Mo composites was examined by tensile creep tests at 900 -1200 °C with an initial tensile stress of 80-220 MPa in a materials testing machine (Schenk Hydropuls PSB250). The specimens solidified at R = 20 mm/h with well aligned Mo fibers (Figures 2 e and f) were chosen for the tests, in which the axis of the tensile specimens (the gage volume is $1 \times 1.3 \times 18.2$ mm^3) is parallel to the growth direction and also the [001] direction of NiAl and Mo fibers.

RESULTS

Microstructure of as-produced NiAl-9Mo composites

The microstructures of as-produced NiAl-9Mo composites with different growth rate during directional solidification are displayed in Figure 2. It is seen that when the growth rate decreased from R = 80 mm/h to R = 20 mm/h Mo fibers aligned themselves gradually from initially irregular distribution (Figures 2a-b) to finally approximately parallel to the growth direction (Figures 2e-f), where the <001> direction of Mo fibers and NiAl matrix are parallel to each other and to the growth direction [6], and therefore, a quasi-single crystal NiAl-9Mo composite with unidirectionally reinforced long Mo fibers was eventually produced

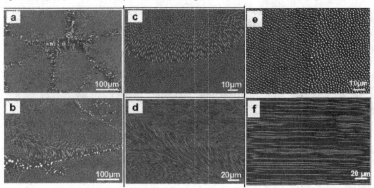

Figure 2. Microstructure of as-produced NiAl-9Mo composites at different growth rate, (a, b) R=80 mm/h. (c, d) R=40 mm/h. (e, f) R=20 mm/h. (a, c, e) parallel to the growth direction. (b, d, f) perpendicular to the growth direction.

RT Fracture toughness of NiAl-9Mo

The measured K_{IC} of as-produced NiAl-9Mo composites with different growth rate and thus, different Mo fiber alignment is listed in Table I. For comparison, the fracture toughness of [001] single crystal NiAl [7] is also given in the table. It can be seen that the fracture toughness of as-produced NiAl-9Mo composites in the current study is apparently higher than that of binary NiAl. Furthermore, in as-produced NiAl-9Mo composites the fracture toughness increased steadily when Mo fibers aligned from irregular to parallel to the growth direction. The specimens solidified at R = 20 mm/h possessed well aligned Mo fibers (Figures 2e-f) and then the highest fracture toughness of 14.5 MPa·m$^{1/2}$ (Table I), almost a factor of two higher than NiAl. The fractography of as-produced NiAl-9Mo composites solidified at different growth rates after flexure tests are shown in Figure 3. It can be seen that as the growth rate reduced and Mo fibers aligned from irregular to parallel to the DS direction, the major fracture mode evolved also from transversal interface debonding (Figure 3a) through crack bridging with partially transversal interface debonding (Figure 3b) and finally reached to the crack trapping (Figure 3c). A quantitative analysis of the contribution to the fracture toughness improvement by different toughening mechanisms was carried out by considering the interface structure and property as well as the mismatch of Young's modulus and K_{IC} between fiber and matrix, the fracture energy of the composite components, and the intrinsic toughening by crack tip blunting in NiAl [6].

Table I. Fracture toughness data for NiAl-9Mo composites

Alloy	K_{IC} (MPa·m$^{1/2}$)
NiAl-9Mo (as-cast, *irregularly aligned Mo fibers as in Figure2 a and b*)	8.7 ± 0.6
NiAl-9Mo (DS, *R*=40 mm/h, *quasi-irregularly aligned*)	10.5 ± 0.4
NiAl-9Mo (DS, *R*=20 mm/h, *well aligned*)	14.5 ± 0.3
[001] single crystal NiAl [7]	8

Figure 3. Fractography of NiAl-9Mo after flexure tests at RT. (a) transversal interface debonding mainly observed in irregularly aligned as-cast specimens (as-produced microstructure similar to Figure 2 a and b), (b) fiber bridging/necking in DS specimens with R = 40 mm/h, (c) crack trapping with fiber cleavage in well aligned specimens with R = 20 mm/h.

Creep resistance of NiAl-9Mo

Figures 4a-d show the creep response of DS NiAl-9Mo composites (R=20 mm/h), plotted as the true strain ε against the time t at different temperatures and initial tensile stresses. By derivation of the ε-t curves over the quasi-steady-state region, the steady creep rate was determined. The steady creep rate $\dot{\varepsilon}$ and the applied tensile stress σ can be described by the Mukherjee-Bird-Dorn Equation [8]:

$$\dot{\varepsilon} = \frac{AEb}{kT} D_0 exp\left(-\frac{Q_c}{RT}\right)\left(\frac{\sigma}{E}\right)^n \qquad (1)$$

where A is a temperature-dependent material constant, E the elastic modulus of NiAl-9Mo that can be evaluated by the rule of mixture with E-modulus of NiAl (E(GPa) = 204.9-0.041*(T+273), T in Celsius [9]) and Mo(E(GPa) =312-0.0418*T, T in Celsius [10]), the calculated composites E-modulus is 173GPa at 900 °C, 169GPa at 1000 °C, 165GPa at 1100 °C and 161GPa at 1200 °C respectively, b is the Burger's vector of NiAl, T the temperature in Kelvin, D_0 is the pre-exponential factor for the self-diffusion constant of Ni in NiAl, Q_c the creep activation energy of NiAl-9Mo, n the stress exponent and σ the applied tensile stress. The parameters n and Q_c of NiAl-9Mo for creep at different temperatures and initial tensile stresses were quantitatively calculated by determining the slopes of $ln\dot{\varepsilon}$ - $ln\frac{\sigma}{E}$ and $ln\left(\frac{\dot{\varepsilon}T}{E}\right)$ - $\frac{1}{RT}$ plots from the curves shown in Figures 4 a-d. Accordingly, the stress exponent of n = 3.5 ~ 5 (see Figure 4e) and the creep activation energy of Q_c = 291±19 kJ/mol (see Figure 4f) are obtained for the creep deformation of well aligned NiAl-9Mo composites at the temperatures between 900 and 1200 °C under the tensile stress of 80 ~ 220 MPa.

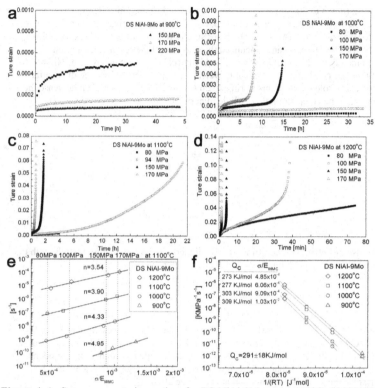

Figure 4. (a-d) true creep strain vs. time for DS NiAl-9Mo at (a) 900°C, (b)1000°C, (c)1100°C and (d)1200°C under initial stresses varying from 80 to 220 MPa. (e) steady creep rate vs. normalized stress. (f) Temperature dependence of the steady creep rate of DS NiAl-9Mo.

DISCUSSION

The creep resistance of well aligned NiAl-9Mo composites measured at 1273 K and 1373 K in the current study was compared with the ones of both single- and polycrystalline NiAl at 1300 K [9], irregularly aligned (or as-cast, microstructure similar to Figure 2a and b) NiAl-9Mo alloy at 1223 K [11], and single crystalline Ni-based superalloy NASAIR100 at 1300 K [12] in Figure 5. It is seen that the creep resistance of well aligned NiAl-9Mo has been remarkably improved compared to either the binary NiAl or the irregularly aligned NiAl-9Mo and is comparable to the Ni-based superalloy. Moreover, Figure 5 shows that around 1373 K under 150MPa tensile stress the well aligned NiAl-9Mo owns a steady creep rate of about $1 \times 10^{-6} \text{s}^{-1}$, which is corresponding to the creep rate of binary NiAl at a similar temperature under, however, a tensile stress of 20MPa [9]. This suggests that during creep in well aligned NiAl-9Mo composites, Mo fibers beard the majority of the tensile stress that decreased drastically the stress level in NiAl matrix (down to 20MPa, e.g., in above mentioned case). This inference is supported by the calculated stress exponent of n = 3.45 ~ 5 and creep activation energy of Q_c = 291±19 kJ/mol for well aligned NiAl-9Mo which are similar to those of binary NiAl (n = 5-6 and Q_c = 290 kJ/mol [9]). Therefore, it can be deduced that the creep kinetics of well aligned NiAl-9Mo is controlled mainly by the creep kinetics of NiAl matrix. The creep behavior of NiAl-9Mo under the current testing conditions is a typical power-law creep governed by edge dislocation climb in the matrix. The role of Mo fibers during creep is to bear the load and then decrease the strain rate and increase the creep strength of whole material.

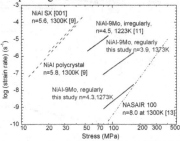

Figure 5. Comparison of creep strength of regularly aligned NiAl-9Mo in current study with the irregularly aligned NiAl-9Mo [11], binary NiAl [9] and a superalloy [13] at HT.

CONCLUSIONS

NiAl composite reinforced by well aligned Mo fibers has been successfully produced by specially controlled directional solidification, leading to the maximized material's mechanical properties along the fiber direction. Such NiAl-9Mo composite with a combination of moderate toughness at RT and remarkably enhanced creep resistance at HT is regarded as a potential candidate material for the structural application at HT above 1000 °C.

REFERENCES

1. S. M. Joslin, Ph.D. Thesis, University of Tennessee, Knoxville, TN (1995).
2. A. Misra, Z.L. Wu, M.T. Kush, and R. Gibala, Philos. Mag. A, 78, 533 (1998).
3. F.E. Heredia, M.Y. He, G.E. Lucas, A.G. Evans, H.E. Dève and D. Konitzer, Acta Metall. Mater., 41, 505 (1993).
4. H. Bei and E.P. George, Acta Mater, 53, 69 (2005).
5. W.F. Brown and J.E. Srawley, ASTM special Publication No.410, ASTM, 13 (1966).
6. L. Hu, W. Hu, G. Gottstein, S. Bogner, S. Hollad and A. Bührig-Polaczek, Mater. Sci, Eng. 539, 211 (2012).
7. A. Misra and R. Gibala, Intermetallics, 8, 1025 (2000).
8. A.K. Mukherjee, J.E. Bird and J.E. Dorn, Trans. ASM., 62, 155 (1964).
9. R.D. Noebe, R.R. Bowman and M.V. Nathal, Int. Mater. Rev., 38, 192 (1993).
10. R.Farraro and Rex. B. McLellan, Metall. Trans., 8, 1563(1977).
11. W. Ren, J. Guo, G. Li and J. Wu, Mater. T. JIM, 45 1731 (2004).
12. M. V. Nathal and L. J. Ebert, Metall. Trans., 16A, 427 (1985).

B2, L1$_0$ and Laves Phase

Mater. Res. Soc. Symp. Proc. Vol. 1516 © 2012 Materials Research Society
DOI: 10.1557/opl.2012.1565

The Effect of Li on Intermetallic Fe-Al Alloys

Xiaolin Li[1,2], Frank Stein[2] and Martin Palm[2]

[1] IEHK, RWTH Aachen, Germany
[2] Max-Planck-Institut für Eisenforschung GmbH, Düsseldorf, Germany

ABSTRACT

A couple of FeAl alloys containing up to 1.4 at.% Li have been produced by vacuum induction melting. Though previous reports indicated a significant effect of Li on the properties of FeAl, no marked changes with respect to binary FeAl are observed. Specifically, no decrease of the lattice constant and no significant increase in ductility are found by alloying with Li. If at all, there is a slight increase of the lattice constant.

INTRODUCTION

Fe-Al intermetallic alloys based on Fe_3Al or FeAl show an excellent corrosion and wear resistance, have relatively low densities compared to other Fe based alloys, and are cost effective as alloys – if at all – do contain only little amounts of strategic elements and they can be produced and processed by standard techniques [1-3]. However, low ductility at room temperature has limited their use to niche applications so far [4]. Because brittleness is not an inherent property of Fe-Al alloys but is caused by environmental embrittlement [2,5], numerous approaches have been made to enhance ductility through alloying (see e.g. [6, 7]). According to a couple of studies, alloying with up to 5 at.% Li considerably improves the mechanical properties of FeAl [8-10] and Fe_3Al [11, 12]. Specifically for FeAl alloyed with Li a decrease of the lattice constant and the hardness [8], and increases of the compressive ductility [10], elongation and ultimate tensile strength [9] have been reported. However, no evidence showing that Li is actually in solid solution in the Fe-Al intermetallic alloys has been presented, e.g. by wet chemical analysis.

It is difficult to alloy Fe-Al intermetallic alloys with Li, because of the low melting point (180.5 °C) and boiling point (1342 °C) of Li. Hence in this study, Fe-Al-Li alloys have been prepared in two different ways, and their compositions, microstructures and basic mechanical properties have been studied.

EXPERIMENT

Three alloys were produced by melting in an induction furnace under argon. The intended composition of all three alloys was Fe-45Al-2Li (all compositions in at.% except noted). Alloys 1 and 3 were produced by melting an Al-Li pre-alloy (Al-9.4 Li) and then adding chips of Fe (99.9 wt.%). Alloy 2 was produced by melting a mixture of Fe and Al (99.99 wt.%) and then adding the Al-Li alloy. All alloys were cast into a cold copper mould with diameters of 29 mm for alloys 1 and 2 and of 17 mm for alloy 3.

Alloys were cut by electrical discharge machining (EDM). Microstructures were characterised by light optical microscopy (LOM) and scanning electron microscopy (SEM;

JEOL JSM-6490). To this end, specimens were mechanically polished to 1 μm grade diamond powder finish and etched by "Ti$_2$ solution" (68 vol.% glycerin, 16 vol.% 70% HNO$_3$, 16 vol. % 40% HF). Wet-chemical analysis was complemented by energy dispersive spectrometry (EDS; EDAX Genesis V5.11) at 15 kV. Phases were identified by powder X-ray diffraction (XRD; Seifert ID 3003) using Co-Kα radiation and scanning the 2θ range 20 to 130 °. For differential thermal analysis (DTA; Setaram SETSYS-18) samples of 3 mm in diameter and 2 or 3 mm in length were investigated in alumina crucibles with a heating rate of 10 K/min under an argon atmosphere. Microhardness measurements (LECO M-400-H1) were performed using a Vickers indenter and a load of 300 g. Compression tests on rectangular samples of 5 x 5 x 10 mm^3 up to 900 °C were performed at a strain rate of 10^{-4} s^{-1} in air on a universal spindle-driven testing machine (Schenck Trebel).

RESULTS

The results of wet chemical analysis and EDS are shown in Table 1. Because Li can not be detected by EDS, the results for EDS have been recalculated (* in Table I) using the data for Li determined by wet chemical analysis. The results show that all alloys contain Li, however less than the intended composition of 2 at. % Li.

Table I. Compositions of the alloys established by wet chemical analysis and EDS (*).

alloy	Al (at. %)	Fe (at. %)	Li (at. %)
#1	54.6	44.0	1.4
#2	46.4 (*46.1)	53.1 (*53.4)	0.5
#3	48.3 (*48.5)	51.1 (*50.9)	0.6

The microstructural and XRD investigations reveal that alloy 2 and 3 are single-phase FeAl while alloy 1 consists of FeAl and FeAl$_2$ (Figure 1). The single-phase alloys have a relatively coarse grain size of about 200 μm, which is typical for FeAl alloys [13]. The microstructure of alloy 1 consists of primary FeAl dendrites surrounded by FeAl + FeAl$_2$ eutectoid. In addition, some FeAl$_2$ precipitates formed in the FeAl dendrites during cooling. It is noted that by alloying with Li apparently no ternary phase has formed and the two-phase structure of alloy 1 is in accordance with the binary Fe-Al phase diagram [14].

DTA analysis of alloys 1 and 2 (Table II) revealed that the transformation temperature from ordered FeAl (B2) to disordered A2 (αFe,Al) and the temperature of the eutectoid reaction ε ↔ FeAl + FeAl$_2$ are not affected by the addition of Li. In contrast, solidus and liquidus temperatures of both alloys are markedly lower than those of binary alloys of the same Al content [14], thus giving further evidence that Li is in solid solution in FeAl. Also, the decrease in solidus and liquidus temperatures is larger for alloy 1, which has a higher Li content.

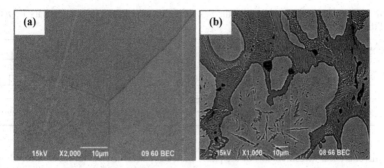

Figure 1. SEM micrographs (back-scattered electron (BSE) contrast) of alloy 2 (a) (Fe-46.4Al-0.5Li) and alloy 1 (b) (Fe-54.6Al-1.4Li). Alloy 2 (Fe-46.4Al-0.5Li) is single-phase FeAl while alloy 1 (Fe-54.6Al-1.4Li) shows primary FeAl dendrites surrounded by FeAl + FeAl$_2$ eutectoid.

Table II. Transformation temperatures of alloys 1 and 2 as determined by DTA compared to those of binary Fe-Al alloys [14] of the same Al content.

alloy	Al (at.%)	$\varepsilon \leftrightarrow$ FeAl + FeAl$_2$ (°C)	solidus (°C)	liquidus (°C)
binary Fe-Al [14]	54.6	1095	1231	1285
# 1	54.6	1093	1205	1257

alloy	Al (at.%)	B2 $\leftrightarrow$ A2 (°C)	solidus (°C)	liquidus (°C)
binary Fe-Al [14]	46.4	1318	1307	1356
# 2	46.4	1319	1287	1339

A decrease of the lattice parameter of FeAl by alloying with Li has been reported [12], which has been explained by the small atomic radius of Li. Table III shows the lattice parameters of FeAl in the three Fe-Al-Li alloys and for binary FeAl of the same Al content [15]. Compared with binary FeAl of the same composition, the lattice parameters of alloy 1 and 2 are slightly higher while the lattice parameter of alloy 3 is within the range of observed lattice parameters. It has been shown that the lattice parameter of FeAl is crucially affected by prior annealing and cooling conditions [15] (Table III). However, after a heat treatment at 400 °C for 120 hours the lattice parameter of alloy 2 shows a value similar to that in the as-cast condition. It can be concluded, that according to the present observations no decrease of the lattice parameter of FeAl is observed through alloying with Li.

Table III. Lattice parameter a_0 of FeAl of the Fe-Al-Li alloys compared to that of binary Fe-Al alloys of the same Al content [15].

alloy	Al (at.%)	a_0 FeAl + Li (nm)		a_0 FeAl (nm)	
		as cast	Annealed (400 °C/120 h)	quenched < 500 °C [15]	quenched from 1000 °C [15]
#1	51.9	0.29086(9)		0.29067	0.29030
#2	46.4	0.29084(8)	0.29093(7)	0.29040	0.29015
#3	48.3	0.29036(5)		0.29078	0.29033

Microhardness measurements of the phase FeAl have been carried out for all three alloys in the as-cast condition, measuring in precipitate-free areas of the matrix in case of alloy 1. It is noted, that no cracks originated from the corners of the Vickers indents. Figure 2 shows the microhardness of FeAl containing Li compared to that of binary FeAl [16]. Because for comparable Al contents comparable values for the microhardness are observed, it can be concluded, that the addition of Li to FeAl does not affect the microhardness. This is in contrast to an earlier report, where a decrease of the microhardness by 30% through addition of nominally 1 at. % Li has been found [10]. Presumably the decrease in microhardness reported in [10] is due to the fact that according to the micrographs shown in [10] the FeAl alloy used for comparison contained considerable amounts of Al_2O_3 and numerous fine precipitates of perovskite carbide Fe_3AlC.

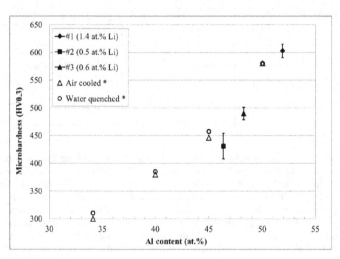

Figure 2. Microhardness in dependence of Al content for binary FeAl (open symbols) and Fe-Al + Li (filled symbols) [16].

The compressive yield strength has been established up to 900 °C. Because of the large grain size, yielding occurs smoothly. However, at temperatures above 500 °C no further work hardening is observed when the strain reaches 2 – 3%. This is similar to iron-rich Fe-Al alloys, which at temperatures above about 400 °C, after an initial period of work hardening at low strain, show no further work hardening [17].

Figure 3 shows the compressive 0.2% yield stress of as-cast alloy 2 in dependence on temperature. At intermediate temperatures, an increase of the yield strength is observed, a phenomenon known as yield strength anomaly (YSA). The YSA is typically observed at higher strain rates in Fe-Al alloys. For the present Fe-Al-Li alloys the maximum of the YSA is observed at about 450 °C. This temperature is about the same temperature at which the YSA is observed in coarse grained binary FeAl alloys with Al contents of 40 – 45 at.% [7].

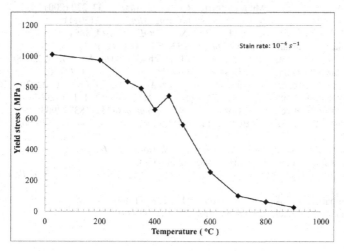

Figure 3. Compressive 0.2% yield of as-cast alloy 2 (Fe-46.4Al-0.5Li) at a strain rate of 10^{-4} s^{-1}.

CONCLUSIONS

From the above results the following conclusions can be drawn:

- It is possible to produce Fe-Al-Li alloys by vacuum induction melting using an Al-Li pre-alloy.
- A marked solid solubility of at least 1.4 at.% Li in FeAl exists.
- No ternary Fe-Al-Li phases were detected in the present investigation.
- As would be expected, addition of Li decreases liquidus and solidus temperatures.
- By the addition of Li no marked decrease of the lattice parameter of FeAl is observed, as has been reported in [12].

- No effect of Li on the microhardness could be found, contrary to a 30% decrease of the microhardness through addition of nominally 1 at.% Li reported in [16].
- The compressive 0.2% yield as well as the temperature of the maximum of the yield stress anomaly (YSA) are comparable to binary FeAl.

REFERENCES

1. D. Hardwick and G. Wallwork: *Rev. High-Temp. Mat.* **4**, 47 (1978).
2. C.G. McKamey, in: N.S. Stoloff and V.K. Sikka, editors: *Physical Metallurgy and Processing of Intermetallic Compounds.* Chapman & Hall, New York, 351 (1996).
3. D.G. Morris and M.A. Munoz-Morris: *Rev. Metal. Madrid*, **37**, 230 (2001).
4. D.G. Morris and M.A. Munoz-Morris: *Adv. Eng. Mat.*, **13**, 43 (2011).
5. C.T. Liu, C.G. McKamey and E.H. Lee: *Scr. Metall. Mat.*, **24**, 385 (1990).
6. R, Balasubramaniam: *J. Alloys Compd.*, **253-254**, 148 (1997).
7. C.T. Liu, E. P. George, P.J. Maziasz and J.H. Schneibel: *Mat. Sci. Eng.* **A258**, 84 (1998).
8. M. Salazar, R. Pérez and G. Rosas: *ATM Int. Mat. Res. Congr.*, **3**, 7 (2001).
9. G. Rosas, R. Esparza, A. Bedolla and R. Pérez: *Mater. Manuf. Process.*, **22**, 305 (2007).
10. M. Salazar, A. Albiter, G. Rosas, R. Pérez: *Mat. Sci. Eng.*, **A351**, 154 (2003).
11. M. Salazar, R. Pérez and G. Rosas: *Mat. Sci. Forum*, **426-432**, 1837 (2003).
12. G. Rosas, R. Esparza, A. Bedolla-Jacuinde and R. Pérez: *J. Mat. Eng. Perform.*, **18**, 57 (2009).
13. D. Li, A. Shan, Y. Liu and D. Lin: *Scr. Metall. Mater.*, **33**, 681 (1995).
14. F. Stein and M. Palm: *Int. J. Mat. Res.*, **98**, 580 (2007).
15. Y.A. Chang, L.M. Pike, C.T. Liu, A.R. Bilbrey and D.S. Stone: *Intermetallics*, **1**, 107 (1993).
16. P. Nagpal and I. Baker: *Metall. Trans.*, **21A**, 2281 (1990).
17. H. Xiao and I. Baker: *Scr. Metall. Mat.*, **28**, 1411 (1993).

Mater. Res. Soc. Symp. Proc. Vol. 1516 © 2012 Materials Research Society
DOI: 10.1557/opl.2012.1566

Effect of Dislocation Sources on Slip in Fe₂Nb Laves Phase with Ni in Solution

N. Takata [1], H. Ghassemi-Armaki [2], Y. Terada [1], M. Takeyama [1] and S. Kumar [2]
[1] Department of Metallurgy and Ceramics Science, Graduate School of Science and Engineering, Tokyo Institute of Technology, 2-12-1, Ookayama, Meguro-ku, Tokyo, 152-8552, Japan
[2] School of Engineering, Brown University, Providence, RI 02912, U.S.A.

ABSTRACT

We have examined the compression response of a ternary Fe₂Nb Laves phase by deforming micropillars with a diameter of ~2 μm produced by focused ion beam milling from a two-phase Fe-15Nb-40Ni (at.%) ternary alloy consisting of the Laves phase and γ-Fe. The Laves phase micropillars exhibit high strength of about 6 GPa (of the order of the theoretical shear strength of the material), followed by a burst of plastic strain and shear failure on the basal plane. If dislocation sources are introduced on a non-basal plane in the micropillars by nanoindentation prior to compression, yielding occurs at a significantly lower stress level of about 3 GPa and plastic deformation by slip proceeds on a pyramidal plane close to (-1-122). Furthermore, if regenerative dislocation sources for basal slip are present in the micropillar, the Laves phase can be continuously plastically deformed in a stable manner to at least 5% strain at a significantly lower stress of 800 MPa. We thus demonstrate the plastic deformation of this ternary Laves phase at the micron-scale at room temperature when sufficient dislocation sources are present.

INTRODUCTION

Our extensive effort has revealed the phase equilibria between γ-Fe (fcc) and Fe₂M Laves phase (C14) in Fe-M-Ni ternary systems (M: Nb, Ti, Mo) at elevated temperature [1-5], based on our design concept for the development of a new type of austenitic heat resistant steel strengthened by TCP (topologically close-packed) Fe₂M Laves phase and GCP (geometrically close-packed) Ni₃Mphase [6]. In Fe-Nb-Ni ternary system, the homogeneity region of Fe₂Nb Laves phase extends towards the equi-Nb concentration direction up to 44 at.% Ni [1,2]. This indicates that more than two thirds of the Fe sublattice sites in Fe₂Nb can be replaced by Ni atoms. The Fe-rich Fe₂Nb Laves phase with over 30 at. %Ni in solution (in equilibrium with γ-Fe) includes numerous basal planar faults extending to several micrometers [7]. One of the interesting findings in the Fe-Nb-Ni ternary Laves phase is that solute Ni atoms significantly soften the Fe-rich Laves phase [1]. This suggests that Ni in solution may enhance the plastic deformability of the Laves phase, resulting in solid solution softening.

In this study, we have examined mechanical properties of the ternary Fe₂Nb Laves phase (in equilibrium with γ-Fe) by uniaxial compression of micropillars (~2 μm diameter) and demonstrate that in certain conditions, it is possible to obtain plastic deformation at this scale at room temperature.

EXPERIMENTAL PROCEDURE

In this study, Fe-15Nb-40Ni alloy (at.%) was prepared by arc-melting and equilibrated at 1473 K/240 h, followed by water quench. The experimental methods were described in detail elsewhere [2, 5]. The sample surface was mechanically polished, followed by the electro-polishing with a Cr-saturated phosphoric acid solution at 35V, 0.5A for 30s at 333 K. Cylindrical micropillars with a diameter of about 2 μm were fabricated by focus ion beam (FIB) milling. Some of them were pre-indented by nanoindentation with a Berkovich triangular pyramid indenter at a maximum load of $1x10^{-2}$ N in order to introduce dislocation sources locally underneath the indent. A cross-sectional TEM sample underneath a nanoindent was prepared by the FIB lift-off technique. These micropillars were compressed using a HYSTRON Triboindenter equipped with a flat punch with a diameter of 10 μm at a constant strain rate of 1.0 x 10^{-3} s^{-1}. The compression direction is [6-3-32] for the C14 hexagonal Laves phase, which was determined by EBSD. The deformed micropillars were subsequently observed by SEM.

RESULTS

Figure 1 shows a back scattering electron image (BEI) showing (a) microstructure of the Fe-15Nb-40Ni alloy and (b) representative micropillars prepared on Laves phase. The Fe-15Nb-40Ni alloy exhibits a two phase microstructure of a ternary Laves phase (ε) with a composition of Fe-27Nb-33Ni (at.%) [2, 5] and γ-Fe. The coarsened eutectic microstructure includes the Laves phase with a width ranging from 10 to 40 μm. There are no cracks within the Laves phase in this alloy. Laves phase micropillars show a diameter of about 2 μm and an aspect ratio (height/diameter) varied from 2 to 4. Note that γ-Fe on the sample surface was preferentially dissolved by the prior electro-polishing.

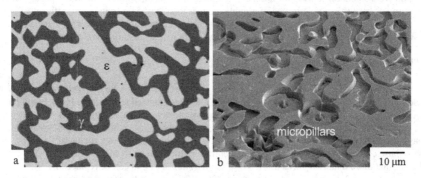

Figure 1. SEM images showing (a) microstructure of Fe-40Ni-15Nb alloy and (b) the location of Laves phase micropillars.

Figure 2 shows a representative load-displacement curve (P-δ) for the pre-nanoindentation to the micropillars, together with a calculated curve for Hertzian contact on an elastic solid [8] assuming that used Berkovich indenter tip is blunt with a radius, R of 400 nm. These curves exhibit a complete overlap till 1.6 x10^{-3} N in load, at which point, the experimental curve exhibits a large displacement excursion (pop-in), a feature that is often associated with the onset of plasticity [9]. This suggests that dislocation nucleation would occur underneath nanoindentation. These dislocations underneath the nanoindent were observed by TEM, as shown in Figure 3. Two inclined sets of slip traces were observed below the indent (Fig. 3a).

One slip trace (the one inclined towards the left side of the image) consists of a number of dislocation loops truncated in the TEM thin specimen (Fig. 3b, c). The electron diffraction pattern indicates these dislocation loops lie on a non-basal plane. These results demonstrate that non-basal dislocation loops are included in the pre-indented micropillars, especially localized close to the top surface of the micropillars.

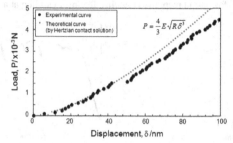

Figure 2. Load-displacement curve for the pre-indentation to the micropillars 2 series (showing pop-in behavior), together with a theoretical curve calculated by Hertzian contact solution assuming an indenter tip radius of 400 nm.

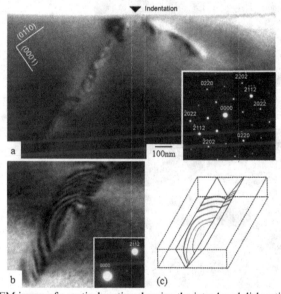

Figure 3. (a) TEM image of a vertical section showing the introduced dislocations underneath a nanoindent, observed from the $[2\bar{1}\bar{1}3]$ direction, (b) an enlarged image for dislocation loops observed under two beam condition of $g = 2\bar{1}\bar{1}2$, (c) a corresponding schematics for a geometry of dislocation loops present in a TEM thin specimen.

271

Compression stress-strain curves for micropillars are illustrated in Figure 4. Micropillars without a nanoindent (micropillar 1 series) are elastically deformed till reaching a maximum nominal stress of 6.0-6.5 GPa, prior to a rapid burst of plastic strain and unloading. In the pre-indented micropillars (micropillars 2 series), stress-strain curves show clear yielding at lower stress of about 3 GPa, followed by a burst of plastic strain. Another micropillar (micropillar 3) exhibits a very different deformation response (Fig. 4b). It shows yielding at a significantly

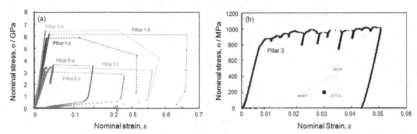

Figure 4. Uniaxial compression stress-strain curves for the Laves phase micropillars: (a) micropillars 1, 2 and (b) micropillar 3.

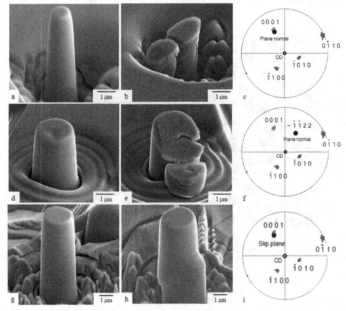

Figure 5. SEM images showing (a, d, g) initial micropillars and (b, e, h) compressed micropillars and (c, f, h) corresponding pole figures showing the possible slip plane normals: (a-c) micropillar 1-b, (d-f) micropillar 2-b and (g-i) micropillar 3.

lower stress level of about 800 MPa, followed by continuously plastic deformation (about 4 % strain) with a negligible work hardening and repeated stress drops.

Figure 5 shows SEM images of the micropillars before and after the compression test, together with pole figures indicating the activate slip plane. Compressed micropillar 1b shows shear failure surface and its two fracture surfaces are besides each other (Fig. 5b). An analysis of the inclination of the shear failure surface confirms basal slip on the shear surface (Fig. 5c). Note that the multiple variants of slip steps appear on the cylindrical side surface of the micropillar. The pre-indented micropillar (micropillar 2b) also shows shear fracture surfaces and the top piece of the micropillar is detached from its bottom piece (Fig. 5e). A crack is also observed on the fracture surface. It is interesting that the shear fracture plane is not the basal plane but rather (-1-122) pyramidal type plane (Fig. 5f). In the compressed micropillar 3, extensive parallel slip traces are present on the cylindrical surface of the micropillar. A single slip band with a width of about 1.2 μm appear to accommodate most of the plastic strain (Fig. 5h). These indicate the operation of a single slip system. A pole figure analysis (Fig. 5i) confirms the slip plane is the basal plane.

DISCUSSION

One of the important findings of our study is high strength level (about 6.2 GPa) of Laves phase micropillars can be reduced substantially by introducing the dislocation loops by pre-nanoindentation. If (0001)<0-110> type dislocations (syncro-Shockely partial dislocations [10, 11]) can be activated on a basal plane in micropillars 1, the corresponding Schmid factors for (0001)[10-10] and (0001)[1-100] slip systems can be calculated to be 0.39 and 0.37, respectively. Calculation of the resolved shear stress gives a value of 2.4 GPa, which corresponds to $G/33$ assuming the basal shear modulus, G of 79 GPa, which was determined by quantum mechanical calculation [12]. This strength level would be close to the theoretical shear strength ($G/6$-$G/30$), which suggests a lack of glissile dislocation source starvation [13] of micropillar 1. Thus, introducing mobile dislocation sources in the form of dislocation loops by nanoindentation results in an occurrence of yielding at a lower stress level of about 3 GPa.

Another important finding is stable plastic deformation of micropillar 3 at a significantly lower stress of 800 MPa. This suggests that the sufficient dislocation sources included in micropillar 3 would enable the stable slip by dislocation multiplication. A careful observation of the slip traces on the micropillar surface confirms the local discontinuity of single slip traces and multi-slip traces, as shown in Figure 6. These features suggest the presence of low-angle grain boundary networks in micropillar 3; such sub-boundaries do not intersect the top of surface of the pillar and hence precludes their identification by EBSD analysis for the sample surface. This low angle boundary would provide a regenerative dislocation source for the dislocation multiplication resulting in the stable plastic deformation.

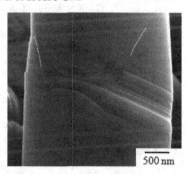

500 nm

Figure 6. SEM image showing the local discontinuity of single slip traces and multi slip traces on the surface of micropillar 3.

Our study demonstrates that in the micrometer-scale, the Fe-Nb-Ni ternary Laves phase can be plastically deformed if sufficient mobile dislocations are available. However, it is not clear whether millimeter-sized single crystal Laves phase specimens will exhibit similar plasticity at room temperature, as do the micron-sized specimen. If the specimen is scaled up, there may be locations in tensile stress in the specimen even under global compression loading. Therefore, extrapolating this plasticity to macroscopic specimens might not be valid.

CONCLUSIONS

We provide a direct evidence for plastic deformation in a ternary Fe_2Nb Laves phase micropillar (in equilibrium with γ-Fe) at room temperature by basal slip and non-basal slip when glissile dislocation sources are present. Introducing dislocation sources can substantially reduce the required stress for slip and change the slip deformation behavior of the Laves phase micropillars. Importantly, this approach is effective to study fundamental plasticity parameters in traditionally brittle materials.

ACKNOWLEDGMENTS

The alloy design and microstructure analysis were performed at Tokyo Institute of Technology, supported in part by the Grant-in-Aid (JY220215) on Advanced Low Carbon Technology Research and Development Program (ALCA), Japan Science and Technology Agency (JST). FIB milling, mechanical testing of pillars were performed at Brown University using the Experimental Shared Facilities that is supported by the NSF-MRSEC on Micro- and Nano-Mechanics of Structural and Electronic Materials (NSF Grant DMR-0520651).

REFERENCES

1. N. Gomi, S. Morita, T. Matsuo, M. Takeyama, Report of JSPS 123rd Committee on Heat Resisting Materials and Alloys **42**, 157 (2004).
2. M. Takeyama, N. Gomi, S. Morita, T. Matsuo, Mater. Res. Soc. Symp. Proc. **842**, 461 (2005).
3. S. Ishikawa, T. Matsuo, M. Takeyama, Mater. Res. Soc. Symp. Proc. **980**, 517 (2007).
4. T. Sugiura, S. Ishikawa, T. Mastuo, M. Takeyama, Mater. Sci. Forum **561-565**, 435 (2007).
5. Y. Hasebe, K. Hashimoto, M. Takeyama, J. Japan Inst. Metals, **75**, 265(2011).
6. I. Tarigan, K. Kurata, N. Takata, T. Matsuo, M. Takeyama, Mater. Res. Symp. Proc. **1295** 317 (2011).
7. N. Takata, S. Ishikawa, T. Matsuo, M. Takeyama, Mater. Res. Soc. Symp. Proc. **1128**, 475 (2009).
8. A. C. Fischer-Cripps, "Contact Mechanics", Nanoindentation, (Springer, 2010) pp.1-20.
9. D. F. Bahr, D. E. Kramer, W. W. Gerberichk, Acta Mater. **49**, 3605 (1998).
10. M. F. Chisholm, S. Kumar and P. Hazzledine, Science **307**, 701 (2005).
11. P. M. Hazzledine, P. Pirouz, Scripta Metall. Mater. **28** 1277 (1993).
12. M. Friak, D. Ma, J. Neugebauer, D. Raabe, Max-Planck-Institut für Eisenforschung GmbH, Private Communications.
13. J. R. Greer, W. D. Nix, Phys. Rev. B **73**, 245410 (2006).

Mater. Res. Soc. Symp. Proc. Vol. 1516 © 2013 Materials Research Society
DOI: 10.1557/opl.2013.142

A Study of Quaternary Cr-Cr$_2$Ta Alloys - Microstructure and Mechanical Properties

Varun Choda[1], Ayan Bhowmik[1], Ian M. Edmonds[2], C. Neil Jones[2] and Howard J. Stone[1]
[1]Department of Materials Science and Metallurgy, Pembroke Street, University of Cambridge, Cambridge CB2 3QZ, United Kingdom
[2]Rolls-Royce plc, P. O. Box 31, Derby DE24 8BJ, United Kingdom

ABSTRACT

Six alloys based on Cr-10Ta-7Si (by at.%) with quaternary additions of 0.5Ag, 5Ti, 1Hf, 3Mo, 3Al, or 3Re (by at.%) substituted for Cr were produced by vacuum arc-melting. The microstructures of the alloys were found to predominantly consist of a eutectic mixture of an A2 Cr-based solid solution and a C14 Cr$_2$Ta Laves phase along with proeutectic Cr$_2$Ta dendrites. Microstructural macro- and micro-scale inhomogeneities were observed in all alloy ingots, which were attributed to the non-equilibrium arc-melting process. The measured lattice parameters of the constituent phases and the elemental partitioning behaviour between the phases have been correlated with the respective covalent atomic radii. The bulk hardnesses of the alloys, along with the hardness of individual phases, have also been reported.

INTRODUCTION

Recently there has been renewed interest in developing Cr based alloys for high temperature structural applications, for example as turbine blades in gas turbine engines [1-3]. This interest has arisen as a result of the higher melting temperatures, higher elastic moduli, and lower densities offered by these alloys in comparison to nickel-base superalloys currently used for these applications [4].

In this study, two-phase Cr-Cr$_2$Ta alloys based on Cr-10Ta have been considered. These are hypereutectic alloys, close to the eutectic composition of Cr-9.6Ta [5]. In these alloys, the Cr$_2$Ta Laves phase possesses a high melting point and considerable high temperature strength. However, it also exhibits unacceptably low fracture toughness at room temperature. The presence of the A2 Cr-solid solution offers some ductile phase toughening; however, this is limited at lower temperatures due to interstitial embrittlement. The high Cr contents in these alloys also offer the potential of forming a protective chromia scale at elevated temperatures, although improvements are still required for these alloys to be competitive with nickel-base superalloys. Previous work on these alloys has therefore focused on improving their room temperature fracture toughness and elevated temperature oxidation resistance [6-10].

EXPERIMENT

Vacuum arc-melting and characterisation

Solid pieces of high purity elements (99.9 wt.%+) were arc-melted in a water-cooled copper hearth into cylindrical ingots of approximately 40g mass. The details of the melting process have been outlined elsewhere [11, 12]. The nominal compositions of the alloys investigated are given in Table I. Ingots were then encapsulated in quartz tubes under an Ar-atmosphere, and heat treated at 1100°C for 72 hours to reduce micro-segregation and to relieve residual solidification stresses induced during processing. The impurity contents of interstitial N and O in the alloys were determined in triplicate by combustion LECO analysis. The alloys were found to contain 0.01-0.04 wt.% O and 1-15 wt. ppm N, which were within acceptable limits when compared to literature [13]. The densities of each alloy were also determined through Archimedes' principle.

Metallographic samples were prepared to investigate the microstructure and composition of the alloys. Bulk chemical mapping was performed using electron dispersive spectroscopy

(EDS) on JEOL 5800-LV and CamScan MX 2600 scanning electron microscopes (SEM), with pure nickel as a calibration standard. Precise compositions were deduced from wavelength dispersive spectroscopy (WDS) using a Cameca SX100 electron probe microanalyser (EPMA). Pure metal standards were used and zircon as a Si standard. For bulk compositional analyses, a broad beam, approximately 50μm in diameter, was used to sample from 10-12 random locations throughout each specimen. For compositional analysis of individual phases a beam size of 1μm was used. Bulk analyses used an operating voltage of 20kV with a 50nA operating current, whereas phase analyses used two beam conditions with a beam current of 40nA for detection of Si and a 10nA beam current for all other elements. Three detectors were also used to pick up the Si-Kβ signals.

Phase identification was additionally performed by X-ray diffraction (XRD) for which ingot pieces of the order of 1g were crushed and powdered in a pestle and mortar to eliminate any textural effects. The powders were mounted on an off-axis cut Si crystal disc and placed in a Bruker D8 ADVANCE X-ray diffractometer. Measurements were performed using Cu Kα (λ = 1.54056 Å) radiation, a 40kV operating voltage, a 40mA current, and a step-size of 0.05°. The precise lattice parameters of the constituent phases were obtained using Rietveld refinement with the GSAS computer program.

Hardness

Macrohardness measurements were performed with a Vickers Pyramid Hardness Testing Machine. Rows of indents, approximately two indent widths apart, were made at 10kg and higher loads, chosen to ensure the scale of the indents was comparable to that of the alloy microstructure. Hardness readings were averaged over 8-12 indents along cross-sectional axes parallel and perpendicular to the arc-melting direction, in an effort to encompass the full range of microstructural inhomogeneities seen under the SEM.

To obtain individual phase hardnesses, mounted samples were first etched by immersion in a solution of 15g KOH, 15g $K_3[Fe(CN)_6]$ and 90ml water. This enabled the constituent phases to be visually identified, with preferential dissolution of the Cr-rich phase. Nanoindentations were made using a MTS NanoIndenter XP with a Berkovich tip under continuous stiffness mode. Nanohardness readings were taken from 6-8 random regions throughout each sample.

Table I. Comparison of nominal and measured compositions for test alloys. Also shown are the compositions of constituent phases.

| Alloy | Bulk composition/ at.% | | Phase composition/at.% | |
	Nominal	Actual	Laves Phase, Cr_2Ta	Cr-solid solution
QAV-1	82.5Cr-10Ta-7Si-0.5Ag	83.7Cr-6.3Ta-10Si-0.05Ag	57.4Cr-24Ta-18.6Si	91.3Cr-1.7Ta-7Si
QAV-2	78Cr-10Ta-7Si-5Ti	70.6Cr-11.5Ta-12.5Si-5Ti	58.7Cr-23.2Ta-14.6Si-3.6Ti	82.7Cr-1.4Ta-10.2Si-5.7Ti
QAV-3	82Cr-10Ta-7Si-1Hf	78.5Cr-8.9Ta-11.7Si-0.8Hf	57.5Cr-24.9Ta-16.4Si-1.2Hf	88.1Cr-1.5Ta-10.4Si-0Hf
QAV-4	80Cr-10Ta-7Si-3Mo	77.7Cr-9Ta-10.5Si-2.8Mo	61.2Cr-24.4Ta-11.6Si-2.9Mo	87.7Cr-1.5Ta-8.4Si-2.4Mo
QAV-5	80Cr-10Ta-7Si-3Al	81.3Cr-6.8Ta-9.1Si-2.8Al	58.7Cr-28.3Ta-11.9Si-1.0Al	86.1Cr-1.9Ta-8.3Si-3.7Al
QAV-6	80Cr-10Ta-7Si-3Re	78.6Cr-9.6Ta-8.4Si-3.4Re	52.7Cr-27.4Ta-13.2Si-6.7Re	87.2Cr-2.8Ta-6.4Si-3.7Re

RESULTS AND DISCUSSION
Alloy composition and phase analysis

Earlier studies have revealed significant differences between nominal and actual compositions after arc-melting of Cr-based refractory metal alloys [14]. The actual bulk compositions of the alloys determined from the EPMA data are summarised in Table I. The phase compositions obtained are also shown in the same table. The results show that QAV-2 is particularly Ta-rich with a bulk composition of 11.5 at.% Ta. QAV-5 on the other hand is deficient in Ta at only 6.8 at.%. Alloy Si concentrations are also generally higher than nominal. Thus, any systematic comparisons of the effect of quaternary additions on ternary Cr-Ta-Si should be tentatively made based on the actual compositions of the alloy. An interesting observation from the concentrations of Si and Ta in the Cr-rich phase is that it is actually a solid solution with primarily Si rather than Ta. Table I also reveals the partitioning behaviour of the alloying elements: Si, Re and Hf partition preferentially to the Laves phase, Ti and Al to the solid solution and Mo to both phases.

X-ray diffraction traces of the alloys are shown in Figure 1. As expected, each alloy reveals peaks characteristic of both the C14 Cr_2Ta Laves phase and an A2 Cr-rich solid solution. The stability and retention of the C14 polytype of the Laves phase has been previously investigated in some detail in binary and Si-containing ternary alloys [15, 17]. The lattice parameters of the phases obtained for each alloy are given in Table II.

By comparing the lattice parameters of the quaternary alloys on test with the lattice parameters of ternary Cr-Ta-Si from the literature, a correlation can be drawn with partitioning behaviour from EPMA data. If the radii of substitutional alloying elements are bigger or smaller than those they replace, the corresponding unit cell will also grow and shrink. Table III, reproduced from [16], gives the relevant atomic radii.

Table II. Lattice parameters for test alloys obtained through Rietveld refinement. Also included are values from two ternary Cr-Ta-Si alloys for comparison from [17].

Alloy	Lattice Parameter/Å	
	A2 Cr	C14 Cr_2Ta
Cr-Ta-5Si	a=2.8825(3)	a=b=4.8929(5), c=7.9975(10)
Cr-Ta-10Si	a=2.8846(1)	a=b=4.8899(4), c=7.9934(10)
QAV-1	a=2.88161(6)	a=b=4.8535(2), c=8.0587(2)
QAV-2	a=2.88659(2)	a=b=4.9091(3), c=8.0185(8)
QAV-3	a=2.88317(4)	a=b=4.8909(4), c=8.0658(7)
QAV-4	a=2.88402(1)	a=b=4.8758(7), c=8.0658(7)
QAV-5	a=2.88769(4)	a=b=4.8812(1), c=7.9972(4)
QAV-6	a=2.88682(6)	a=b=4.9077(2), c=8.0852(7)

Table III. Atomic radii of the elements based on covalent bonding considerations from [16].

Element	Cr	Ta	Si	Ag	Ti	Al	Mo	Hf	Re
Atomic radius/Å	1.39	1.70	1.11	1.45	1.60	1.21	1.54	1.75	1.51

In a study of Cr-Ta-Si alloys by Bhowmik et al. [17], it was found that Si partitioned strongly to the topologically close packed Laves phase, as the small Si atom was easily accommodated on the Cr sublattice, shrinking the unit cell of Cr_2Ta with increasing Si addition. As the alloys investigated in this study possess Si contents in the region of 10 at.%, the measured lattice parameters may be most appropriately compared against the 10Si alloy in [17]. QAV-1, 4, and 5 display C14 Cr_2Ta a lattice parameters smaller than that of the 10Si alloy, whereas the c lattice parameter is larger. QAV-2, 3, and 6 have larger C14 a and c parameters than the 10Si alloy. The A2 Cr unit cells of QAV-2, 4, 5, and 6 alloys are larger than that of 10Si alloy. These observations suggest the following effects upon site occupancy in the Laves phase:

1. Mo may substitute onto the Ta sublattice in the unit cell.
2. Hf being bigger than Ta and Cr possibly preferentially substitutes onto the C14 Ta sublattice. This would also be consistent with the fact that Hf forms an analogous Cr_2Hf Laves phase [3].
3. Re increases the lattice parameters of the C14 phase, possibly by substitution for Cr.

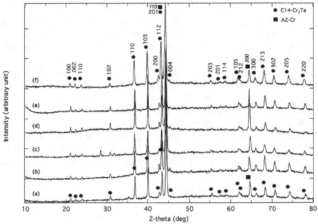

Figure 1. Diffraction patterns for test alloys with (a) QAV-1, (b) QAV-2, (c) QAV-3, (d) QAV-4, (e) QAV-5, and (f) QAV-6.

Microstructure

Figure 2 shows representative back-scattered electron images (BSEI) of all alloys after annealing at 1100°C for 72 hours. The microstructures were commonly comprised of a two-phase Cr-rich solid solution (dark phase) and Cr_2Ta Laves phase (light phase), with the contrast arising from differences between the electron densities of these phases. The eutectic Cr-Cr_2Ta lamellae are labelled **A** while Cr_2Ta Laves phase dendrites correspond to **B**. The alloys have different volume fractions of Laves phase, this is most apparent in QAV-2, to a lesser extent in QAV-5, and QAV-6, all in contrast with QAV-1, QAV-3, and QAV-4. The higher volume fractions of Cr_2Ta dendrites in QAV-2, 5 and 6 are consistent with the higher Ta contents in these alloys compared to QAV-1, 3 and 5 (Table I). In general, at higher magnifications, fine sub-micron precipitates of Cr_2Ta were seen within Cr-rich ribbons in the eutectic, which become supersaturated in Ta as solidifying eutectic colonies impinge. Laves phase dendrites seem to also solidify, when intact, in a hexagonally symmetric fashion, with six dendrite limbs apparently growing simultaneously. QAV-1 and QAV-2 showed similar Laves phase dendrites whereas in QAV-3, 4, 5, and 6 the dendrites displayed a generally higher aspect ratio, appearing as plates (rods in 2D).

In general, microstructural heterogeneities were observed along the thickness of all the alloys. The bulk of the alloy ingot exhibited a hypereutectic microstructure whereas the bottom, having undergone greater undercooling in contact with the water-cooled Cu-hearth, usually displayed columnar eutectic colonies with finer interlamellar spacings.

QAV-1 – 82.5Cr-10Ta-7Si-0.5Ag

The microstructure of QAV-1 was found to primarily comprise of an elongated eutectic mixture and proeutectic Laves phase dendrites. EDS showed the Laves phases as enriched in Ta, relatively depleted in Cr, and enriched in Si. Only 10% of the nominal Ag composition was retained in the final composition, due to the low boiling point of Ag in comparison to high melting point of Ta (Table I). At such a low concentration of Ag, the partitioning behaviour of this element in the alloy could not be reliably observed from the EDS data.

QAV-2 – 78Cr-10Ta-7Si-5Ti

278

The volume fraction of Laves phase in QAV-2 is significantly larger than that observed in QAV-1, with fragmented eutectic colonies and coarser Laves phase dendrites. Compositional analysis also shows Ti enrichment in the Cr-solid solution in comparison to the Laves phase. Of all the quaternary additions Ti in fact shows the highest solubility in the solid-solution phase (Table I). In addition, Ti and Si-rich regions were observed to form networks, not corresponding to either of the two major phases (Figure 3a). It is speculated that these regions may correspond to discrete areas of Ti and Si super-saturation within the Cr-rich phase, for example inter-dendritic boundaries, since XRD analysis did not reveal any peaks corresponding to a third phase.

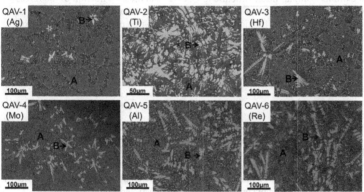

Figure 2. From left to right, top: BSEI of QAV-1, 2, 3 and bottom: QAV-4, 5, and 6.

QAV-3 – 82Cr-10Ta-7Si-1Hf

The microstructure of QAV-3 shows a higher volume fraction of the Cr-rich phase than QAV-1 but appreciably lower than that observed in QAV-2. The eutectic exists in the form of discrete globular colonies with finer inter-lamellar spacing compared to QAV-2. EDS area maps suggest Hf partitions preferentially into the Cr_2Ta Laves phase and this correlates with the EPMA compositional results in Table I, with negligible solubility in the Cr-rich phase.

QAV-4 – 80Cr-10Ta-7Si-3Mo

The microstructure of QAV-4 is generally composed of equiaxed eutectic colonies and proeutectic Laves phase. Higher magnifications reveal the extremely fine nature of the eutectic lamellae and solid state precipitates of Cr_2Ta growing from the supersaturated inter-eutectic Cr-rich phase. Hexagonally symmetric dendrites of the Laves phase exhibit an aspect ratio in between that displayed in QAV-1 and QAV-3, and EDS maps show a preference for Mo to partition to Cr-rich solid solution.

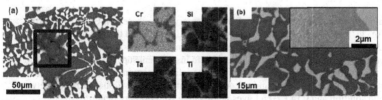

Figure 3. (a) EDS area map showing solute partitioning in QAV-2: note particularly the preferential co-partitioning of Si and Ti. (b) BSEI showing extensive Cr_2Ta precipitation from supersaturated solid solution in QAV-6 with, inset, magnified secondary electron image of same region.

QAV-5 – 80Cr-10Ta-7Si-3Al

In QAV-5, the eutectic colonies in certain regions display a lower packing density of the lamellae than in QAV-1 or 4. There is also a slight distortion of the eutectic, losing the tight disclinated texture seen in the other alloys, with whorls of Cr-rich phase amongst the eutectic sheaves. In other regions, there is a significant fraction of the alloy at the ideal eutectic composition with no hypereutectic Cr_2Ta being seen. This highlights the extent of macro-segregation in the arc-melted $Cr-Cr_2Ta$ systems under investigation.

QAV-6 – 80Cr-10Ta-7Si-3Re

QAV-6 exhibits a similar microstructure to QAV-5, and the area fraction of the Laves phase is comparable to that of eutectic. The eutectic sheaves are distorted and web through the microstructure around a significant fraction of inter-eutectic Cr-rich phase. Table I shows Re partitions strongly to the Laves phase. A faint contrast in the back-scattered mode within the solid solution phase was observed in this alloy as seen in Figure 3b. High-magnification revealed this to be a large extent of sub-micron Cr_2Ta precipitation within the solid solution. A relatively precipitate-free area was observed around the cluster of these precipitates near the periphery of the Cr-rich inter-eutectic pockets.

Hardness

The macrohardnesses of the alloys are shown in Table IV along with the nanohardnesses for the constituent phases. All the quaternary alloys were found to possess an improved hardness compared to the ternary Cr-Ta-Si alloys, which was reported to be ~ 446 VPN [11]. Unsurprisingly, considering the high volume fraction of Laves phase dendrites, QAV-2 exhibits the highest bulk hardness. However, nanohardnesses of the Cr-rich phase suggest some solid solution strengthening of this phase occurs, particularly upon Mo-addition. Given the hypereutectic nature of all the alloys, it is believed that the volume fraction of the hard Laves phase intermetallic has a greater bearing upon the overall hardness than other sources of hardening. The high deviations between the phase hardnesses are attributed to the fine scale of the microstructure, and difficulty pinpointing phases under low magnification optical microscopy. For example the higher hardness of the solid solution in QAV-6 might be due to interference from the fine Cr_2Ta precipitates within the phase. Note also the apparent extreme hardnesses of the Laves phase, particularly in QAV-3, 4, and 6.

Table IV. Macrohardness measurements for test alloys and nanohardnesses for constituent phases.

Alloy	Bulk Macrohardness/VPN	Laves Phase Nanohardness/GPa	Cr-solid solution Nanohardness/GPa
QAV-1	462±10	11.0±0.8	8.1±1.7
QAV-2	564±16	11.0±1.6	8.2±0.8
QAV-3	488±16	19.6±1.0	8.4±1.9
QAV-4	510±20	21.0±1.4	8.3±0.3
QAV-5	517±15	17.0±4.9	6.7±1.2
QAV-6	518±7	21.0±2.0	7.9±1.1

SUMMARY

The microstructures and hardnesses of six quaternary alloys based on Cr-10Ta-7Si with additions of Ag, Ti, Mo, Al, Hf, and Re have been studied. It was found that the arc-melting process introduced macro-scale heterogeneities across the ingots. This was thought to arise from incomplete remelting of ingots during processing, due to insufficient superheat to keep the refractory mixture molten leading to differential rates of solidification in different ingot sections. The microstructure of all alloys was found to comprise primarily of a two-phase eutectic between A2 Cr-rich solid solution and C14 Cr_2Ta Laves phase along with primary dendrites of proeutectic Laves phase.

Lattice parameters of the constituent phases were rationalised based on geometric principles of atomic radii. The alloys retained a metastable C14 hexagonal crystal structure, likely due to the Si-stabilising effect [18, 19].

The highest hardness was obtained with QAV-2, which would be attributed to a high Laves phase fraction in this alloy. Ag additions resulted in the lowest macrohardness. Data characterising the fracture behaviour of these alloys will be communicated in a further work.

ACKNOWLEDGMENTS

The authors acknowledge the financial support provided under the EPSRC and Rolls-Royce Strategic Partnership (EP/H500375/1) to carry out the work. The assistance of Dr. Iris Buisman, Department of Earth Sciences, University of Cambridge, with the EPMA experiments is also duly regarded.

REFERENCES

[1] K.S. Kumar, C.T. Liu, Acta Metallurgica 45 (1997) 3671-3686.
[2] M.P. Brady, C.T. Liu, J.H. Zhu, P.F. Tortorelli, L.R. Walker, Scripta Materialia 52 (2005) 815-819.
[3] Y.F. Gu, H. Harada, Y. Ro, Journal of Metals, 56 (2004) 28-33.
[4] W.D. Klopp, Journal of the Less-Common Metals, 42 (1975) 261-278.
[5] A. Bhowmik, H.J. Stone, Metallurgical and Materials Transactions A, 43 (2012) 3283-3292.
[6] M.P. Brady, J.H. Zhu, C.T. Liu, P.F. Tortorelli, L.R. Walker, Intermetallics, 8 (2000) 1111-1118.
[7] T. Takasugi, S. Hanada, M. Yoshida, Materials Science & Engineering A, 192/193 (1995) 805-810.
[8] T. Takasugi, K.S. Kumar, C.T. Liu, E.H. Lee, Materials Science & Engineering A, 260 (1999) 108-123.
[9] M. Takeyama, C.T. Liu, Materials Science & Engineering A, 132 (1991) 61-66.
[10] F. Laves, H. Witte, Metall-Wirtschaft,-Wissenschaft und -Technik, 14 (1935) 645-653.
[11] A. Bhowmik, C.N. Jones, I.M. Edmonds, H.J. Stone, Journal of Alloys and Compounds, 530 (2012) 169-177.
[12] A. Bhowmik, K.M. Knowles, H.J. Stone, Intermetallics, 31 (2012) 34-47.
[13] C.L. Briant, K.S. Kumar, N. Rosenberg, H. Tomioka, International Journal of Refractory Metals and Hard Materials, 18 (2000) 9-11.
[14] Y.F. Gu, Y. Ro, H. Harada, Metallurgical and Materials Transactions A, 35 (2004) 3329-3331.
[15] K.S. Kumar, P.M. Hazzledine, Intermetallics, 12 (2004) 763-770.
[16] B. Cordero, V. Gomez, A.E. Platero-Prats, M. Reves, J. Echeverria, E. Cremades, F. Barragan, S. Alvarez, Dalton Trans., issue 21 (2008) 2832-2838.
[17] A. Bhowmik, H.T. Pang, S. Neumeier, H.J. Stone, I. Edmonds, MRS Proceedings, 1295 (2011) 1-6.
[18] A. Bhowmik, H.J. Stone, Journal of Materials Science, *accepted* (2013).
[19] R.C. Mittal, S.K. Si, K.P. Gupta, Journal of the Less Common Metals, 60 (1978) 75-82.

Mater. Res. Soc. Symp. Proc. Vol. 1516 © 2013 Materials Research Society
DOI: 10.1557/opl.2012.1752

Micropillar Compression Deformation of Fe-Zn Intermetallic Compounds in the Coating Layer of Galvannealed Steel

Norihiko L. Okamoto, Daisuke Kashioka and Haruyuki Inui

Department of Materials Science and Engineering, Kyoto University, Sakyo-ku, Kyoto 606-8501, Japan

ABSTRACT

The deformation behavior of two of the five Fe-Zn intermetallic phases (Γ, Γ_1, δ_{1k}, δ_{1p} and ζ), which are formed in the coating layer of galvannealed steel, has been investigated through uniaxial compression tests for single-phase polycrystalline micropillars. The ζ phase is ductile to some extent while the Γ_1 phase is brittle. These results are consistent with the Peierls stress estimated from the crystal structures by assuming the primitive Peierls-Nabarro model.

INTRODUCTION

Zinc-coated (galvanized) steel is widely used in applications in automotive and building industries. Zinc is coated to improve the aqueous corrosion resistance of steel by a shielding mechanism called galvanic protection, in which the substrate steel is cathodically protected by the sacrificial corrosion of the zinc coating because zinc is less noble (more electronegative) than iron. The galvanized steel is sometimes further heat-treated (galvannealed: GA) to alloy the zinc coating with the substrate iron by diffusion, resulting in improved coating adhesion and weldability. The coating layer of galvannealed steel consists of a lamellar series of intermetallic compounds in the Fe-Zn system; Γ (Fe_3Zn_{10}), Γ_1 ($Fe_{11}Zn_{40}$), δ_{1k} ($FeZn_7$), δ_{1p} ($FeZn_{10}$) and ζ ($FeZn_{13}$) in decreasing order of the iron content (see figures 1(a,b)) [1]. The deformation and fracture behavior of these intermetallic compounds influences the press formability response of the galvannealed steel. During press forming, zinc coating occasionally fails as a result of decohesion at the coating/substrate interface (flaking) and/or particle formation by intracoating failure (powdering). The coating failure occurs more significantly with both increasing Fe content and thickness of the coating layer. It has been reported in review articles that the δ_1 (δ_{1k}/δ_{1p}) phase is the most ductile and the Γ (Γ/Γ_1) and ζ are brittle [2], whereas the ζ phase is the most ductile and the δ_1 and Γ are brittle [3]. However, the coating failure has been only phenomenologically understood and still under discussion. No mechanical properties of each phase have been investigated except for micro-Vickers hardness, partly because of difficulties in preparing single-phase specimens in bulk form. If any, the neighboring phases (δ_{1k}/δ_{1p} as well as Γ/Γ_1) have not been always distinguished because the layers are as thin as a single micrometer [4-7]. However, recent advances in fabrication processes with precise control of material dimensions down to nanometer level have made it possible to investigate mechanical properties at these scales [8-10]. In the present paper, we investigate compression deformation behavior of two of the five intermetallic phases through compression tests of micrometer-sized specimens prepared via the focused ion beam (FIB) method (figure 1(c)).

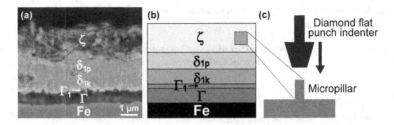

Figure 1. (a) Cross-sectional SEM backscattered electron image of GA steel. The Fe-Zn intermetallic compounds, Γ, Γ_1, δ_{1k}, δ_{1p} and ζ, form a lamellar structure in the coat layer. (b) Schematic of the layer structure of the GA coating. Micropillar specimens are machined from each intermetallic layer for compression tests. (c) Schematic of micropillar compression testing.

CRYSTAL STRUCTURES AND PREDICTED PLASTICITY

Basically all the five Fe-Zn compounds possess rather complicated crystal structures as shown in figure 2. The Γ phase possesses a cubic unit cell which contains 52 atoms and can be regarded as consisting of $3\times3\times3$ bcc-based cells, including two atomic vacancies at the corner and body center of the unit cell [11-12]. The Γ_1 phase possesses a $2\times2\times2$ supercell of the Γ phase including 408 atoms per unit cell [13]. While the δ_{1p} phase comprises a huge hexagonal lattice including 556 atoms per unit cell, in which Fe atoms form clusters [12], almost nothing is known about the crystal structure of the δ_{1k} phase except that the δ_{1k} phase has a superlattice structure of the δ_{1p} phase [14]. The ζ phase has a base-centered monoclinic unit cell in which Zn_{12} icosahedra encapsulating a Fe atom are linked to one another [15].

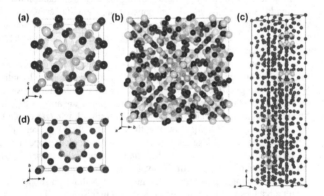

Figure 2. Crystal structures of intermetallic compounds in the Fe-Zn system. (a) Γ [12], (b) Γ_1 [13], (c) δ_{1p} [12], and (d) ζ [15-16] phases. Black: Zn, light gray: Fe, dark gray: mixed occupation of Fe and Zn.

Within the Peierls-Nabarro (PN) model [17-18], shear stress required for the motion of edge dislocations (PN stress), τ_p, is given by

$$\tau_p = \frac{2\mu}{1-v}\exp\left(-\frac{2\pi d}{(1-v)b}\right),\qquad(1)$$

where μ, v, d and b stand for the shear modulus, Poisson's ratio, atomic plane distance and magnitude of Burgers vector, respectively. Since the d/b ratio in the exponential largely dominates the magnitude of the PN stress, we compare the d/b ratios for the Fe-Zn intermetallic compounds as well as pure iron, as tabulated in Table I. At first, the shortest translational vector is chosen for the slip (Burgers vector b) vector. Then, the plane with the largest atomic plane distance that includes the selected slip vector is chosen for the slip plane. It is to note here that the selection of the slip vector and plane might not be correct and other slip systems might occur in each phase. The d/b ratios estimated for the Fe-Zn intermetallic compounds are considerably small (0.099–0.276) compared to that for iron (0.817), indicating that the expected magnitude of τ_p for those compounds is very high. Among the Fe-Zn compounds, the ζ phase has the largest d/b ratio of 0.276, suggesting it may exhibit ductility, whereas the Γ_1 phase has the smallest d/b ratio of 0.099, suggesting its poorest ductility.

Table I. Crystal structure parameters, shortest Burgers vectors, and largest atomic plane distances for the Fe-Zn intermetallic compounds and pure iron.

	Composition	Pearson Symbol	Lattice Constant (nm)	Burgers Vector, b	Atomic Plane Distance, d	d/b
Γ [19]	Fe_3Zn_{10}	$cI52$	a=0.9018	1/2<111> 0.7810 nm	{110} 0.1349 nm	0.173
Γ_1 [13]	$Fe_{11}Zn_{40}$	$cF408$	a=1.7963	1/2<110> 1.2702 nm	{111} 0.1260 nm	0.099
δ_{1k}^{*}	$FeZn_7$	-	-	-	-	-
δ_{1p} [12]	$FeZn_{10}$	$hP556$	a=1.2787 c=5.7222	1/3<11$\bar{2}$0> 1.2787 nm	(0001) 0.1298 nm	0.102
ζ [16]	$FeZn_{13}$	$mC28$	a=1.0862 b=0.7608 c=0.5061 β=100.32°	[001] 0.5061 nm	{110} 0.1395 nm	0.276
Fe [20]	-	$cI2$	a=0.2867	1/2<111> 0.2483 nm	{110} 0.2027 nm	0.817

[*]The crystal structure of the δ_{1k} phase has not been determined yet although it has been reported that it possesses a superlattice structure of the δ_{1p} phase [14].

EXPERIMENTAL PROCEDURES

Galvannealed mild steel with a coating thickness of approximately 10 μm was soaked in a pure zinc bath at 450°C for 30 min to develop the coating layer (figure 3(a)). The heat-treated GA was subsequently immersed into a salt bath at 450°C for 24 h (figure 3(b)). Single-phase square columnar specimens 3.0−4.0 μm on a side with an aspect ratio of 1:3−1:4 were machined from the heat-treated GA steels by using a JEOL JIB-4000 FIB at an operating voltage of 30 kV and a beam current of 100−300 pA. Uniaxial compression tests were conducted with a flat punch indenter tip in a Shimadzu MCT-211 micro compression tester at room temperature (figure 1(c)). The compression tests were performed with a constant stress rate of 1−3 MPa/s, which corresponds to a nominal strain rate of $3 \times 10^{-5} - 3 \times 10^{-4}$ s^{-1} in the elastic deformation region. Before and after the compression tests, the specimen surface was observed by a JEOL JSM−7001FA scanning electron microscope (SEM).

RESULTS AND DISCUSSION

Figure 4 shows SEM secondary electron images as well as stress-strain curves obtained for the Γ_1 and ζ single-phase polycrystalline micropillar specimens. While the micropillar of the Γ_1 phase is fractured without exhibiting any plastic deformation at a stress level as high as 1,650 MPa (figure 4(b)), that of the ζ phase exhibits ductility to some extent as exemplified by slip lines observed on the orthogonal surfaces of the micropillar specimens (figure 4(c)). It is not surprising that the Γ_1 phase with the smallest d/b ratio exhibits no plastic deformation while the ζ phase with the largest d/b ratio exhibits plasticity. To statistically examine the brittleness/ ductility of the Γ_1 and ζ phases, additional compression tests were carried out for several polycrystalline micropillars. For the Γ_1 phase, all the micropillars investigated were fractured exhibiting no plastic deformation. For the ζ phase, some micropillars were fractured soon after yielding at a strain level of only around 0.5%. However, our preliminary experiments of

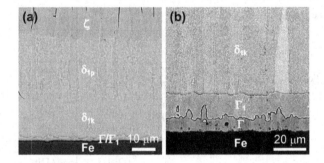

Figure 3. Cross-sectional SEM backscattered electron images of post annealed GA steel sheet. (a) Immersed in a pure zinc bath at 450°C for 30 min (Specimen A). (b) Subsequently immersed into a salt bath at 450°C for 24 hrs (Specimen B). Micropillar specimens for the Γ_1 phase are machined from Specimen B, while those for the ζ phase are machined from Specimen A.

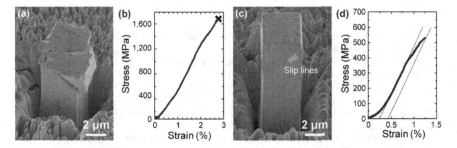

Figure 4. SEM secondary electron images of single-phase polycrystalline micropillar specimens after compression tests for (a) Γ_1 and (c) ζ phases. Stress-strain curves obtained for the micropillar compression tests for (b) Γ_1 and (d) ζ phases.

single-crystalline micropillar compression indicate that the ζ phase exhibits considerably large ductility in a single-crystalline form [21]. This discrepancy between the polycrystalline and single-crystalline forms infers that the ζ phase possesses at least one soft slip mode but does not satisfy the von Mises criterion, according to which five independent slip systems are required for a crystal grain to undergo an arbitrary imposed plastic deformation [22-23]. Owing to the low symmetry of the monoclinic lattice (space group: $C2/m$) [15-16], many non-equivalent slip planes have to operate to satisfy the von Mises criterion. In order to clarify the reason for the discrepancy in the ductility between polycrystalline and single-crystalline forms, the deformation modes in single crystals of the ζ phase are now under survey [21].

CONCLUSIONS

Compression deformation behaviour of two of the five Fe-Zn intermetallic phases (Γ_1 and ζ) has been investigated through uniaxial compression tests of single-phase polycrystalline micropillars which are fabricated by the FIB technique. The ζ phase is ductile to some extent while the Γ_1 phase is brittle. These results are consistent with the Peierls stress estimated from the crystal structures by assuming the primitive Peierls-Nabarro model.

ACKNOWLEDGMENTS

This work was supported by JSPS KAKENHI grant number 24246113 and the Elements Strategy Initiative for Structural Materials (ESISM) from the Ministry of Education, Culture, Sports, Science and Technology (MEXT) of Japan, and in part by Advanced Low Carbon Technology Research and Development Program (ALCA) from the Japan Science and Technology Agency (JST). This work was also supported by Research Promotion Grant from ISIJ and Grants for Technical Research from JFE 21st Century Foundation.

REFERENCES

1. P. Villars, *Pearson's Handbook: Crystallographic Data for Intermetallic Phases* (ASM International, Amsterdam, 1997).
2. J. Mackowiak and N. R. Short, *Inter. Metals Rev.* **24,** 1 (1979).
3. A. R. Marder, *Prog. Mater Sci.* **45,** 191 (2000).
4. M. A. Ghoniem and K. Lohberg, *Metall* **26,** 1026 (1972).
5. G. F. Bastin, F. J. J. Vanloo, and G. D. Rieck, *Z. Metallkd.* **65,** 656 (1974).
6. P. J. Gellings, E. Willemdebree, and G. Gierman, *Z. Metallkd.* **70,** 312 (1979).
7. C. E. Jordan and A. R. Marder, *J. Mater. Sci.* **32,** 5593 (1997).
8. D. M. Dimiduk, M. D. Uchic, and T. A. Parthasarathy, *Acta Mater.* **53,** 4065 (2005).
9. M. D. Uchic, P. A. Shade, and D. M. Dimiduk, *Annu. Rev. Mater. Res.* **39,** 361 (2009).
10. J. R. Greer and J. T. M. De Hosson, *Prog. Mater Sci.* **56,** 654 (2011).
11. M. H. Hong and H. Saka, *Philos. Mag. A* **74,** 509 (1996).
12. C. H. E. Belin and R. C. H. Belin, *J. Solid State Chem.* **151,** 85 (2000).
13. A. S. Koster and J. C. Schoone, *Acta Crystallogr. Sect. B: Struct. Sci.* **37,** 1905 (1981).
14. M. H. Hong and H. Saka, *Scripta Mater.* **36,** 1423 (1997).
15. R. Belin, M. Tillard, and L. Monconduit, *Acta Crystallogr. Sect. C: Cryst. Struct. Commun.* **56,** 267 (2000).
16. P. J. Gellings, E. Willemdebree, and G. Gierman, *Z. Metallkd.* **70,** 315 (1979).
17. R. Peierls, *Proc. Phys. Soc.* **52,** 34 (1940).
18. F. R. N. Nabarro, *Proc. Phys. Soc.* **52,** 90 (1940).
19. J. K. Brandon, R. Y. Brizard, P. C. Chieh, R. K. Mcmillan, and W. B. Pearson, *Acta Crystallogr. Sect. B: Struct. Sci.* **30,** 1412 (1974).
20. R. Kohlhaas, P. Dunner, and Schmitzp.N, *Z. Angew. Phys.* **23,** 245 (1967).
21. M. Inomoto, N. L. Okamoto, and H. Inui, *Mater. Res. Soc. Symp. Proc.* submitted.
22. R. von Mises, *Z. Angew. Math. Mech.* **8,** 161 (1928).
23. G. W. Groves and A. Kelly, *Philos. Mag.* **8,** 877 (1963).

Silicides and LPSO Phases

Mater. Res. Soc. Symp. Proc. Vol. 1516 © 2013 Materials Research Society
DOI: 10.1557/opl.2013.17

Crystal Structures of Long-Period Stacking-Ordered Phases in the Mg-TM-RE Ternary Systems

Kyosuke Kishida, Hideyuki Yokobayashi, Atsushi Inoue and Haruyuki Inui
Department of Materials Science and Engineering, Kyoto University,
Sakyo-ku, Kyoto 606-8501, JAPAN

ABSTRACT

Crystal structures of long-period stacking-ordered (LPSO) phases in the Mg-TM (transition-metal)-RE(rare-earth) systems were investigated by atomic resolution high-angle annular dark-field scanning transmission electron microscopy (HAADF-STEM). The 18R-type LPSO phase is constructed by stacking 6-layer structural blocks, each of which contains four consecutive close-packed planes enriched with TM and RE atoms. Formation of the TM_6RE_8 clusters with the $L1_2$ type atomic arrangement is commonly observed in both Mg-Al-Gd and Mg-Zn-Y LPSO phases. The difference between the crystal structures of Mg-Al-Gd and Mg-Zn-Y LPSO phases can be interpreted as the difference in the in-plane ordering of the TM_6RE_8 clusters in the structural block. The Mg-Al-Gd LPSO phase exhibits a long-range in-plane ordering of Gd and Al, which can be described by the periodic arrangement of the Al_6Gd_8 clusters with the $L1_2$ type atomic arrangement on lattice points of a two-dimensional $2\sqrt{3}a_{Mg} \times 2\sqrt{3}a_{Mg}$ primitive hexagonal lattice, although the LPSO phase in the Zn/Y-poor Mg-Zn-Y alloys exhibits a short-range in-plane ordering of the Zn_6Y_8 clusters.

INTRODUCTION

Mg alloys containing ternary Mg-TM(Transition-metal)-RE(Rare-earth) phases with long-period stacking-ordered (LPSO) structures have received a considerable amount of attention in recent years [1-3]. Although reasons why these alloys can simultaneously exhibit high strength and high ductility have been remained largely unsolved, ternary LPSO phases have been believed to play important roles in endowing them with excellent mechanical properties. Mg-Zn-RE LPSO phases are reported to consist of structural blocks with five to eight close-packed atomic planes, forming various polytypes with different numbers of the close-packed atomic planes in the structural blocks and with different stackings of the structural blocks [4]. In the absence of the in-plane long-range ordering of the constituent atoms (as usually assumed in most studies in Mg-TM-RE LPSO phases), polytypes expressed as 10H, 14H, 18R and 24R polytypes are reported to form, among which 14H and 18R polytypes are the most dominantly observed ones [1,2]. The Mg-TM-RE LPSO structures have been generally characterized by periodic occurrence of stacking faults within the HCP stacking of parent Mg and also by enrichment of TM and RE atoms in two atomic layers adjacent to the stacking fault. Figure 1 shows a typical example of an 18R-type LPSO phase observed in Zn/Y-poor Mg-Zn-Y alloys [5]. An atomic resolution image taken by high-angle annular dark-field scanning electron microscopy (HAADF-STEM) (figure 1a) indicate that the LPSO phase has 18 layers in the hexagonal unit cell and stacking faults exist every six layers in a hcp stacking. In addition, the selected area electron

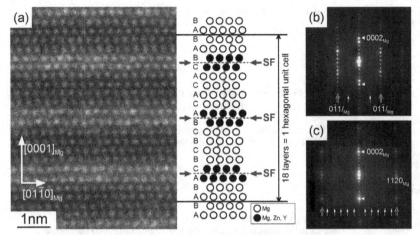

Figure 1. (a) HAADF-STEM image and (b,c) SAED patterns taken from an 18R-type LPSO phase in a Zn/Y-poor Mg-Y-Zn alloy in (a),(b) [2$\bar{1}\bar{1}$0] and (c) [1$\bar{1}$00] projections. Arrows in (b) and (c) indicate the positions of diffuse streaks extending along [0001] [5].

diffraction (SAED) patterns exhibit diffuse scattering (marked by arrows in figures 1b and 1c), which implies the occurrence of a short-range ordering of Zn and Y in the enriched layers. Although the atomic resolution HAADF-STEM imaging has great advantages in determining the ordered arrangements with the strong Z-dependence of the contrast [6,7], it is still difficult to observe the short range ordering in the Mg-Zn-Y LPSO phase partly because of relatively low concentrations of Zn and Y. Thus, the details of the in-plane arrangement in the enriched layers have not been fully clarified yet.

We have very recently investigated the crystal structure of both the 18R- and 14H-type LPSO phase newly found in the Mg-Al-Gd system by atomic resolution HAADF-STEM and transmission electron microscopy (TEM) and successfully determined the in-plane ordered arrangement of Gd and Al in the enriched layers [5,8-10]. Because of the in-plane ordering natures, the LPSO phase in the Mg-Al-Gd system cannot be described as an 'LPSO' phase any longer in a strict crystallographic sense but as an order-disorder (OD) intermetallic phase with a so-called OD structure, which have been reported in many minerals but not in intermetallic compounds so far. In this paper, we will review the details of the crystal structures of the OD/LPSO phases in Mg-TM-RE alloys on the basis of the OD theory especially focusing on the difference in the in-plane ordering natures

EXPERIMENT

Ingots of Mg-Al-Gd and Mg-Zn-Y ternary alloys with nominal compositions of Mg - 3.5 at.%Al - 5.0 at.%Gd and Mg - 1.0 at.%Zn - 2.0 at.%Y were produced by high-frequency induction-melting in an argon atmosphere. The ingots of the Mg-Al-Gd alloy were homogenized at 550 °C for 2 hours and then heat-treated at 400 °C for 10 hours. Microstructures of both alloys were examined by transmission electron microscopy and scanning transmission electron microscopy with JEOL JEM-2000FX and JEM-2100F electron microscopes operated both at 200 kV. Chemical compositions were analyzed by energy dispersive x-ray spectroscopy (EDS) in the STEM. Specimens for TEM and STEM observations were cut from heat-treated ingots, mechanically polished, and electropolished in a solution of perchloric acid (60 %), n-butyl alcohol and methanol (3:30:130 by volume) with 0.2 M of LiCl under 17 V at -55 °C.

RESULTS AND DISCUSSION

Crystal structure of 18*R*-type Mg-Al-Gd OD phase

HAADF-STEM images of the LPSO phase taken along $[2\bar{1}\bar{1}0]$ and $[1\bar{1}00]$ directions are shown in figures 2a and 2b, respectively. In the atomic resolution STEM-HAADF images, bright spots directly correspond to the position of the atomic columns with the strong Z-dependence of the contrast [6,7]. Since Gd has the largest atomic number (#64) among the three component elements of Mg-Al-Gd LPSO phase (Mg : #12 and Al: #13), the brighter and the darkest spots should correspond to the atomic columns containing higher amount of Gd atoms and those of pure Mg columns, respectively. Enrichment of Al in Mg is practically impossible to detect from the brightness variations in STEM-HAADF images because of very small atomic number difference between Al (#13) and Mg (#12). In addition to the bright spots corresponding to pure Mg atomic columns, there are bright spots with two different brightness in the HAADF-STEM image of the $[2\bar{1}\bar{1}0]$ incidence (figure 2a), indicating the different levels of the Gd enrichment occurring for these atomic columns in the corresponding projection. Of significance to note is that the Gd enrichment occurs in the four consecutive close-packed planes, instead of two which has been reported for other Mg-TM-RE LPSO phases [11]. The level of the Gd enrichment is obviously higher in the inner two layers than in the outer layers in the quadruple Gd-enriched

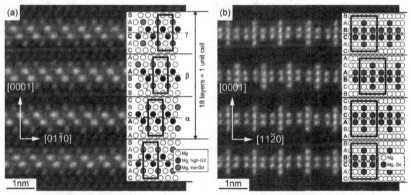

Figure 2. Atomic resolution HAADF-STEM images of the Mg-Al-Gd LPSO phase taken along (a) $[2\bar{1}\bar{1}0]$ and (b) $[1\bar{1}00]$.

layers. In the HAADF-STEM image of the [1$\bar{1}$00] incidence (figure 2b), the Gd enrichment is also confirmed to occur in the four consecutive close-packed planes. In this case, however, the level of the Gd enrichment seems not to depend on atomic column, but the spacing of these enriched atomic columns in the inner two layers is smaller (basically one third) than that in the outer layers. The brighter spots corresponding to the Gd-enriched atomic columns are observed to regularly arrange along [01$\bar{1}$0] and [11$\bar{2}$0] directions respectively in the HAADF-STEM images taken along [2$\bar{1}\bar{1}$0] and [1$\bar{1}$00] directions with the periods twice (for both inner and outer layers) and six times (for the outer layers) those of the interplanar distances of (01$\bar{1}$0) and (11$\bar{2}$0) planes of the hcp structure. These observations clearly indicate that the Gd atoms actually exhibit long-range ordered arrangements in-plane in the quadruple enriched close-packed planes. If we ignore the in-plane ordering of the Gd atoms in the quadruple enriched layers, the stacking sequence of the present LPSO phase in the Mg-Al-Gd system is identified to be the same as that previously reported for other 18R-type Mg-TM-RE LPSO phases [4,5,11]. Three structural blocks, each of which consists of six close-packed atomic planes, ABACBC-, BCBACA- and CACBAB-blocks, are identified and are designated α-, β- and γ-blocks, respectively. These three structural blocks are confirmed to be identical in terms of the atomic arrangement.

The in-plane arrangements of the Gd atoms in the inner and outer layers of the quadruple Gd-enriched layers were successfully determined based on the results of the HAADF-STEM observations. Although the ordering of the Al atoms is quite difficult to deduce solely from the contrast variation in the HAADF-STEM images, the in-plane arrangements of the Al atoms can be inferred from the observed local contraction of intercolumnar distances occurring in the close vicinity of the Gd-enriched atomic column, which is easily recognized in the HAADF-STEM image of the [1$\bar{1}$00] incidence (figure 2b) [8]. Atomic arrangements in each of the six atomic planes are schematically illustrated in figure 3 for the α-block with the stacking of the ABACBC-type. The in-plane long-range ordered arrangement of the Gd- and Al in each structural block can be described as a periodic arrangement Al_6Gd_8 clusters with the $L1_2$-type atomic arrangement on

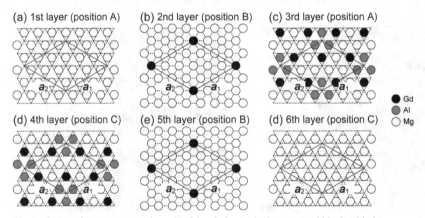

Figure 3. Atomic arrangements in each of the six layers in the structural block with the stacking of ABACBC (α-block).

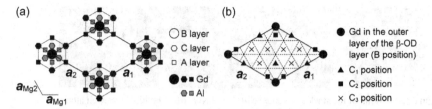

Figure 4. (a) Periodic arrangement of Al_6Gd_8 clusters with the $L1_2$-type atomic arrangement in the quadruple layers projected along [0001]. (b) Possible stacking positions of the β-block stacked on the α-block. Positions of α- and β-blocks are indicated with those of the Gd atoms in the outer layers in the B and C positions, respectively.

lattice points of a two-dimensional $2\sqrt{3}a_{Mg} \times 2\sqrt{3}a_{Mg}$ primitive hexagonal lattice, where a_{Mg} is referred to the length of the unit vector along the a-axis of Mg (figure 4a). The ideal chemical formula of the structural block is calculated to be $Mg_{29}Al_3Gd_4$ (Mg - 8.3 at.%Al - 11.1 at.%Gd), which coincides well with the experimental result of EDS analysis of Mg - 6.8 at.%Al - 10.8 at.%Gd.

We now investigate how 6-layer structural blocks stack each other by taking an example of the stacking of the β-block on top of the α-block. We take the position of the Gd atoms in the outer layer in the quadruple Gd-enriched layers in investigating the stacking of structural blocks, since the difference in the stacking in several different (non-equivalent) stacking positions is best described with them. Figure 4b shows the arrangement of Gd atoms in the outer layer in the B position of the α-block together with possible stacking positions for the outer layer in the C position of the β-block. Because of the 12 times larger in-plane unit cell of the structural block, 12 different positions are generated from a single C position. The symmetry of the structural block classifies the 12 different stacking positions into 3 types (C_1, C_2 and C_3), each of which has 3, 3 and 6 crystallographically equivalent positions, respectively (figure 4b). The preference of these three stacking positions can be inspected by checking the relative shifts of the positions of the Gd-enriched columns in the outer layers between neighboring structural blocks observed in the <1$\bar{1}$00>-projected HAADF-STEM images. The shifts are either 0 or 1/3 for the C_1 position, 0 or 1/6 for the C_2 position, and 1/6, 1/3 or 1/2 for the C_3 position, when referred to the unit of the distance between neighboring Gd-enriched columns in the outer layers (figure 4b). Inspection of the experimental HAADF-STEM image of the [1$\bar{1}$00] incidence indicates that the shifts of either 0 or 1/3 occur preferentially (figure 5), indicating that the C_1 stacking positions are preferred. Of importance to note in figure 5 is that the shifts of 0 and 1/3 between neighboring structural blocks occur randomly without any regularity. Thus, the crystal structure of the Mg-Al-Gd LPSO phase should be described as one-dimensionally disordered structure, which can be best described with the crystallographic concept of the order-disorder (OD) structure [8,9,12,13].

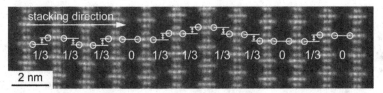

Figure 5. Variation of the stacking sequence of the 6-layer structural blocks viewed along [1$\bar{1}$00]. The relative shifts of the stacking position for the Gd atoms in the outer layers occurring between neighboring structural blocks are indicated in the unit of the projected unit cell dimension along [$\bar{1}$2$\bar{1}$0].

Structure description with the OD theory and polytypes with the maximum degree of order (MDO polytypes)

The theory of the OD structure (OD theory) was originally developed to describe crystal structures with one-dimensional stacking disorder observed in many minerals and some ceramics, such as wollastonite, brochantite, hillebrandite, SiC and so on [12,13]. In the OD theory [8,9,12,13], a crystal structure is described with two sets of partial operations (POs) of symmetry, namely λ-POs and σ-POs [9,12,13]. The λ-POs correspond to POs that transform an OD layer into itself. The set of λ-POs of a particular OD layer forms one of the 80 layer groups, which are the groups of symmetry operations of a structure assembled with three-dimensional objects on the two-dimensional lattice, and therefore, lacking periodicity in the third direction. The second type of PO, σ-PO transforms an OD layer into an adjacent one above it. For a given set of λ-POs, i.e. a given layer group, a set of σ-POs can be derived based on the symmetry of the layer group. A whole family of the derivative structures described with a complete set of POs is called an OD-groupoid family. In general, there exist two or more geometrically equivalent stacking configurations of two adjacent OD layers according to the symmetry of the OD layer, and therefore infinite possible stacking sequences including ordered and disordered structures can be described with a complete set of POs in the concept of the OD structures.

For the present LPSO phase in the Mg-Al-Gd system, the 6-layer structural block corresponds to the OD layer and the layer group, corresponding to the λ-POs for the OD layer, is the trigonal-type of $P(\bar{3})1m$ [13,14]. Three different OD-groupoid families can be derived for the OD phase depending on stacking positions [8] and are characterized by translation vectors correlating two neighboring OD layers as follows:

$$\text{Family } C_1 : t_i = -\frac{1}{3}a_i + h \quad (i = 1 \sim 3) \tag{1},$$

$$\text{Family } C_2 : t_i = \frac{1}{6}a_i + h \quad (i = 1 \sim 3) \tag{2},$$

$$\text{Family } C_3 : t_i = \frac{1}{2}a_i + \frac{1}{6}a_j + h \quad (i \neq j, i, j = 1 \sim 3) \tag{3},$$

where the vectors a_1, a_2, a_3 (= - (a_1 + a_2)) and h corresponds to three in-plane unit vectors and one unit vector along the stacking direction of the OD layer. Symmetry elements for the layer group of $P(\bar{3})1m$ corresponding to the λ-POs in the OD layer for the Mg-Al-Gd LPSO phase and the σ-POs transforming an OD layer (L_0) into an adjacent one (L_1) for the three OD-groupoid families are schematically illustrated in figure 6.

Among three possible OD-groupoid families, the $18R$-type OD intermetallic compound in the Mg-Al-Gd system is confirmed to take the OD-groupoid family C_1 that described as follows using the so-called OD-groupoid symbol proposed by Dornberger-Schiff [8,9,12,13].

$$P \quad 1 \quad 1 \quad 1 \quad (\bar{3}) \quad \frac{2}{m} \quad \frac{2}{m} \quad \frac{2}{m}$$

$$\left\{ 1 \quad 1 \quad 1 \quad \begin{pmatrix} \bar{3} \\ 3_3 \end{pmatrix} \quad \frac{2_{1/3}}{n_{1/3,2}} \quad \frac{2_{-1/3}}{n_{1/3,2}} \quad \frac{2}{n_{-2/3,2}} \right\} \qquad (4).$$

The first and second lines represent the λ-POs and σ-POs, respectively. The notations used in the OD-groupoid symbol are similar to those defined in the international tables for crystallography A [15]. The readers are referred to [8,9,12,13] for the detailed meaning of the OD-groupoid symbols.

In any OD-groupoid family, there exist some simple polytypes, designated as polytypes with the maximum degree of order (MDO polytypes) [11,14]. Three MDO polytypes, namely $1M$ (space group: $C2/m$), $2M$ (space group: $C2/c$) and $3T$ (space group: $P3_112$ and $P3_212$) in the Ramsdell notation, are found to be derived for the OD-groupoid family described by equation (4). These three polytypes are expressed with combinations of the stacking vectors of equation (1) as follows.

$$\text{MDO1} \ (1M, C2/m): \ \boldsymbol{t}_i \ (i = 1, 2, 3) \qquad (5a)$$

$$\text{MDO2} \ (2M, C2/c): \ \boldsymbol{t}_i + \boldsymbol{t}_j \ (i = 1, 2, 3, \ j = 1, 2, 3, \ i \neq j) \qquad (5b)$$

$$\text{MDO3} \ (3T, P3_112): \ \boldsymbol{t}_1 + \boldsymbol{t}_2 + \boldsymbol{t}_3 \ (\text{equivalent to } \boldsymbol{t}_2 + \boldsymbol{t}_3 + \boldsymbol{t}_1 \text{ and } \boldsymbol{t}_3 + \boldsymbol{t}_1 + \boldsymbol{t}_2) \qquad (5c)$$

$$\text{MDO3'} \ (3T, P3_212): \ \boldsymbol{t}_1 + \boldsymbol{t}_3 + \boldsymbol{t}_2 \ (\text{equivalent to } \boldsymbol{t}_2 + \boldsymbol{t}_1 + \boldsymbol{t}_3 \text{ and } \boldsymbol{t}_3 + \boldsymbol{t}_2 + \boldsymbol{t}_1) \qquad (5d)$$

The crystallographic parameters of MDO polytypes for the three OD groupoid familes are summarized in Table 1.

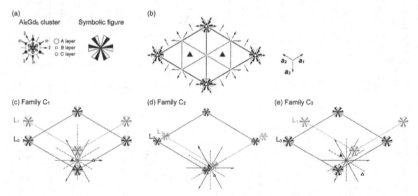

Figure 6. (a) atomic arrangement of the Al_6Gd_8 cluster characterizing the symmetry of the motif projected along the stacking direction and a corresponding symbolic figure for the motif, (b) diagram of symmetry elements for the layer group of $P(\bar{3})1m$ corresponding to the λ-POs in the OD layer for the Mg-Al-Gd LPSO phase and (c-e) schematic illustrations of the σ-POs transforming an OD layer (L_0) into an adjacent one (L_1) for three OD-groupoid families.

Table 1. The OD groupoid symbols for the OD groupoid families generated by three different types of preferential block stacking of types C_1 to C_3 in the 18R-type Mg-Al-Gd LPSO phase and crystallographic parameters for MDO polytypes derived for each of the OD groupoid families [8,9].

Stacking type	OD groupoid symbol	Ramsdell notation for the MDO polytype	Space group for the MDO polytype	Lattice parameters for the MDO polytype
C_1	$P\ 1\ 1\ 1\ (\bar{3})$ $\dfrac{2}{m}\ \dfrac{2}{m}\ \dfrac{2}{m}\ 2$ $\left\{1\ 1\ 1\begin{pmatrix}\bar{3}\\3_3\end{pmatrix}\right\}$ $\dfrac{2_{1/3}}{n_{1/3,2}}\ \dfrac{2_{-1/3}}{n_{1/3,2}}\ \dfrac{2}{n_{-2/3,2}}$	$1M$	$C2/m$ (#12)	$a = 1.12$ nm, $b = \sqrt{3}a$, $c = 1.62$ nm $\alpha = 90°$, $\beta = 103.29°$, $\gamma = 90°$
		$2M$	$C2/c$ (#15)	$a = 1.12$ nm, $b = \sqrt{3}a$, $c = 3.18$ nm $\alpha = 90°$, $\beta = 96.74°$, $\gamma = 90°$
		$3T$	$P3_112$ (#151) $P3_212$ (#153)	$a = b = 1.12$ nm, $c = 4.74$ nm $\alpha = \beta = 90°$, $\gamma = 120°$
C_2	$P\ 1\ 1\ 1\ (\bar{3})$ $\dfrac{2}{m}\ \dfrac{2}{m}\ \dfrac{2}{m}\ 2$ $\left\{1\ 1\ 1\begin{pmatrix}\bar{3}\\3_3\end{pmatrix}\right\}$ $\dfrac{2}{n_{1/3,2}}\ \dfrac{2_{-1/6}}{n_{-1/6,2}}\ \dfrac{2_{1/6}}{n_{-1/6,2}}$	$1M$	$C2/m$ (#12)	$a = 1.12$ nm, $b = \sqrt{3}a$, $c = 1.59$ nm $\alpha = 90°$, $\beta = 96.74°$, $\gamma = 90°$
		$2M$	$C2/c$ (#15)	$a = 1.12$ nm, $b = \sqrt{3}a$, $c = 3.17$ nm $\alpha = 90°$, $\beta = 93.38°$, $\gamma = 90°$
		$3T$	$P3_112$ (#151) $P3_212$ (#153)	$a = b = 1.12$ nm, $c = 4.74$ nm $\alpha = \beta = 90°$, $\gamma = 120°$
C_3	$P\ 1\ 1\ 1\ (\bar{3})$ $\dfrac{2}{m}\ \dfrac{2}{m}\ \dfrac{2}{m}\ 2$ $\left\{1\ 1\ 1\begin{pmatrix}\bar{3}\\3_3\end{pmatrix}\right\}$ $\dfrac{2_{1/6}}{n_{5/6,2}}\ \dfrac{2_{-1/2}}{n_{-1/6,2}}\ \dfrac{2_{1/3}}{n_{-2/3,2}}$	$1A$	$P\bar{1}$ (#2)	$a = b = 1.12$ nm, $c = 1.66$ nm $\alpha = 93.23°$, $\beta = 106.37°$, $\gamma = 120°$
		$2M_1$	$C2/c$ (#15)	$a = 1.12$ nm, $b = \sqrt{3}a$, $c = 3.29$ nm $\alpha = 90°$, $\beta = 106.45°$, $\gamma = 90°$
		$2M_2$	$C2/c$ (#15)	$a = 1.12$ nm, $b = \sqrt{3}a$, $c = 3.25$ nm $\alpha = 90°$, $\beta = 103.29°$, $\gamma = 90°$
		$2M_3$	$C2/c$ (#15)	$a = 1.12$ nm, $b = \sqrt{3}a$, $c = 3.17$ nm $\alpha = 90°$, $\beta = 93.38°$, $\gamma = 90°$
		$3T$	$P3_1$ (#144) $P3_2$ (#145)	$a = b = 1.12$ nm, $c = 4.74$ nm $\alpha = \beta = 90°$, $\gamma = 120°$

298

Identification of the OD character with electron diffraction patterns

The characteristics of the OD structure are clearly appeared in selected area electron diffraction patterns [8,9,16]. For the crystal belongs to an OD groupoid family, reflections are classified into two types. The first ones are so-called OD family reflections that are commonly appeared in SAED patterns taken from all possible polytypes in the OD family [16]. The second ones are termed as characteristic reflections that appear at different positions depending on polytypes. Since two or more crystallographically equivalent stacking relations exist for an OD structure, the OD phase generally exhibits one-dimensionally disordered nature derived by random selection of these three equivalent stacking relations in stacking OD layers. Then, only the reciprocal lattice lows of the characteristic reflections exhibit sharp streaks extending along the stacking direction, while those of the OD-family reflections contain discrete spots. Typical selected area electron diffraction (SAED) patterns taken from the Mg-Al-Gd OD intermetalic phase in some low-indexed projections along $[2\overline{1}\overline{1}0]$ and $[1\overline{1}00]$ as well as those calculated for the three MDO polytypes are summarized in figures 7 and 8. The reciprocal lattice rows of OD family reflections and characteristic reflections for the OD-groupoid family described by Eq. (1) are marked by two-headed arrows and one-headed arrows, respectively. If these OD family reflections together with characteristic reflections are observed, the phase should be identified as one of the OD phase

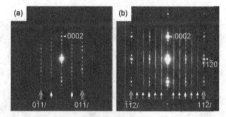

Figure 7. SAED patterns in the $[2\overline{1}\overline{1}0]$ and $[1\overline{1}00]$ projections of the Mg-Al-Gd LPSO phase.

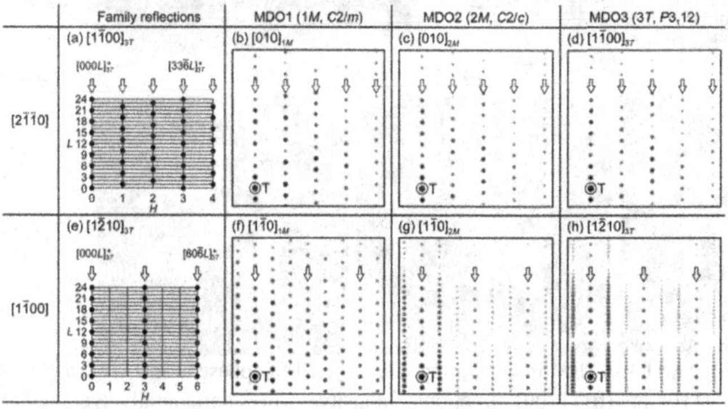

Figure 8. SAED patterns in the $[2\overline{1}\overline{1}0]$ and $[1\overline{1}00]$ projections calculated for (a,e) the OD family structure and (b-d,f-h) three MDO polytypes in the OD-groupoid family C_1.

Crystal structure of 18*R*-type Mg-Zn-Y LPSO phase

In order to see the similarity and difference in the 'LPSO' structure in the Mg-Al-Gd and other Mg-TM-RE systems, the crystal structure of an 18*R*-type LPSO phase formed in the Y/Zn-poor Mg-Zn-Y alloy was reinvestigated by HAADF-STEM imaging and SAED analysis, as shown in figure 9. Intensity profiles horizontally averaged over the HAADF-STEM images of the [$2\bar{1}\bar{1}0$] and [$1\bar{1}00$] incidences (figure 9a and 9b) clearly indicate the occurrence of the enrichment of Y and Zn atoms in the consecutive four layers in each of six-layer structural blocks with the higher enrichment occurring in the inner two layers. This is consistent with what is observed for the 'LPSO' structure in the Mg-Al-Gd system. Although the in-plane long-range ordering of Y and Zn atoms in the consecutive four layers is not readily evident, characteristic double-dagger patterns (corresponding to the distribution of Y and Zn atoms) that indicate the ordering to form Zn_6Y_8 clusters with the $L1_2$-type atomic arrangement is seen here and there in the consecutive four layers, as some of them are indicated with a frame in the HAADF-STEM image of the [$1\bar{1}00$] incidence (figure 9b). The RE enrichment in the consecutive four layers accompanied by the TM_6RE_8 cluster formation is likely to be a common characteristic to any Mg-TM-RE LPSO phases, although the periodicity of the TM_6RE_8 cluster formation may depend on alloy system. Diffuse streaks in the SAED patterns (marked by arrows in figure 9c-9e) are considered to correspond to short-range ordering of the Zn_6Y_8 clusters with the $L1_2$-type atomic

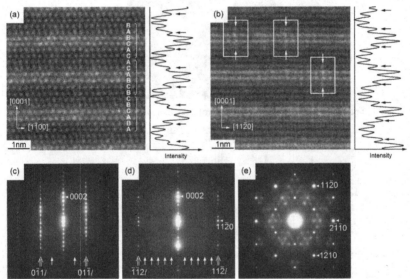

Figure 9. (a), (c) HAADF-STEM images of the 18*R* type LPSO phase in the Mg-Zn-Y system taken along (a) [$2\bar{1}\bar{1}0$] and (b) [$1\bar{1}00$] directions. Intensity profiles horizontally averaged over the images are also shown. SAED patterns taken along [$2\bar{1}\bar{1}0$], [$1\bar{1}00$] and [0001] are shown in (c) – (e), respectively [8].

300

arrangement. If we assume that the structure of the structural block is described as the matrix-Mg phase being embedded with many TM_6RE_8 clusters, then the stability and size of the TM_6RE_8 clusters in the Mg matrix could be considered as the controlling factors for the formability of the LPSO phase and periodicity of their distribution, i.e. the extent of long- (short-) range ordering in the quadruple layers, respectively. Controlling factors for the formation and ordering of TM_6RE_8 clusters are currently under survey in our group.

CONCLUSIONS

The 18R-type Mg-Al-Gd LPSO phase is composed of 6-layer structural blocks with fully-ordered atomic arrangement. The enrichment of RE (and TM) atoms occurs in four consecutive close-packed atomic planes in each structural block and the long-range atomic ordering involving a periodic arrangement of Al_6Gd_8 clusters of the $L1_2$ type atomic arrangement occurs in the four consecutive atomic planes. However, the stacking sequence of the 6-layer structural blocks does not exhibit any long-range order along the stacking direction. Because of these characteristics, the LPSO phase in the Mg-Al-Gd system cannot be described as an 'LPSO' phase any longer in a strict sense but as an order-disorder (OD) intermetallic phase with a so-called OD structure. In the case of the Mg-Y-Zn LPSO phase, the Zn_6Y_8 clusters are also confirmed to be formed but without exhibiting any long-range in-plane ordering. Thus, the formation of the TM_6RE_8 clusters is found to be a common characteristic for the Mg-TM-RE LPSO phase. Depending on whether the in-plane atomic ordering in the quadruple layers is in the long-range or short-range, the LPSO phase is described as one of the OD structures or LPSO structure.

ACKNOWLEDGMENTS

This work was supported by Grant-in-Aid for Scientific Research from the Ministry of Education, Culture, Sports, Science and Technology (MEXT), Japan (No. 23360306, and No. 23109002) and in part by the Elements Strategy Initiative for Structural Materials (ESISM) from the MEXT, Japan.

REFERENCES

1. Y. Kawamura, K. Hayashi, and A. Inoue, Mater. Trans., **42**, 1171 (2001).
2. Y. Kawamura, T. Kasahara, S. Izumi, and M, Yamasaki, Scripta Mater., **55**, 453 (2006).
3. Y. Kawamura and M. Yamasaki, Mater. Trans., **48**, 2986 (2007).
4. M. Matsuda, S. Ii, Y. Kawamura, Y. Ikuhara and M. Nishida, Mater. Sci. Eng. A, **393**, 269 (2005).
5. H. Yokobayashi, K. Kishida, H. Inui, M. Yamasaki, and Y. Kawamura in *Intermetallics-Based Alloys for Structural and Functional Applications*, edited by B. Bewlay, M. Palm, S. Kumar, and K. Yoshimi (Mater. Res. Soc. Symp. Proc., **1295**, Warrendale, PA, 2011), pp. 267-272.
6. S.J. Pennycook, A.R. Lupini, M. Varela, A.Y. Borisevich, Y. Peng, M.P. Oxley and M.F. Chisholm, in *Scanning Transmission Electron Microscopy for Nanostructure Characterization*, edited by W. Zhou and Z.L.Wang, (Springer, New York, 2006) p.152.

7. K. Kishida and N.D. Browning, Physica C **351**, 281 (2001).
8. H. Yokobayashi, K. Kishida, H. Inui, M. Yamasaki, and Y. Kawamura, Acta Mater., **59**, 7287 (2011).
9. K. Kishida, H. Yokobayashi, H. Inui, M. Yamasaki, and Y. Kawamura, Intermetallics., **31**, 55 (2012).
10. K. Kishida, H. Yokobayashi, H. Inui, M. Yamasaki, and Y. Kawamura in *Mg2012: 9th International Conference on Magnesium Alloys and their Applications*, edited by W.J. Poole and K.U. Kainer (ICMAA2012, Vancouver, BC, 2012), pp. 429-434.
11. E. Abe, Y. Kawamura, K. Hayashi and A. Inoue, Acta Mater., **50**, 3845 (2002).
12. K. Dornberger-Schiff, Acta Cryst., **9**, 593 (1956).
13. K. Dornberger-Schiff and K. Fichtner, Krist. Tech., **7**, 1035 (1972).
14. V. Kopský and D.B. Litvin DB (eds.), *International Table for Crystallography, Vol. E*, second ed., (John Wiley & Sons, Ltd., West Sussex, 2010).
15. Th. Hahn (ed.), *International Table for Crystallography, Vol. A*, fifth ed., (Springer, Dordrecht, 2005).
16. S. Merlino, in *Modular aspects of minerals / EMU Notes in Mineralogy*, Vol. 1, edited by S Merlino., (Eötvös University Press, Budapest, 1999), p.29.

Mater. Res. Soc. Symp. Proc. Vol. 1516 © 2012 Materials Research Society
DOI: 10.1557/opl.2012.1684

Multiphase Mo-Si-B alloys processed by directional solidification

Manja Krüger[1], Georg Hasemann[1], Iurii Bogomol[2], Petr I. Loboda[2]

[1] Otto-von-Guericke University Magdeburg, Institute for Materials and Joining Technology, P.O.
Box 4120, D-39016 Magdeburg, Germany
[2] National Technical University of Ukraine "Kiyv Polytechnic Institute", Kiev, Ukraine

ABSTRACT

Multiphase Mo-Si-B alloys are potential candidates for applications in the aerospace and power generation industry due to their enhanced creep and oxidation resistance at ultra-high temperatures. It is documented that the microstructure and the resulting properties of Mo-based alloys are heavily influenced by their fabrication procedure. In this study we investigate different multiphase Mo-Si-B alloys processed by zone melting (ZM) starting from cold pressed elemental powders. Microstructural characterization of zone melted alloys based on SEM investigations shows elongated arrangements of phases parallel to the growing direction as well as homogeneously distributed phases in the cross-section for some of the alloys investigated. First compression creep tests were performed at about 1100°C. In comparison to the creep resistance of powder metallurgically (PM) processed alloys the behaviour of ZM materials was found to be substantially improved. Hence, targeted application temperatures of around 1200°C to 1300°C may become feasible. Furthermore, the oxidation behaviour was found to be influenced by the volume fraction of the Mo solid solution phase since the volatilization of the Mo solid solution phase leads to a mass loss of the compound.

INTRODUCTION

While Ni-based superalloy turbine blade materials already operate at very high homologous temperatures, new metallic materials that can withstand surface temperatures higher than 1100°C would be desirable in order to increase the thermodynamic efficiency of gas turbines. Three phase Mo-Si-B alloys, consisting of Mo solid solution (α-Mo) and the intermetallic phases Mo_3Si and Mo_5SiB_2, have been the subject of intensive research because of both promising mechanical properties at ambient as well as at high temperatures and oxidation resistance. Previous research clarified that ingot metallurgical processes like arc-casting lead to inhomogeneous and coarse grained microstructures with intermetallic matrices. Materials processed in this manner show brittle behaviour at low temperatures [1,2]. On the other hand, processes like He gas atomization [3] or mechanical alloying [4] and subsequent consolidation of the powders result in fine-grained microstructures with a Mo solid solution matrix and a homogeneous distribution of the individual intermetallic phases. These materials provide decreased brittle-to-ductile-transition temperatures (DBTT) and, therefore, a good balance of the properties at ambient and elevated temperatures. However, the ultra-fine grained microstructure of powder metallurgically (PM) processed materials was superplastically deformed at temperatures in excess of 1300°C [5]. This limits the application to 1200°C or less.

The zone melting (ZM) technique offers great potential for processing materials with improved creep resistance due to the possibility of manufacturing longitudinally elongated microstructures in analogy to state-of-the-art directionally solidified and single crystalline nickel-base superalloys [6] or near–eutectic Cr-Cr₃Si [7] and NiAl-Mo alloys [8]. The possibility to form directionally grown microstructures in multiphase Mo-Si-B alloys will be proven in the present work. As demonstrated in our previous work [9] similar results regarding the BDTT can be obtained for ZM and PM processed alloys. In the present study we will investigate the creep behaviour of ZM alloys, since their coarser grained microstructure suggest a great potential for improved creep resistance. Additionally, the oxidation behaviour of different alloy compositions will be investigated at 1100°C (which may be the targeted application temperature) because oxidation plays an important role for structural applications of these alloys in air.

EXPERIMENTAL

Elemental powders of Mo, Si and B of 99.95, 99.6 and 98% purity, respectively, were mixed and cold pressed to produce ternary Mo-Si-B compositions, namely Mo-6Si-8B, Mo-9Si-8B and Mo-9Si-15B (in at.%). In Figure 1a the nominal compositions of these alloys are shown in the liquidus projection reported by Yang and Chang [10]. Moreover, mixtures of elemental powders with stoichiometric compositions of MoSi₂ and MoB₂ were annealed to produce the corresponding compounds, which subsequently were again mixed in a second step to prepare the alloy MoSi₂-10MoB₂ (in wt.%), which was reported to be the eutectic alloy composition by Kryklyva et al. (see Fig. 1b [11]).

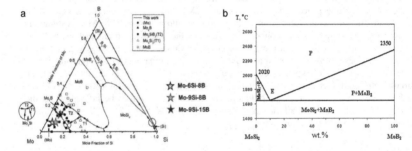

Figure 1. a) Liquidus projection of the Mo-Si-B system [10], the compositions chosen for this study are marked with stars; b) schematic binary phase diagram of MoSi₂-MoB₂ system [11].

Vertical zone melting using induction heating was carried out in a protective He atmosphere in a Kristall 206 device starting from the green samples to produce cylindrical specimens with typical dimensions of 6 mm diameter and 100 mm length using either a solidification rate of 3 mm/min (for Mo-6Si-8B, Mo-9Si-8B, Mo-15Si-8B) or 1 mm/min for MoSi₂-10MoB₂. The microstructures of the consolidated samples were characterized by scanning electron microscopy (FEI ESEM XL30 FEG equipped with EDX). Compression creep tests at constant stresses in the range between 50 MPa and 300 MPa were carried out using rectangular samples with a cross section of 3 x 3 mm² and a height of 5 mm (taken in direction of crystal growth) which were prepared by electro-discharge machining and grinding. All tests were carried out using a Zwick electromechanical testing device equipped with a Maytec furnace under a protective atmosphere of flowing Ar. The oxidation performance of ZM samples was evaluated after annealing cylindrical samples with 5 mm diameter and 4 mm height at 1100°C for selected periods of time in air. Prior to the oxidation experiments the specimens were ground and polished with SiC paper down to 1200 grit and ultrasonically cleaned in ethanol. Cross-sections of selected samples were investigated utilizing SEM after grinding and polishing.

RESULTS AND DISCUSSION

Figure 2 shows micrographs of ZM alloys Mo-6Si-8B, Mo-9Si-8B Mo-9Si-15B and near-eutectic MoSi$_2$-MoB$_2$ alloy, respectively, taken in the direction of crystal growth and perpendicular to solidification direction. Alloys Mo-6Si-8B and Mo-9Si-8B consist of a two-phase intermetallic matrix, i.e. a continuous network composed of Mo$_5$SiB$_2$ (dark gray) and Mo$_3$Si (light gray), as well as large, partially dendritic islands of α-Mo phase (bright phase), which is primary solidifying phase (Fig. 1a). In both cases the microstructure morphology shows no significant dependence on the zone melting direction as demonstrated in Figures 2a-d. By contrast, alloy Mo-9Si-15B exhibits an anisotropic microstructure. As shown in Figure 2f this microstructure possesses, in addition to large α-Mo particles, eutectic regions and intermetallic regions transverse to the zone melting direction. The eutectic structure shows elongated, fiber-like morphologies in the direction of crystal growth (Fig. 2e), which illustrates the feasibility of processing fibrous structures in ternary Mo-Si-B alloys using the zone melting technique.

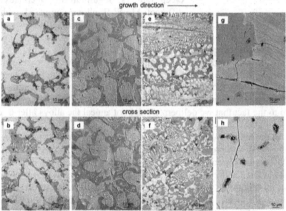

Figure 2. Microstructure of alloys Mo-6Si-8B (a,b), Mo-9Si-8B (c,d), Mo-9Si-15B (e,f) and MoSi$_2$-MoB$_2$ (g,h) parallel and transverse to the crystal growth direction.

Eutectic alloy compositions are promising materials for successful directional solidification via the zone melting method, but there is only little information about ternary eutectic Mo-Si-B alloys available in literature. However, Kryklyva and co-workers [11] provided a schematic phase diagram of the binary system MoSi$_2$-MoB$_2$, wherein a eutectic point at about 10 wt.% MoB$_2$ was represented (Fig. 1b). Our first attempt to process this presumed eutectic alloy lead to the expected anisotropy of the microstructure as demonstrated in the micrographs in Figures 2g and h, in which the individual phases have globular morphology in the transverse cross section and elongated structures in the direction of crystal growth. However, the microstructure is comparatively coarse and it was found that the alloy consists of at least three additional phases Mo_3Si, Mo_5Si_3 and Mo_2B (detected by XRD measurements) in addition to $MoSi_2$ and MoB_2 as anticipated on the basis of the phase diagram. More detailed information about the specific phase composition and the properties of this alloy will be given elsewhere [12].

Alloy composition [at.%]	Fraction of phases [%]		
	α-Mo	Mo₃Si	Mo₅SiB₂
Mo-6Si-8B	64	10	26
Mo-9Si-8B	43	24	33
Mo-9Si-15B	39	18	43

Table I. Phase composition of different ZM alloys.

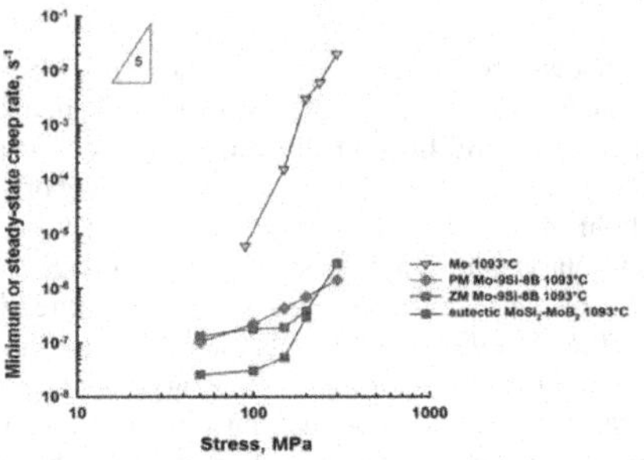

Figure 3. Logarithm of minimum or steady-state creep rate vs. logarithm of creep stress for different Mo-Si-B alloys compared with pure Mo (data taken from [15]).

The phase compositions of the three phase ZM alloys investigated in this study, which were evaluated on the basis of SEM micrographs combined with EDX and XRD measurements, are given in Table I. The volume fractions of the obtained phases differ from the predictions of the 1800°C isothermal section of the ternary phase diagram [13], which is in contrast to PM Mo-Si-B alloys with the same nominal compositions we investigated in our previous studies [4,14]. This may be due to the fact that we deliberately did not anneal the alloys after directional solidification.

The creep performance of the new ZM Mo-Si-B alloys is assessed in Figure 3 in comparison to a PM Mo-9Si-8B alloy (for details on processing and the resulting microstructure see [4,14]). The test temperature is 1093°C which is currently a typical service temperature of structural alloys used in gas turbines. As a reference the lower creep performance of pure Mo [15] is presented in the Norton-plot shown in Figure 3 which illustrates the strengthening effect due to the intermetallic phases which are formed in ternary Mo-Si-B alloys. In PM Mo-9Si-8B alloy the combined volume fraction of the intermetallic phases Mo_3Si and Mo_5SiB_2 amounts to around 50%, whereas the fraction of these intermetallic phases in the ZM alloy with the same nominal composition is around 57%. At stress levels between 50 MPa and 300 MPa the creep performance of the PM alloy is comparable to the behaviour of the ZM alloy. This is surprising at first because the PM alloy possesses a much finer microstructure with grain sizes of around 1.5 μm. However, as FIB tomography studies [16] show, the intermetallic particles which are much more creep resistant than the α-Mo phase [17] are homogeneously distributed in the microstructure due to the specific PM processing route and may be responsible for the satisfactory creep behaviour of this alloy. The stress exponents (expressed as the slope $n = \Delta \log \dot{\varepsilon} / \Delta \log \sigma$ assuming power law creep) of PM and ZM alloy Mo-9Si-8B are found to be

2.5 and 1.7, respectively. Compared to the before-mentioned alloys a creep resistance improved by around one order of magnitude was observed for eutectic $MoSi_2$-MoB_2 which can be explained by the absence of the α-Mo phase with its inferior creep resistance compared to the intermetallic phases.

Furthermore, the oxidation behavior of ZM alloys becomes an important issue at the targeted high application temperatures. In Figure 4 the formation of the protective glass layer is shown using the example of alloy Mo-9Si-8B. The cross-sectional micrograph shown in Figure 4 c demonstrates that a dense and well-sticking borosilicate glass layer which protects the surface from continuing oxidation forms during annealing the material 100 h at 1100°C in air. The fissured surface of the substrate material below the protective SiO_2-B_2O_3 layer (marked by arrows in Fig. 4c) indicates a selective oxidation process at the beginning of the exposure to high temperatures as shown in our previous work [9]. Both the inhomogeneous phase distribution and the comparatively coarse microstructure of ZM materials lead to only partial formation of the protective glass layer during the first stage of annealing. Hence, the localized loss of material occurs due to Mo oxide volatilization [18,19] before the viscous glass layer can protect the surface completely. As expected, the mass loss measured after oxidation tests at 1100°C was increased with increase in the volume fraction of the α-Mo phase in the alloys investigated in this work (Fig. 4 d, mass loss shown for 100 hours of exposure in air).

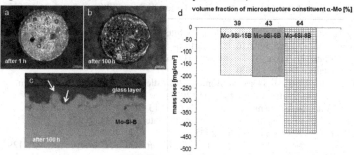

Figure 4. a) and b) Oxidized specimen of alloy Mo-9Si-8B after annealing at 1100°C, c) cross section showing the protective borosilicate glass layer that formed after 100 h of exposition and d) dependence of mass loss on the volume fraction of α-Mo in different Mo-Si-B alloys.

SUMMARY AND CONCLUSIONS

Multiphase Mo-Si-B alloys with different compositions that were produced via a zone melting (ZM) technology were comparatively assessed. The microstructure of the alloys Mo-6Si-8B and Mo-9Si-8B with an intermetallic matrix as well as large and partially dendritic α-Mo phase regions shows no clear dependence on the zone melting direction. However, in the case of materials Mo-9Si-15B and $MoSi_2$-MoB_2 an anisotropic microstructure having globular phase regions in the transverse section and elongated structures in the direction of crystal growth were observed. The creep performance at 1093°C was found to be comparable for either PM and or ZM processed alloy Mo-9Si-8B. Improved creep behavior by around one order of magnitude was observed in alloy $MoSi_2$-MoB_2 which is exclusively composed of intermetallic phases, whereas alloy Mo-9Si-8B exhibits a high fraction of around 43% of the well-deformable α-Mo phase.

In addition, the oxidation performance at 1100°C was found to be mainly dependent on the fraction of the α-Mo phase, since this phase suffers from the problem of significant mass loss due to volatilization. Nonetheless, our investigations show that a protective glass layer was formed after 100 h of annealing at 1100°C. Therefore, in future work pre-oxidation experiments will be carried out to form the protective glass layer before cyclic or continuous oxidation tests will be carried out. While it is known that high initial mass loss during heating can be avoided by homogeneous distribution of the oxidation resistant intermetallic phases, a more fundamental approach is needed to influence the phase distribution by different processing parameters during zone melting, e.g. varying the solidification velocity or repeated zone melting.

ACKNOWLEDGMENTS

The support of this work by a visiting scientists program of the Deutscher Akademischer Austauschdienst (DAAD) is gratefully acknowledged. We appreciate partial funding by the Methodisch-Diagnostisches Zentrum für Werkstoffprüfung e.V., Magdeburg.

REFERENCES

1. J. H. Schneibel, R. O. Ritchie, J. J. Kruzic, P. F. Tortorelli, Metall. Mater. Trans. **36** A, 525 (2005)
2. H. Choe, D. Chen, J. H. Schneibel, R. O. Ritchie, Intermetallics **9**, 319 (2001)
3. D. M. Berczik, U.S. Patent 5,595,616, East Hartford, United Technologies Corp. (1997)
4. M. Krüger, S. Franz, H. Saage, M. Heilmaier, J.H. Schneibel, P. Jéhanno, M. Böning, H. Kestler, Intermetallics **16**, 933 (2008)
5. P. Jéhanno, M. Heilmaier, H. Saage, M. Böning, H. Kestler, J. Freudenberger, S. Drawin, Materials Science and Engineering A **463**, 216 (2007)
6. F. L. VerSnyder and M. E. Shank, Mater. Sci. Eng. **6**, 213 (1970)
7. H. Bei, E. P. George, E. A. Kenik, G. M. Pharr, Acta Mater. **51**, 6241 (2003)
8. H. Bei, E. P. George, Acta Mater. **53**, 69 (2005)
9. M. Krüger, M. Heilmaier, V. Shyrska, P. I. Loboda, Mater. Res. Soc. Symp. Proc. Vol. 1295, DOI 10.1557 (2011)
10. Y. Yang, Y. A. Chang, Intermetallics **13**, 121 (2005)
11. I. Kryklyva, A. Dudka, M. Heilmaier, Physics and Chemistry of Solid State **12**, 365 (2011), in Ukraine
12. G. Hasemann, M. Krüger, I. Bogomol, P. I. Loboda, manuscript in preparation.
13. S.-H. Ha, K. Yoshimi, K. Maruyama, R. Tu, T. Goto, Mater. Sci. Eng. A **552**, 179 (2012)
14. M. Heilmaier, M. Krüger, H. Saage, J. Rösler, D. Mukherji, U. Glatzel, R. Völkl, R. Hüttner, G. Eggeler, C. Somsen, T. Depka, H.-J. Christ, B. Gorr, and S. Burk, JOM **61**, 7, 61 (2010)
15. H. J. Frost, M. F. Ashby, Pergamon Press, ISBN 0-08-029338-7.1987.
16. O. Hassomeris, G. Schumacher, M. Krüger, M. Heilmaier, J. Banhart, Intermetallics **19**, 470 (2011)
17. A. P. Alur, N. Chollacoop, K. S. Kumar, Acta Mater. **52**, 5571 (2004)
18. T. A. Parthasarathy, M. Mendiratta and D. M. Dimiduk, Acta Mat. **50**, 1857 (2002)
19. S. Burk, B. Gorr, V. B. Trindade and H.-J. Christ, Oxid Met **73**, 163 (2010)

Mater. Res. Soc. Symp. Proc. Vol. 1516 © 2013 Materials Research Society
DOI: 10.1557/opl.2013.360

Phase-Field Simulation of Lamellar Structure Formation in MoSi2/NbSi2 Duplex Silicide

Yuichiro Koizumi[1], Toshihiro Yamazaki[1,2], Akihiko Chiba[1], Koji Hagihara[3], Takayoshi Nakano[4], Koretaka Yuge[5], Kyosuke Kishida[5] and Haruyuki Inui[5]
[1] Institute for Materials Research, Tohoku University, 2-1-1 Katahira, Sendai, Miyagi 980-8577, Japan
[2] Department of Materials Processing, Tohoku University, 6-6-02 Aoba Aramaki, Aoba-ku, Sendai, Miyagi 980-8579, Japan
[3] Department of Adaptive Machine Systems, Osaka University, 2-1 Yamadaoka, Suita, Osaka 565-0871, Japan
[4] Division of Materials and Manufacturing Science, Osaka University, 2-1 Yamadaoka, Suita, Osaka 565-0871, Japan
[5] Department of Materials Science and Engineering, Kyoto University, Yoshida Honmachi, Sakyo-ku, Kyoto 606-8501, Japan

ABSTRACT

We conducted phase-field simulations of microstructural evolution in C11$_b$-MoSi$_2$ / C40-NbSi$_2$ dual phase alloy with and without Cr-addition to examine the factors responsible for the formation and stability of the lamellar structure on the basis of thermodynamics, micromechanics and first-principles calculations. The first principles calculation was used for evaluating the interfacial energy, segregation energy of solute Cr-atoms and lattice parameters of imaginary disilicides for estimating the effects of solute distribution on the lattice misfit. When both of lattice misfit and the anisotropy of interfacial energy is taken into account, a lamellar structure similar to that observed experimentally is formed. In the absence of Cr-addition, the straightness of lamellar structure decreased slightly. When an isotropic interfacial energy is assumed, lamellar structure is not formed. Instead, a microstructure with habit planes parallel to $\{1\,0\,\bar{1}\,1\}$ plane of C40-phase is formed. Thus, the anisotropy of interfacial energy is crucial for the lamellar structure formation rather than the elastic energy due to lattice misfit.

INTRODUCTION

MoSi$_2$-based materials are promising candidates for ultrahigh-temperature structural applications of ultrahigh efficiency gas-turbine power generation system [1-3]. Recently, oriented lamellar structure formed in C11$_b$-MoSi$_2$/C40-NbSi$_2$ duplex-silicide has been found to improve high temperature strength and room temperature toughness [4]. More recently, Hagihara et al. [5] demonstrated that Cr-addition is effective to improve the thermal stability of the lamellar structure, which is associated with Cr segregation at the lamellar interface. In practical use, however, the stability of the lamellar structure needs to be further improved. In order to improve the thermal stability of the lamellar structure, the deep understanding of the factors that govern the stability of microstructure is important. On the other hand, the phase-field method has become very powerful tool for simulating microstructural evolution in alloys and examining the factors which affect the morphology of microstructures [6]. In the present study, the effects of (i) lattice misfit between C40-marix phase and C11$_b$-precipitate, (ii) anisotropy of interfacial energy on the morphology of the dual phase structure and the effect of Cr-addition are examined by using the phase-field simulation.

THEORY AND METHOD

In the lamellar structure of $C11_b$- and C40-phases in Mo-Nb-Si-based alloy system, the crystals of the two phases are aligned so that one of {1 1 0} planes of $C11_b$-phase are parallel to the (0 0 0 1) plane of C40-phase, and the [0 0 1] direction of $C11_b$-phase is parallel to one of <1 0 $\bar{1}$ 0> directions of C40-phase. The atomic arrangement within a (0 0 0 1) plane of C40-phase and a (1 1 0) plane of $C11_b$-phase are nearly identical, and the only difference between the two crystal structures is the sequence of stacking of the layers. Specifically, C40-structure has an ABDABD-type structure and $C11_b$-structure has an ACAC-type stacking, and coherent lamellar interfaces are formed parallel to these planes. Hereafter the indices of plane and direction in C40-phase are denoted as (h k i l)$_{C40}$ and [u v t w]$_{C40}$, respectively, and those in $C11_b$-phase are (h k l)$_{C11b}$ and [u v w]$_{C11b}$, respectively. The coordinate is defined as below.

x-axis // $[0\ 0\ 0\ 1]_{C40}$ // $[1\ 1\ 0]_{C11b}$

y-axis // $[\bar{1}\ 2\ \bar{1}\ 0]_{C40}$ // $[1\ \bar{1}\ 0]_{C11b}$

z-axis // $[1\ 0\ \bar{1}\ 0]_{C40}$ // $[0\ 0\ 1]_{C11b}$.

The order parameter φ is defined so that $\varphi = 0$ and $\varphi = 1$ represent C40-structure and $C11_b$-structure, respectively. The interface is defined as a region where the φ varies from 0 to 1 continuously. First, the energies of C40-phase, f_{C40}, and $C11_b$-phase, f_{C11b}, is calculated as a function of temperature and composition using the CALPHAD method [7], and then the gradient-independent part of free energy density f_{chem} is defined as

$$f_{chem} = f_{C40}h(\phi) + f_{C11_b}(1 - h(\phi)) + W \cdot g(\phi) \quad (1)$$

where $h(\varphi)$ is the interpolation function and $g(\varphi)$ is the double-well function which penalizes the intermediate state between the two structures; W is the parameter which determines the magnitude of the energy penalty for the intermediate state. The details of $h(\varphi)$ and $g(\varphi)$ are described in the previous paper of [8]. The value of W was derived from the interfacial energy evaluated by the first principles calculation. The detail is described in the companion paper [9].

The coarse grained phase-field (CGPF) simulation of the microstructural evolution is conducted for the space of 128 × 128 × 128 μm cubic by using a 128 × 128 × 128 grids. Periodic boundary condition is employed at the edges of the simulation box. In the CGPF, the thickness of the interface is assumed to be 13 μm, which is 10000 times larger than the real thickness of the intermediate layer where the stacking sequence is that of neither of C40- nor $C11_b$-structure (approximately 1.3 nm). The interface can be then described as the variation of φ from 0 to 1 over several grid points. Gradient energy coefficient for the gradient in x-direction (i.e. the direction perpendicular to the lamellar interface), $\kappa_{\varphi,x}$, can be derived to be 2.66×10^{-11} J/m from the ab-initio value of the interfacial energy, 27.2 mJ/m², and the thickness of lamellar interface, 1.3 nm, by using the relationship for equilibrated interface [10] expressed by Eq. (2).
$\kappa_{\varphi,z} = 3\gamma_s/4.$ (2)
In the CGPF of this study, the 10000^2 times larder value of 2.66×10^{-3} J/m is used for compensating the decrease in the gradient due to the coarse graining. The energy penalty W is

set to be the $1/10000^2$ of the values used in the microscopic phase-field simulation focused on the interfacial segregation [8]. The gradient energy is described as below,

$$\kappa_{\phi,x}\left(\partial\phi/\partial x\right)^2+\kappa_{\phi,y}\left(\partial\phi/\partial y\right)^2+\kappa_{\phi,z}\left(\partial\phi/\partial z\right)^2 \qquad (3)$$

For examining the effect of the anisotropy of interfacial energy, the values of $\kappa_{\phi,Y}$ and $\kappa_{\phi,Z}$ are changed variously, while keeping the value of $\kappa_{\phi,x}$ is fixed to be 2.66×10^{-3} J/m.

A first principles calculation demonstrated that the $(1\,\overline{2}\,1\,0)_{C40}/(1\,\overline{1}\,0)_{C11b}$ interface has the lowest interfacial energy among the various C40/C11$_b$ interfaces perpendicular to $(0\,0\,0\,1)_{C40}$, and its value is 1760 mJ/m^2 (0.110 eV/Å^2).

The total free energy of the system is given by integrating the summation of the chemical free energy density and gradient energy density and elastic strain energy density due to the lattice misfit. The details of the calculation of free energies are described in the separated paper [9]. As initial condition, random number of φ between 0 and 1 was assigned to each grid so that C11$_b$-phase nucleation events occur without giving additional fluctuation focusing on the morphology of the microstructure and its stability rather than the kinetics of initial micro structural evolution. The distribution of solute concentrations and order parameter are evolved at 1673 K, where lamellar structure is formed experimentally, to decrease the total energy by solving Cahn-Hilliard equation [10] and Allen-Cahn equation [11].

RESULTS AND DISCUSSION

Figure 2 shows the distributions of order parameter on the cross sections through the center of the simulation box after evolved for 1000 s' under different conditions of interfacial anisotropy and solute concentrations. It is clearly seen that the morphology of the C40/C11$_b$ duplex structure greatly depends on the anisotropy of interfacial energies. When the value of interfacial energy evaluated by the first principles calculation is used for all the interfaces perpendicular to $(0\,0\,0\,1)_{C40}$ for the Cr-containing system, a lamellar structure with most of interfaces aligned along $(0\,0\,0\,1)_{C40}$ is formed as shown in Figs.1a, d and g. Figures 1b, e and h show the microstructure obtained in the absence of Cr-addition by using the same values of interfacial energies. Compared to the case with Cr-addition, the straightness of the lamellar interface appears to have decreased. Note that some of the interfaces are still not parallel to (0 0 0

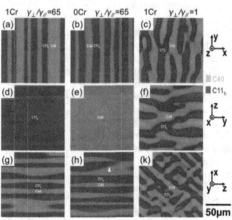

Figure 1. Duplex structures formed in the CGPF simulation using anisotropic (a, b, d, e, g, h) and isotropic (c,f,k) interfacial energies with Cr-addition (a,c,d,f,g,k) and without Cr-addition (b,e,h), observed on **x-y** plane (a-c), **y-z** plane (d-f) and **z-x** plane (g-k).

311

1)$_{C40}$ as indicated by an arrow head in Fig. 1g. This agrees with the experimental fact that the lamellar structure is not permanently stable and collapsed after a very long annealing especially in MoSi$_2$-NbSi$_2$ pseudo-binary system, and a more sharply defined lamellar structure with a higher thermal stability is formed in a Cr-added MoSi$_2$-NbSi$_2$ duplex-disilicide [12]. When the interfacial energy is assumed to be isotropic, a microstructure slightly elongated along y-direction (i.e. $[\bar{1}\,2\,\bar{1}\,0]_{C40}$ or $[1\,1\,0]_{C11b}$) is formed as in Figs.1c and f. However, no lamellar structure is formed, and the microstructure has $\{1\,0\,\bar{1}\,1\}_{C40}$ type habit planes when the microstructure is observed on z-x plane (Fig.1k). This means that the reduction of elastic energy due to the lattice misfit is not responsible for the formation of lamellar structure, and indicates that the anisotropy of interfacial energy is requisite for the formation of lamellar structure. This confirms that that the anisotropy of C40/C11$_b$ interface is the major origin of the formation of lamellar structure.

Figures2a, b and c are the concentration profiles of constituent transition metals in [0 0 0 1]$_{C40}$ direction (i.e. x-direction) along the lines of the bottom edges in Figs. 1a,b and c, respectively, and the profiles of order parameter φ along the same lines are also indicated in Fig.2d, e and f, respectively. In all the cases, C40-phases are enriched with Nb and Cr, and C11b-phases are Mo-rich. In the Cr-containing system, the profiles of Nb-concentration appear proportional to those of Cr-concentration. There is no significant difference in the compositions of both phases for the cases of isotropic and anisotropic interfacial energy. Also, it is seen that the values of order parameter φ for C40-phase and C11$_b$-phase are deviated from the originally assigned values (i.e. 0 and 1 respectively). Such a deviation of order parameter has never been seen in the case where elastic energy is neglected (not shown in this paper), and therefore it is attributed to the effect of elastic energy. This is probably because the lattice misfit is too large to be accommodated only by elastic deformation. The deviation can be interpreted as a result of the large reduction of elastic energy by forming the intermediate state between C40-phase and C11$_b$-phase, which is unfavorable when the elastic energy is neglected, is preferred.

Figure 3 shows the distribution of internal stress in the same microstructure as in Fig.1. In each image, the intensity of stress in vertical direction (i.e. y-direction on x-y plane, z-direction on y-z plane and x-direction on z-x plane) is indicated in color scale. The C40-phases are compressed and the C11$_b$-phases are expanded in y-direction (Figs.3a,b) and z-direction (Figs.3d,e) in the

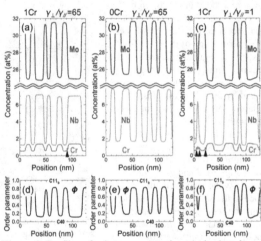

Figure 2. Profiles of solute concentrations (a, b, c) and order parameter (d, e, f) along the bottom edges of Fig.2a, Fig.2b and Fig.2c, respectively.

lamellar structures, while the C40-phased are expanded and C11$_b$-phases are compressed in **x**-direction in all the cases (Figs.3g-i). The internal stresses exerted on the C40-phase in the lamellar structure with Cr-addition (Figs.3a,g) are lower than those in the case without Cr-addition (Figs.4b,h). However, the internal stress on C11$_b$ phase is higher as seen in Fig. 3a. This may be related with the fact that the lamellar structure can be stabilized by Cr-addition. However, the internal stress is still as high as 6 GPa, which is extremely higher than the fracture stress of the lamellar disicilide deformed parallel to the lamellar interface [13] and the yield stress of C40-NbSi$_2$ with various additional atoms [14]. This is probably responsible for the deviations of the values of order parameters for C40-phase and C11b-phase (Fig. 2).

These high values of internal stress suggest that the lattice misfit is too large to be compensated by the changes in lattice parameter of the constituent phases associated with the solute concentration change and Cr-addition. This qualitatively agrees with the discussion by Nakano et al. [4] that the lattice misfit cannot be eliminated completely only by the changes in the lattice parameters by the formation of solid solution. Hagihara et al [5] suggested that the stabilization of lamellar structure by Cr-addition is due to the Cr-segregation at lamellar boundaries which they detected by analytical transmission electron microcopy (TEM). However, no significant segregation of Cr was formed in the simulation of the present study. Although a slight spike at a lamellar interface is observed in the Cr-concentration profile as indicated by an arrowhead in Fig.2a, it is much less significant compared to that measured experimentally. The slight segregation is attributed to the chemical interaction between the solute Cr and the lamellar interface in the companion paper [9]. The study indicate also the possibility of the segregation by the pile-up of Cr-atom expelled from the growing C11$_b$ phase for the case where the diffusion of Cr is assumed to be much slower than those of Mo and Nb. But the diffusivities of Nb and Cr are unknown, and they are assumed to be equal to that of Mo in MoSi$_2$ in the present study, and such a pile-up is not formed in the simulation of present study. If there are misfit dislocations on the lamellar interfaces they will be able to relax the lattice misfit and decrease the internal stress. However, few intrinsic misfit dislocations have been observed on the lamellar interface of C11$_b$-MoSi$_2$/C40-NbSi$_2$ alloy. Instead, Moiré contrast indicating the lattice misfit of 1-2% were observed on the lamellar interface [12]. Besides, Wei et al. [15] observed dislocations with both C11$_b$-phase and C40-phase in the lamellar structure of MoSi$_2$-15mol%TaSi$_2$ alloy which were not subjected to any plastic deformation. The dislocations are considered to be generated during solidification and/or annealing for lamellar structure formation. Probably, the internal stress due to the lattice

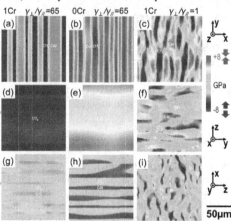

Figure 3. Stress distributions in the C40-C11$_b$ duplex structure of MoSi$_2$-NbSi$_2$ with and without Cr-addition simulated under different conditions of interfacial energy anisotropy. The location of each image is same as in Fig.1.

misfit is responsible for the formation of the dislocations. A new model that takes into account the plastic deformation will be useful for evaluating more precisely the stability of the microstructure. In any case, the results of the present study have clearly shown that the anisotropy of the interfacial energy is crucial to the formation of lamellar structure.

CONCLUSIONS

The roles of lattice misfit and the anisotropy of interfacial energy in the formation of $C11_b/C40$ lamellar structure in $MoSi_2/NbSi_2$ duplex silicide have been investigated by a phase-field simulation. The main findings are as follows.

• The lattice misfit between the two phases on its own does not give rise to the formation of lamellar structure. A microstructure with $C11_b$ phase elongated along $[1\ \bar{1}\ 0\ 0]_{C40}$ direction and $[1\ \bar{1}\ 0]_{C11b}$, with $\{1\ 0\ \bar{1}\ 1\}$-type habit plane are preferred for reducing elastic misfit energy only.

• The anisotropy of interfacial energy, i.e. the energy of $(0\ 0\ 0\ 1)_{C40}/(1\ 1\ 0)_{C11b}$ lamellar interface which was confirmed to be lower than those of interface perpendicular to the lamellar interface by nearly two order of magnitude is responsible for the lamellar structure formation.

• The addition of Cr can modify the distribution of internal stress due to the lattice misfit slightly. But the remaining internal stress is as high as 6 GPa in the simulation

ACKNOWLEDGMENTS

This research was supported by the Advanced Low Carbon Technology Research and Development Program of the Japan Science and Technology Agency (JST). For conducting the calculation, facilities of the Center for Computational Materials Science of Institute for Materials Research and the Cyberscience Center both belonging to Tohoku University have been utilized.

REFERENCES

1. A. K. Vasudevan, J. J. Petrovic, *Mat. Sci. Eng. A* **155**, 1-17 (1992).
2. K. Ito, H. Inui, Y. Shirai, M. Yamaguchi, *Philos. Mag. A* **72**, 1075-1097 (1995).
3. H. Inui, K. Ishikawa, M. Yamaguchi, *Intermetallics* **8**, 1159-1168 (2000).
4. T. Nakano, Y. Nakai, S. Maeda, Y. Umakoshi, *Acta Mater.* **50**, 1781-1795 (2002).
5. K. Hagihara, T. Nakano, S. Hata, O. Zhu, Y. Umakoshi. *Scripta Mater.* **62**, 613-616 (2010).
6. Y. Z. Wang, J. Li, *Acta Mater.* **58**, 1212-1235 (2010).
7. N. Sundman, J. Ågren, *J. Phys. Chem. Sol*, **42**, 297-301 (1981).
8. Y. Koizumi, T. Nukaya, S. Suzuki, S. Kurosu, Y. Li, H. Matsumoto, K. Sato, Y. Tanaka, A. Chiba, *Acta Mater.* **60**, 2901-2915 (2012).

9.   T. Yamazaki, Y. Koizumi, A. Chiba, K. Hagihara, T. Nakano, K. Yuge, K. Kishida, H. Inui, MRS Fall Meeting (Boston, MA 2012).
10. J. W. Cahn, J. E.Hilliard, *J. Chem. Phys.* **28**, 258 (1958).
11. S. M. Allen, J. W. Cahn, *Acta Metal.* **27**, 1085-1095 (1979).
12. K. Hagihara, Y. Hama, M. Todai, T. Nakano, MRS Fall Meeting (Boston, MA 2012).
13. T. Nakano, K. Hagihara, Y. Nakai, Y. Umakoshi, *Intermetallics* **14**, 1345-1350 (2006)
14. T. Nakano, K. Hagihara, *Scripta Mater.* (in press) doi:10.1016/j.scriptamat.2012.10.053
15. F. G. Wei, Y. Kimura, Y. Mishima, *Intermetallics*, **9**, 661-670 (2001).

Mater. Res. Soc. Symp. Proc. Vol. 1516 © 2012 Materials Research Society
DOI: 10.1557/opl.2012.1655

Influence of microstructure and processing on mechanical properties of advanced Nb-silicide alloys

C. Seemüller[1]*, M. Heilmaier[1], T. Hartwig[2], M. Mulser[2], N. Adkins[3], and M. Wickins[3]

[1] Physical Metallurgy, IAM-WK, Karlsruhe Institute of Technology, Engelbert-Arnold-Str. 4, 76131 Karlsruhe, Germany
[2] Powder Technology, Fraunhofer Institute for Manufacturing Technology and Advanced Materials, Wiener Straße 12, 28359 Bremen, Germany
[3] IRC in Materials Processing, The University of Birmingham, Elms Road, Edgbaston, Birmingham B15 2TT, UK.

*Corresponding author: Tel. +49 721 608 46556, christoph.seemueller@kit.edu

ABSTRACT

In this study different powder metallurgical processing routes, commonly used for refractory metal based materials, were evaluated on their impact on mechanical properties of a multi-component Nb-20Si-23Ti-6Al-3Cr-4Hf (at.%) alloy. Powder was produced by gas-atomization or high energy mechanical alloying of elemental powders and then consolidated either by HIPing or powder injection molding (PIM). The PIM process requires fine particles. In this investigation powder batches of gas-atomized powder (< 25 μm) and mechanically alloyed powder (< 25 μm) were compacted via PIM. Fine (< 25 μm) and coarser (106-225 μm) particle fractions of gas-atomized powder were compacted via HIPing for comparison. Quantitative analysis of the resulting microstructures regarding porosity, phase formation, phase distribution, and grain size was carried out in order to correlate them with the ensuing mechanical properties such as compressive strength at various temperatures.

INTRODUCTION

In the hot sections of today's turbines mostly nickel-base superalloys are used which operate at homologous temperature up to about 0.85 (1100 °C). To improve turbine efficiency the temperature has to be further increased requiring new materials with increased melting temperature. Niobium based silicide (NbSi) composites are a promising materials group to reach that goal. Powder metallurgical processing routes give the opportunity to obtain a fine-grained, homogeneous microstructure. Compaction processes like powder injection molding (PIM), net-shape hot isostatic pressing (HIP), or selective laser melting (SLM) provide the ability to produce near net shape components, requiring a minimum effort in post processing. They also yield potential in cost reduction compared to classical casting techniques. At present, however, NbSi-alloys still remain in the development stage leading to alloys mostly being produced by arc-melting/casting [1-3]. By contrast, powder metallurgical approaches with NbSi-alloys have been scarcely investigated in literature [4], despite being a promising method. Therefore, the goal in the present study is to compare the impact of two different powder production routes, namely mechanical alloying (MA) and gas-atomization (GA), combined with different compaction methods such as hot isostatic pressing and powder injection molding on microstructure and mechanical properties of Nb-Si alloys.

EXPERIMENTAL DETAILS

Powder production

Gas-atomized powders were produced by weighing and plasma melting element pieces of a nominal composition of Nb-20Si-23Ti-3Cr-6Al-4Hf (all compositions given in at.%). The hereby produced rods were subsequently gas-atomized via the electrode induction-melting gas atomization (EIGA) process and separated into different size fractions by sieving from which the < 25 µm and the 106-225 µm fractions were used for later compaction.

Mechanically alloyed powders were produced by milling of elemental powders of a nominal composition of Nb-20Si-23Ti-3Cr-6Al-4Hf with 5 mm diameter hardened steel balls in a Zoz CM01 simoloyer attritor. To reduce ductility, and hence to avoid sticking of the powders to balls and vial walls, the vial was continuously cooled to -15 °C. A time cycle consisting of 30 seconds of milling at 1200 rpm (≈7.3 m/s tip speed) followed by a pause of 30 seconds, ensured proper dissipation of the heat produced by ball/powder collisions. Ball-to-powder weight ratio (BPR) was 10:1. Only the size fraction with particle sizes < 25 µm was used.

Compaction

Powders were compacted using a conventional HIPing process or powder injection molding (PIM). For HIPing powders were filled in a steel can on a vibrating table. After evacuation and gas-tight sealing of the can, HIP took place at 1230 °C for 4 hours at a pressure of 150 MPa. For the PIM process a wax-polymer based binder was used with a powder load of 70 % and 52 % for GA and MA powders, respectively. Before sintering at 1500 °C for 3 hours under vacuum ($1 \cdot 10^{-4}$ mbar), a combination of solvent debinding at room temperature and thermal debinding at 600 °C was done.

The samples used in this work with their respective combinations of powder size batches and compaction processes are described in Table I.

Table I. Powder sizes and compaction process combinations used in this study. Abbreviations as defined in the text.

Sample name	Process	Powder production	Particle size [µm]
HIP GA 106-225	HIP	Gas-atomization	106-225
HIP GA -25	HIP	Gas-atomization	<25
PIM GA -25	PIM	Gas-atomization	<25
PIM MA -25	PIM	Mechanical alloying	<25

Microstructural characterization

Surfaces for the characterization of the microstructure were ground to P4000 grit using SiC paper and then polished with diamond suspension to 1 µm. Finally, the samples were polished with colloidal silica (particle size below 0.05 µm) using MasterPrep® by Buehler. SEM micrographs were taken with a Zeiss LEO1530. X-ray diffraction for phase analysis was performed on a Bruker D8 equipped with an area detector.

Compression testing

Samples were produced by electro discharge machining in case of HIP samples (4 × 4 × 8 mm³) and PIM GA samples (3.5 × 3.2 × 5 mm³) or were manufactured directly as cylindrical compression samples in case of PIM MA (Ø4mm x 5 mm). Subsequently, they were grinded coplanar down to P1000 grit. At room temperature (RT) compression tests were done with a Zwick Zmart.Pro electromechanical testing machine and at elevated temperatures (between 600-1000 °C) a Zwick Z100 with attached vacuum furnace by Maytec was used. All compression tests were conducted with constant engineering strain rate of 10^{-3} 1/s.

RESULTS AND DISCUSSION

Microstructure

XRD measurements of the four samples show similar patterns with the same peaks being present, but with different relative intensities (

Figure 1). Those peaks correspond to niobium solid solution (Nb_{ss}), α-Nb_5Si_3, and γ-Nb_5Si_3 (in what follows designated by the crystallographically identical Ti_5Si_3). The higher relative intensity of the Ti_5Si_3 peaks in the PIM samples than in the HIP samples compared to the Nb_5Si_3 peaks, suggests that the sintering conditions of the PIM process, namely 1500 °C, facilitate the formation of Ti_5Si_3. Nb_3Si seems not to be present at all in this material.

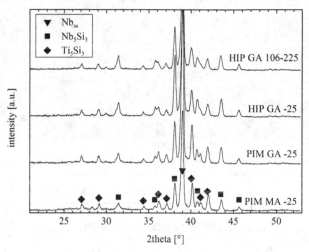

Figure 1. XRD patterns showing the presence of Nb_{ss}, Nb_5Si_3 and Ti_5Si_3 in all four samples

In Figure 2a-d representative micrographs of the four samples in back scatter electron mode are shown. In all cases, white areas correspond to hafnia (HfO_2), which typically forms during powder metallurgy via metallic Hf scavenging oxygen [4]. Light grey areas correspond to Nb_{ss} while darker grey areas consist of the silicides (Nb_5Si_3 and Ti_5Si_3). The two PIM samples have a similar microstructure regarding phase sizes and shapes. However, samples from the coarser gas-atomized powder show some larger regions of silicide. Both HIP samples comprise a much more heterogeneous microstructure with partially globular grains with dendritic structures in between. This would seem to be inherited from the microstructure of the gas-atomized powder. . Gradients inside the phases point to segregations of components during solidification. In the HIP

GA 106-225 sample there are also regions of silicide being more than a factor of 10 larger than for the fine powder fraction due to slower cooling rates of the larger droplets. A quantitative phase analysis was not possible because of the too low backscatter electron contrast.

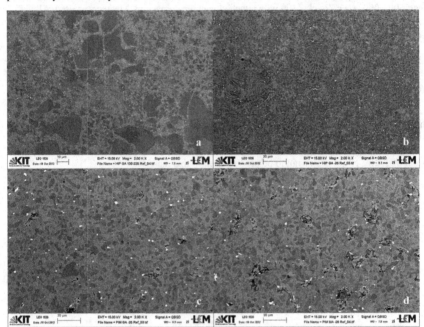

Figure 2. Backscatter electron micrographs of a) HIP GA 106-225, b) HIP GA -25, c) PIM GA -25, and d) PIM MA -25; all micrographs scaled to same magnification.

Table II. Porosity measurements for all 4 manufacturing routes as defined in Table I.

Sample name	Porosity [%]	Standard deviation [%]	Mean pore size [μm]	Standard deviation [μm]
HIP GA 106-225	0.3	0.1	0.4	0.2
HIP GA -25	0.2	0.1	1.6	0.7
PIM GA -25	3.3	0.3	5.1	3.2
PIM MA -25	5.0	0.3	4.2	4.1

Porosity measurements from optical cross-sectional micrographs reveal higher densities for the HIPed powders than for the PIMed powders (Table II, at least three micrographs with magnifications as shown in Fig. 2 were analyzed). Also pore sizes are larger for PIM samples as compared to HIP specimens.

Compression tests

Figure 3a and b show exemplary stress-strain curves at RT and at 1000 °C, respectively, for samples manufactured by the four routes defined in Table I. Most notably, at RT all samples show premature failure in compression. The PIM MA -25 shows no plastic deformation, at all. Probably, this is due to pickup of gaseous impurities during the mechanical alloying process and residual porosity which may act as stress concentrator. Yet, the obtained yield strengths approach (and partly exceed) 2000 MPa. The differences between the HIP GA 106-225, HIP GA -25 and PIM GA -25 can be attributed to grain boundary hardening. The higher cooling rates of the fine fraction, yields a finer microstructure in case of HIP. In case of PIM, the heat treatment leads to a much coarser structure. At temperatures above 800 °C all samples show ductile behavior. However, the observed maximum in strength with a subsequent decay in stress level points to the formation of defects, such as cracks, leading to shearing of the sample at high strains.

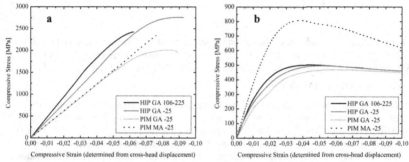

Figure 3. Compressive engineering stress-strain curves at a) RT and b) 1000 °C

The temperature dependence of the maximum strength is displayed in Figure 4. The decrease in compressive strength with temperature is negligible up to 600 °C (between 6-9 %), commonly rationalized by the decrease in Young's modulus. Above that temperature it drops about 75 %, when reaching 1000 °C. The PIM GA -25 samples show the lowest strength in this study. One likely reason for this behavior could lie in the high porosity that resulted from the PIM process. At RT the higher amount and larger size of the pores leads to stress concentrations around the pores. Therefore, the matrix yields at lower external stresses than the denser HIP samples. However, the matrix is not ductile enough to accommodate for larger plastic strains leading to brittle fracture at about 2 % of plastic strain. Also, the finer microstructure of HIP samples may lead to grain boundary hardening. At high temperatures (1000 °C) the plastic deformability of the matrix is sufficient for ductile behavior and the level of maximum strength of the PIM sample is at 95 % of the HIP sample. With a density of 96.7 % and 99.7 % for PIM and HIP samples, respectively, this remaining difference can, to the first order, be approximated taking into account the smaller effective cross-section of the PIM samples. Even though having a lower density of 7.0 g/cm³, compared to CMSX-4 (density of 8.7 g/cm³) and other NbSi-alloys the samples produced in this study lie in the same strength regime at high temperatures, but achieve much higher compressive strengths at lower temperatures [2,5,6].

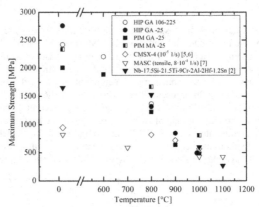

Figure 4. Temperature dependence of the maximum strengths of the four samples HIP GA 106-225, HIP GA -25, PIM GA -25, and PIM MA -25. For comparison literature data of CMSX-4 [5], MASC in tensile mode [6], and Nb-17.5Si-21.5Ti-9Cr-2Al-2Hf-1.2Sn [2] are included.

CONCLUSION AND SUMMARY

A multi-component niobium silicide composite was produced by powder metallurgy. Gas-atomized or mechanically alloyed powders were compacted by hot isostatic pressing or powder injection molding, respectively. The microstructure was more homogeneous for powder injection molded samples albeit densification and compressive strength were better for the hot isostatically pressed ones. Porosity and gaseous impurities seem to have a major influence on the deformation behavior, especially at low temperatures. Therefore, if powder injection molding may be used for future component production the densification step has to be improved. However, all samples showed a high room temperature strength and reasonable high-temperature strength.

ACKNOWLEDGEMENTS

This work was funded through the HYSOP project by the European Commission in the Framework Programme 7.

REFERENCES

[1] W. Ligang, J. Lina, C. Renjie, Z. Lijing, and Z. Hu, Chin. J. Aeronaut. **25** (2), 292–296 (2012).
[2] B.P. Bewlay, M.R. Jackson, J.-C. Zhao, P.R. Subramanian, M.G. Mendiratta, and J.J. Lewandowski, MRS Bulletin. *Sep. 2003*, 646.
[3] I. Grammenos and P. Tsakiropoulos, Intermet. **18** (2), 242–253 (2010).
[4] P. Jehanno, M. Heilmaier, H. Kestler, M. Boning, A. Venskutonis, B. Bewlay, and M. Jackson, Metall. Mater. Trans. A **36A** (3), 515–523 (2005).
[5] M. Wenderoth, S. Vorberg, B. Fischer, Y. Yamabe-Mitarai, H. Harada, U. Glatzel, and R. Völkl, Mater. Sci. Eng. A **483–484**, 509–511 (2008).
[6] C.D. Allan, doctoral thesis, 1995
[7] B.P. Bewlay, M.R. Jackson, and H.A. Lipsitt , Metall. Mater. Trans. A **27** (12), 3801–3808 (1996).

AUTHOR INDEX

SUBJECT INDEX

Printed in the United States
by Baker & Taylor Publisher Services